For ABBY and STEVEN

Contents

Acknowledgments

I am grateful to all the people who offered their expertise and shared their experiences with me about wind power. Many friends have provided me with information, ideas, and assistance during the two-year period in which I wrote this book. I want to particularly thank the following people:

Tony Rau for his crisp technical drawings and illustrations.
Ken Ketner for spending considerable time reviewing the manuscript and adding his insights.
Earle Rich for his thoughtful comments and invaluable help with wind distribution formulas.
Carol Stoner for her foresight and patience.
Dan Wallace for a fine editing job.
Abby and Steven Marier for their support and encouragement.
Martin Jopp for teaching me the basics of wind power.

Many others took the time to offer information or to describe their experiences with wind power, and I thank them all. They are: Dr. Ken Barnett, Bob Bartlett, Irv Benson, Douglas and Flo Brooks,

Chris Chomiak, Dr. Duane Cromack, John D'Angelo, Joel and Sherri Davidson, Richard Davis, Carel DeWinkel, Curt Eggert, Larry Elliot, Gary Grotte, Charlie Hall, Frank Hansen, Bror Hanson, Bruce and David Hilde, Bob Howard, Marcellus Jacobs, Joe Joddock, Dr. C. G. Justus, Dr. Henry Kelly, Joe Laleman, Dr. Louis Liljedahl, John McGeorge, Dave Madsen, Terry Mehrkam, Philip Moore, Dr. Vaughn Nelson, Tom Nichols, Scott Nielsen, William Nissley, Vernon and Marlys Orf, Ken Preston, Derek Schruers, Caleb Scott, Jim Sencenbaugh, Bob Sherwin, William Stuart, Chuck Syverson, Craig Toepfer, Harry Toppings, Robert Wagner, Elizabeth Willey, Henry Wood, and Thomas Zaborski.

Introduction

Imagine using wind power to read by, to pump water, to operate a drill, or to perform any number of other tasks around your home. These thoughts alone may kindle feelings of the satisfaction possible from beating the high cost of utility energy with wind power. As one who uses this free, nonpolluting energy source to generate electricity for my home and shop, I can tell you that energy independence is part of the satisfaction, and that the satisfaction is real. I have talked to many owners of wind power systems over the past several years, and all have spoken of one common recollection, the thrill and awe they experienced when they first watched their wind system turn into the wind to begin spinning. These people commented, too, that using wind power signified a change in their lives. Some became more conscious of how energy is used in all of its forms, while others chose wind power because their perspective on energy use had already changed. I found among most of them a uniform commitment to do something about our energy problems.

My own involvement with wind power began in 1971 when, as an electrical engineer in Chicago, I faced the prospect of being surrounded by several nuclear power plants. I became interested in the possibilities of using renewable energy sources, such as solar and wind power, and began publishing *Alternative Sources of Energy* magazine in that same year. From this vantage point I watched interest in wind power develop from a small group of experimenters into the present growing industry.

Before long, my writing on wind and solar energy and my

engineering background interested me in becoming a "wind experimenter." Along with others I learned that wind power was not a new technology. It had worked for thousands of rural people from the 1920's right up until the 1950's when the Rural Electrification Administration (R.E.A.) finished installing utility lines to provide low-cost electricity for most of rural America.

I also learned about restoring old wind systems left over from the pre-R.E.A. period and was impressed with the quality of work invested in many of them. Among the more inspiring models were the old Jacobs Wind Electric Units, which were considered the Cadillac of wind systems in their day. Made in Minneapolis, Minnesota from 1931 to 1957 by Marcellus and Joe Jacobs, hundreds of these units still existed in the early 1970's. These units provided a ready-made introduction to wind power for many wind experimenters, including myself, at a time when there were few other wind systems available in this country.

Not content simply to write about alternative energy sources, my wife, Abby, and I resolved to use it in our daily lives. We moved to Minnesota to build our own wind-powered, solar-heated, energy-efficient home.

In 1972, I learned that a man named Martin Jopp, known as the "Wind Wizard," lived in rural Princeton, Minnesota and had parts and tools for restoring old Jacobs wind systems. I ended up spending many hours working with this electrical whiz. Soon Mr. Jopp had me redoing the electrical windings for my own wind power generator and casting new aluminum parts in his small foundry. When my rebuilt wind system first began turning and generating power, I felt a great sense of fulfillment in knowing that I had been closely involved in its resurrection.

Since that time, wind power has come into its own as a viable source of energy, with many more options available now than in the pre-R.E.A. era. In Grandpa's day, the only way to store electricity from a wind system was to use storage batteries. This is still the best use of wind power for those who live in remote areas where utility power is not easy to obtain, or for those wishing to be independent of the power company. A real breakthrough in the use of wind power, however, was the development of the utility-connected wind system that generates electricity compatible with regular utility electricity, and does not require batteries for storage. Having originally proved its value by providing power for remote homesites and communications equipment, wind power has now taken its place as a supplementary source of power for anyone with a suitable wind power location who also uses utility power.

With utility rates rising almost annually, wind power is an attractive option for neutralizing the effects of these increases. Once the wind system is installed, the cost of the wind-generated power is fixed, and this fixed cost stands up well against the increasing cost of future utility-generated electricity.

Besides future costs, there is also the question of the availability of energy from centralized power sources in the future. We have already gone through several periods of energy shortages and rapidly increasing prices. The future is anyone's guess, but few predict lower energy prices, and few are willing to guarantee that there will be uninterrupted supplies of fossil fuels and electricity in the future. The move to energy conservation and the use of renewable energy sources such as wind power is no passing fad. It is a commonsense move for anyone interested in protecting himself from future increases and in becoming more energy independent.

Wind power is not something for merely the future, though, as some would like to indicate. It is a practical, environmentally sound source of energy right now. No scientific breakthrough is required to make it feasible. As wind power generating systems become economically beneficial in more areas, there is potential for the use of several million home- and farm-size installations in the United States alone. The only limitations are how fast manufacturers can learn to produce large numbers of reliable systems, and how fast our society wants to adjust its laws and financial structures to pave the way for the use of renewable energy systems.

Because an increasing number of people are considering wind power, I wrote *Wind Power for the Homeowner* to help answer the practical questions that may arise for anyone who is interested in installing a wind system but who is not quite sure where to start or what judgments have to be made. I've attempted to write this book in a straightforward manner with the minimum of necessary technical terms and mathematics. Yet, I've also tried to present enough detail about all aspects of wind power, to give the potential user a firm background in the process of choosing and specifying a wind system. Ordinarily, most people do not design a wind system on their own, but as with any new technology or product, a good background in the concepts and in the decisions that are involved in choosing and operating a wind system may be the best investment you can make.

There are some definite steps to take in deciding if wind power is suitable for you and how you can use it best. The first considerations are to find out if there is sufficient wind in your area, and if your site is suitable for using wind power, considerations discussed in Chapter 1. Although measuring the wind at a site may seem to be an obvious point at which to begin, sometimes unbridled enthusiasm for wind power causes the potential user to overlook such considerations and immediately install expensive equipment. Like all renewable energy sources, wind power is not suited to all regions and all situations. No amount of interest and enthusiasm will change that fact and you should gain peace of mind from *knowing* whether or not your site is suitable for wind power at the start.

The wind is free, of course, but the equipment to capture the power in the wind is not. Just how to evaluate the economics of a wind system is covered in Chapter 2. Purchasing a wind system requires about the same investment as buying a new car. The comparison ends there, however, because the automobile is an energy consumer while the wind system is a producer. Using a renewable energy source such as wind power presents a special situation that we do not generally meet in our consumer society. We are familiar with paying a relatively low price for an electrical, gas, or oil-powered appliance and then paying for the fuel as we use it. With wind power, the fuel—the wind—is always available at the same low price, and it is the cost of the equipment that determines the value of the energy generated.

Besides cost considerations, there are legal and institutional concerns to deal with before a wind system can be installed. These are covered in Chapter 3. Ours is an increasingly regulated society, and so there are zoning and electrical codes as well as utility regulations to be aware of. These need not be major concerns, but becoming familiar with all aspects beforehand can prevent unnecessary difficulties later. On the positive side, there are state and federal tax programs that can help to reduce the initial cost of a wind system.

You do not need to be an engineer to use wind power, but understanding the components of a wind system and how they function is a definite aid in being able to decide upon the equipment to use. The components installed in the air and on the ground are described in Chapters 4 and 5, and in Chapter 6 choosing the proper size system is discussed along with the various ways of using wind power. Chapter 7 describes how to use electricity directly from storage batteries that are charged by a wind system. This option limits the number and types of appliances that can be operated from the wind system, but it is still an attractive option for people who live without utility power, and for those who find that limiting the number of appliances they use fits in with their low-energy consumption lifestyle.

After all of the planning and choosing, the wind system can be installed. As discussed in Chapters 8 and 9, most homeowners will not engage in the actual work of installing their tower and wind system unless they live in remote areas, but being familiar with the process will help the homeowner assess the costs of the method and equipment involved. Also, a wind system involves the user to varying degrees in its upkeep and operation, so I've included maintenance and fine-tuning tips from several experienced wind power users in Chapter 10. Finally, research results in wind power are promising, and in Chapter 11 we take a look at the developments that may well lead to the wind system of the future.

After reading *Wind Power for the Homeowner,* I think you will find, as I have, that although wind power is not a panacea for all of our energy problems, it is a very practical technology that more and more people are using successfully. Perhaps you will be one of them.

Chapter 1

Selecting a Site and Measuring the Wind

The wind blows everywhere, but it blows stronger and more often in some areas than in others. You may be enthusiastic about wind power, but before you rush out to buy any equipment, it would be wise for you to find out how much wind is available in your area. It may turn out that your location does not have sufficient wind to make operating a wind power system practical. If so, it's better to be disappointed at this initial stage than after having spent several thousand dollars on wind equipment.

Of course, before you can find out whether or not you have enough wind in your area to support a wind power system, you need to know how much wind is enough wind. Fundamental principles have been formulated regarding the effect of wind speed variations and the efficiency of certain elements in the design of wind systems. Since these rules of thumb help to define the acceptable range of wind speeds for generating energy, it's helpful to learn a bit about them and the basic components of modern wind systems.

The Power in the Wind

Energy from the wind is a form of solar energy. Winds are turbulent masses of air rushing to even out the differences in atmospheric pressure created when the sun heats the air more in one place than in another. Intuitively, we know that the faster the wind blows, the more power is available—the "stronger" the wind is. Though the wind is invisible, its effects are quite apparent. On some

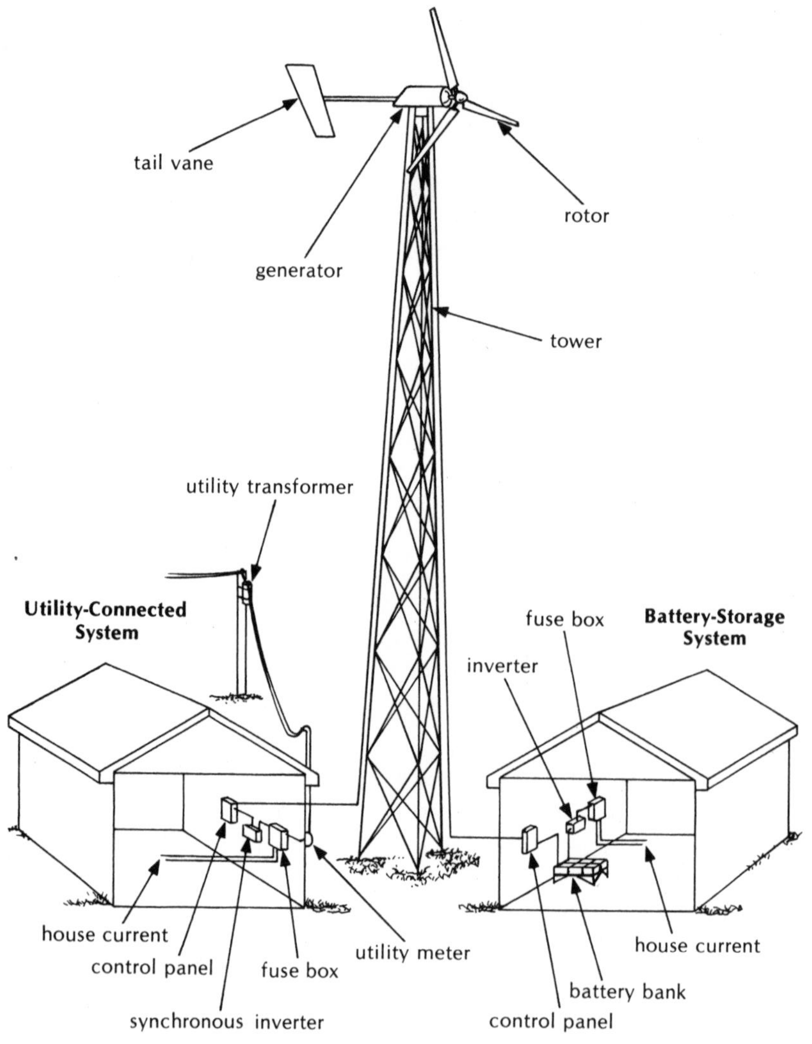

Illus. 1-1 The Basic Components of a Wind System—Components include a rotor, generator, tower, wiring, control panel, and storage batteries for direct current (d.c.) or a synchronous inverter for alternating current (a.c.).

days, in a "light" wind, a kite will barely stay aloft. At other times, trees and even buildings can be damaged or destroyed by the power of the wind. Thus, the wind can be insufficient to support a wind system, but it can also be too powerful for wind equipment to function without damage.

Present-day wind systems consist of a rotor or turbine (which often has blades somewhat similar to those of an airplane propeller) and an electrical generator, both of which are usually mounted upon a tower. To convert wind power into electrical or mechanical power, the wind rotates the turbine or rotor which turns the shaft of the electrical generator or a mechanical device. If the system produces electricity, the electrical power courses through wires to be used immediately or to be stored in battery banks for later use. Inverters, for changing the type of electrical current supplied, and a control panel also make up parts of the typical wind power system. Of all of these components, the characteristics of the rotor and generator and their reaction to varying wind speeds play the most significant roles in determining the power available in the wind.

General Rules of Thumb

Wind power has been used for centuries to pump water and grind grain. Starting with basic observations on the speed of the wind and the length of the windmill blades, Dutch millwrights developed various practical rules of thumb over the years to predict the power of a mill and to develop the best possible design. But it was not until 1927 that Alfred Betz of Göttingen, Germany computed a precise formula for the power that is available from the wind (see App. A). Part of Betz's formula shows that:

> *The power in the wind is proportional to the cube of the wind speed.*

This part of Betz's formula states that the power in the wind increases or decreases as a ratio of the cube of the original wind speed and the new wind speed. For example, if the power in the wind is known at a wind speed of 10 miles per hour (mph), and the wind speed increases to 11 mph, then the increase in the power in the wind can be found as follows:

$$P = \text{power in the wind}$$

$$P \times \frac{11^3}{10^3} = P \times \frac{11 \times 11 \times 11}{10 \times 10 \times 10} = P \times 1.331$$

The example shows that an increase in wind speed from 10 mph to 11 mph, just one mph, or 10 percent, would cause a 33 percent increase in the power in the wind. Similarly, the power in the wind does not just double if the speed of the wind doubles, it increases by a factor of 8 (since 2 cubed is 8). Thus, a 20-mph wind has 8 times as much power as a 10-mph wind. The "cubed law" relationship between the power in the wind and the speed of the wind explains why it is so important to know something about local wind speeds before installing a wind system. A small increase in wind speed produces a large increase in power.

The power in the wind is proportional to the area swept by the rotor.

If a wind rotor sweeps a circle of 100 square feet (sq. ft.) as it turns, then a rotor that sweeps a circle of 200 sq. ft. intercepts twice as much power in the same wind ($200 \div 100 = 2$). The area of a circle is equal to $\pi \times R^2$ where π equals 3.1416 and R is the radius of the circle or the length of one blade. Thus we can say that the power in the wind increases in proportion to the ratio of the blade length (radius) squared. For example, a rotor with a 15-foot radius has an area of 707 sq. ft. ($3.1416 \times 15 \times 15 = 707$) and can produce 9 times as much power as a rotor with a 5-foot radius (78 sq. ft. area) since the ratio of the radii is $15^2 \div 5^2 = 225 \div 25 = 9$.

The power in the wind is proportional to the density of the air.

The density of the air is determined by factors such as the temperature of the air, the barometric reading, and the elevation. The air density, and thus the power in the wind, can vary from 10 to 15 percent throughout the year. At an altitude of 3,500 feet, the air density is about 10 percent less than at sea level. In practice, however, changes in air density are usually not taken into account when estimating wind energy because changes in wind speed have a much greater influence.

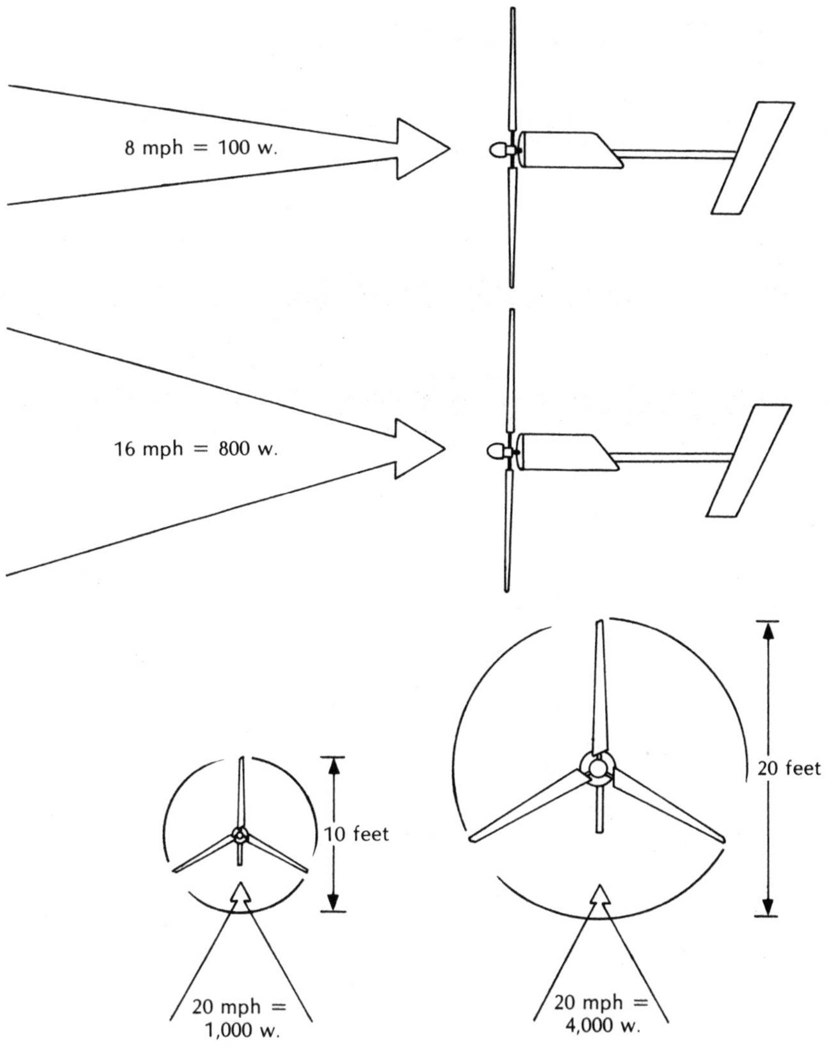

Illus. 1-2 Betz's Law—Doubling the wind speed produces an eightfold increase in the power from the wind. Doubling the rotor diameter produces a fourfold increase in the available power.

A perfect wind system extracts a maximum of only 59.3 percent of the total power in the wind.

This rule refers to what is called the wind's "available power" and represents the maximum amount of power that a wind system can produce at a certain wind speed. In practice, a well-designed wind system can convert only about 50 percent of the available power to mechanical or electrical energy. Furthermore, there are other losses in power resulting from the use of storage batteries or electronic conversion devices. In reality, only about 20 to 50 percent of the available power in the wind is converted to usable electrical or mechanical energy. In this book we are concerned mainly with the conversion of wind energy into electricity. (However, the basic principles are the same for converting wind energy to mechanical power, such as for pumping water.) Although such systems are often referred to as wind generators, a more descriptive phrase is wind-driven generator or wind-powered generator. In many technical journals, they are referred to as wind energy conversion systems (WECS). Here, we will refer to WECS as wind power systems, a convenient and easily identifiable general phrase.

Average Wind Speeds

Two words frequently used to describe the wind are variable and gusty. Indeed, wind speeds change from moment to moment. In fact, sensitive wind measuring instruments show that changes in the wind speed of up to 30 mph or more are possible in ½ second or less! Thus, rather than speak of the speed of the wind, we usually refer to the average wind speed. Wind speeds can be averaged for a second, a minute, a day, a month, a year or more. For home wind power applications, it suffices to know the monthly and yearly average wind speeds.

Generally speaking, areas with wind speed averages of 12 to 16 mph or above are excellent sites for wind systems, and areas with average speeds of 10 to 12 mph are considered moderate to good. Areas with wind speed averages of 8 to 10 mph are adequate for the use of wind power but the power they generate might be marginal in economic value compared to the cost of the system. Such areas require a more detailed assessment of the average wind speeds over a full year. For example, the yearly average wind speed for a certain site may be only 8 or 9 mph, but the average speed for the winter months may be above 10 mph. If the homeowner's greatest power demand occurs during the winter months, then the area could be suitable. Except in rare situations, areas with wind speeds below 8 mph are considered unsuitable for the use of wind power.

Now that you know the range of wind speeds that will support a wind power system, the next thing to do is to determine how your area measures up to these ranges. Easily accessible sources of average wind speed data are found at local airports, universities, meteorological testing stations, agricultural research stations, and pollution monitoring stations. Contact as many of these agencies as you can and find out from them the monthly or yearly wind speed averages in your area.

Several hundred airports in the United States send their weather data to the National Climatic Center in Asheville, North Carolina. This data is included in the Center's *Comparative Climatic Data Manual.* The average wind speed figures are based upon one minute averages taken every three hours and then averaged over several years. Average wind speed data from the *Comparative Climatic Data Manual* is included in Appendix B. Of the 271 stations listed in Appendix B, 73 percent have annual average wind speeds of 8 mph or greater; 33 percent have averages between 8 and 10 mph; and 12 percent have averages above 12 mph. You can use the annual average wind speed for the closest airport to give you an approximate idea of the wind speeds at your site.

As you examine these monthly wind speed averages in Appendix B, you'll notice that wind speeds vary considerably from month to month. Most areas show a decided reduction in average wind speeds during the summer months and the highest wind speeds during the winter. For example, Springfield, Illinois has an average of only 8.0 mph in August but has an average wind speed of 14.1 mph in March. Thus, when you look at the averages, try to be aware of extreme highs and lows throughout the year in monthly averages. Also, think about any time during the year when you may use more energy than usual, and see whether or not you would have to make any adjustments of your energy use according to the average wind speed during those times. In Table 1-1 four examples of the wind speed averages over a year are shown, with their varying monthly averages and their suitability for wind power systems.

Getting in touch with your airport and other agencies will give you a general idea of the wind speeds in your area, and very probably you'll receive encouraging reports of wind averages above the critical 8-mph mark. In fact, in some parts of the midwestern Plains states such reports are so encouraging, registering consistently excellent averages near the 12-mph mark or above, that people in these areas don't bother doing extensive investigations of wind speeds. Instead, they begin immediately to put their wind system components together, and the odds are that they won't get into trouble if they site their systems properly. Other factors, however, come into play in determining an area's suitability for wind power that may supersede the general reports from airports or weather stations.

Airport data is helpful as an initial estimate of local wind

TABLE 1-1
Monthly Variations in Average Wind Speeds for Four Cities

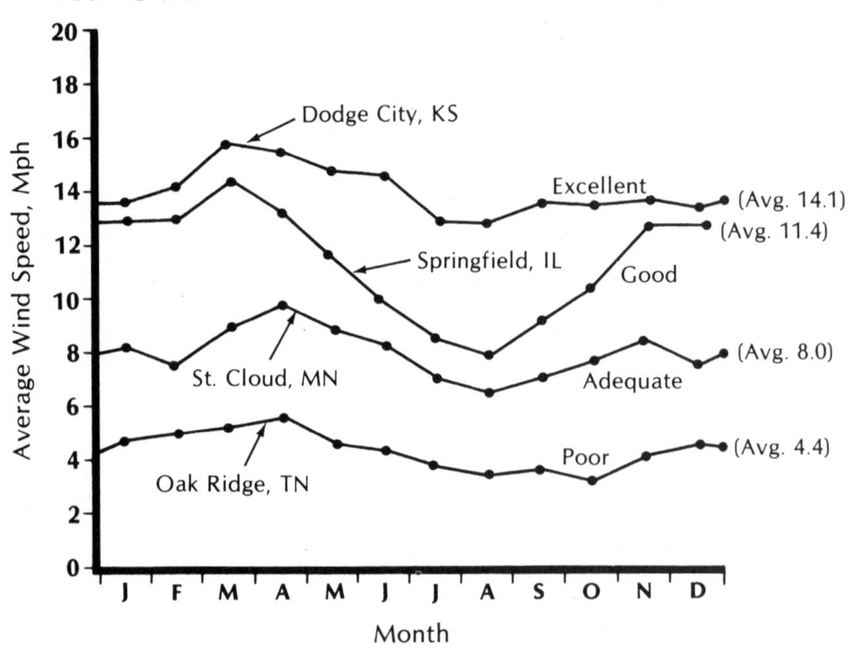

speeds, but using it should not replace actual measurements of wind averages at your site. The reason is that airport wind data is kept primarily for navigational purposes and not for estimating wind energy. In fact, airports are usually located in the least windy locations within the area. Also, airport wind-measuring instruments are sometimes too near the ground to properly represent wind speeds higher up. At some airports, the buildings shield the wind instruments from winds coming from certain directions. You can find out the height and location of the airport's wind instruments either by going there for your own look, or by obtaining the *Local Climatological Data* sheet for your local airport from the National Climatic Center at a nominal cost. These sheets indicate the height of the wind instruments, how long the data has been taken, and the directions in which the wind is blocked from the instruments due to obstructions.

Jack Reed of Sandia Laboratories conducted a study, published in the book *Wind Power Climatology in the United States,* in which the average wind speed data for over 750 wind-reporting airports and stations in the United States and Canada was converted into available wind power data. The available power is computed in watts per square meter (watts/$m.^2$) of area. Most likely, watts/$m.^2$ is an unfamiliar phrase to you. Before we go any further, a brief explanation of the units used to measure wind and power energy would be helpful, since many of the devices used to record such data register it in these units.

The traditional definition of power is the ability to do work (in this case the wind system generator's ability to do work). In electrical terms, power is measured in units of watts (w.) or kilowatts (kw.), one kilowatt being the same as 1,000 watts. Electrical generators for home-size wind systems have power ratings of anywhere from 200 to about 20,000 w. (.2 to 20 kw.). The power rating of the generator in watts is similar to the horsepower rating of a car's engine; both are convenient standards to compare the performances of different machines.

What, then, is the difference between power and energy? Whereas the power rating of a generator is a measure of its ability to do work, energy is a measure of how much work was actually done. Energy is what you pay for when you get an electric bill and energy is what you want to get from a wind-powered generating system. The most commonly used unit for electrical power is the watt, and the unit of measurement for energy is the watt hour (wh.). If a light bulb with 100 w. of power burns for one hour, the energy used is 100 wh. If the bulb is on for two hours, the energy used is 200 wh. A kilowatt hour (kwh.) is the same as 1,000 wh. and kilowatt hours are the energy units generally used when computing electricity bills. A 100-watt light bulb left on for ten hours uses 1,000 wh. of electrical energy or one Kwh.

The amount of energy that a wind system can generate can be estimated if you know the power rating of the wind system and the average wind speed at the site where it will be used. This estimate can be made by using either a wind energy monitor, by the power curve method, or by the standard distribution method, all of which are discussed fully in Chapter 6. But, knowing the basic units of energy will be enough for now.

As stated before, Reed's study relies upon watts/m.^{2} to assess the suitability of an area for wind power (to convert watts /m.^{2} to

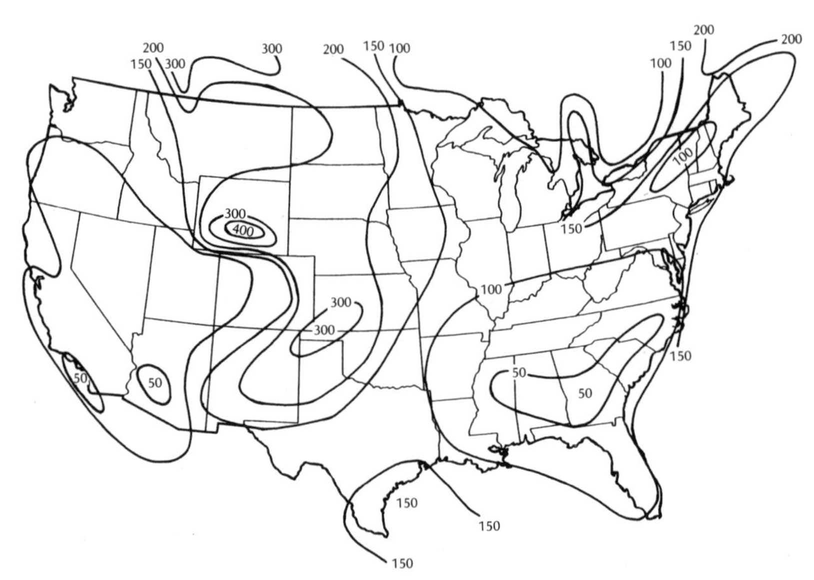

Illus. 1-3 The Available Power in the Wind in the United States—The numbers on the map represent the potential wind power in that area measured in watts/m.^{2}. (Jack Reed, Wind Power Climatology in the United States, *Document #SAND* 74-0348 *[Springfield, Va.: National Technical Information Service, 1975.])*

watts per square foot [watts/sq.ft.], multiply by .09). The map in Illustration 1-3 summarizes the data from that study. This map is useful for illustrating those regions of the United States that are generally suitable for wind power.

Areas with available power of less than 100 watts/m.^{2} are usually not good wind power sites. This includes the extreme southeastern and southwestern United States except for various coastal areas. Areas with available power of 100 to 200 watts/m.^{2} are considered inadequate to moderate wind locations, which include most of the more populated regions of the United States. The Great Plains states and many coastal areas have available power averages of 200 watts/m.^{2} and above, and these are very often good wind locations.

The calculation of watts/m.^{2} should not be used to predict the amount of energy that can be generated at a particular location—it is for comparison purposes only. Also, locations within a low wind region frequently have much higher available average wind speeds, often because of the effects of a body of water.

If the airport records state the case in clear-cut terms (that is, your area's averages are either well beneath the minimally suitable 8 mph or well above the good 12 mph average), your decision on the feasibility of a wind power system might be made at once. However, it's wise not to act too hastily in accepting a poor forecast in this event because it's the gray area between 8 mph and 12 mph that most demands the performance of a more exacting analysis of the wind speeds at your site. Even if the local airport figures show a 10- to 12-mph average, increasing the odds that you will have a good location for using wind power, you should still do a wind site analysis.

A simple wind site analysis can be performed for little expense, and is one of the most important steps in deciding whether or not to install a system, as well as being intrinsic to proper installation itself. A site analysis first involves picking the most favorable, least-obstructed location on your property for a wind system, and then measuring the wind speeds at the site for a minimum of three months up to a year. Measuring average wind speeds for a full year will provide you with a more complete picture of your area's wind patterns, including knowledge of those months that are most windy, and those that are the least windy.

Observing the Wind

People tend to grossly overestimate their local wind speeds simply on the basis of observation. People tend to remember the windy days and forget the calm days. My own experience is typical. In my original enthusiasm to install a wind system, I assumed that it was not necessary to measure wind speeds since I live in a windy midwestern state—Minnesota—where average speeds of 10 and 12 mph are common. It was not until a year after I had installed a wind system that I checked the wind speed data for St. Cloud, Minnesota (see App. B) and found our local average wind speed to be only 8 mph! As it turns out, the winds have been sufficient to power our home, another instance in which the actual wind system site proved to experience better wind averages than the local airport.

Although personal estimates of local average wind speeds are

TABLE 1-2 Beaufort Scale

Description of Wind	Observation	Speed, Mph
Calm	Smoke rises vertically.	0–1
Light air	Smoke drifts slowly.	1–3
Light breeze	Wind felt on face. Leaves rustle.	4–7
Gentle breeze	Leaves and small twigs in constant motion. Flags or streamers extend.	8–12
Moderate breeze	Raises dust. Small branches move.	13–18
Fresh breeze	Small trees begin to sway.	19–24
Strong breeze	Large branches in motion. Umbrellas difficult to hold.	25–31
Moderate gale	Whole trees in motion.	32–38
Fresh gale	Breaks twigs off trees. Difficult to walk.	39–46
Strong gale	Slight structural damage to roofs and signs possible.	47–54
Full gale	Trees uprooted. Considerable structural damage occurs.	55–63
Storm	Widespread damage	64–72

not very accurate, there is general merit in becoming familiar with the wind's effects on the surroundings as physical indicators of particular speeds. The Beaufort Scale, shown in Table 1-2, has been developed as an aid in estimating wind speeds by watching the wind's effect on trees and other objects. Also, once a wind system is installed, the turning rotor itself becomes a good visual indication of the speed of the wind.

Site Selection

After finding out the average wind speeds from your local airport and the other suggested agencies, the next steps are to choose a site for the installation of the wind system and then to test the wind speeds at the site for several months. These results can then be compared to the airport and other agencies' data.

In choosing your site for a possible wind power installation, the effects on the wind due to the local topography can be of prime importance. Hills, trees, valleys, bodies of water, and nearby buildings can affect wind speeds, especially if these obstacles are in the path of winds from the most prevalent wind directions. Normally, wind direction is not a major concern for the homeowner, but impediments to the wind can make an important difference in power output. Thus, you need to inventory such objects on your own and surrounding land to find just the right spot for your system.

Obstacles and Height

In most cases, practical considerations will limit the possible locations for a home-size wind system. If you have a half acre of land, your choices will obviously be more limited than someone with 100 acres of land. Furthermore, locating the wind system over 1,000 feet or so from the house can mean that you'll need to run out some expensive lengths of electrical cable.

Because of these physical limitations, home-size systems must often be located near obstacles such as trees and buildings. These obstacles slow down the wind from certain directions and they also cause the wind to be more turbulent. Reduced wind speeds limit the amount of energy that the wind system can generate, and excessive turbulence can cause undue stress on the wind turbine. The solution is to install the wind system on a tower that is high enough to overcome the effects of such nearby obstacles. The generally accepted rule is to install the wind system on a tower high enough so that the center of the rotor is at least 30 feet higher than any obstacles within 300 feet of the tower. To completely avoid turbulence, the tower should be located at a distance from the obstacle of at least 10 times the height of the obstacle. This is not always possible, but the "30 feet higher" rule should always be considered a minimum requirement.

The importance of the height of the tower in planning a wind system cannot be overemphasized. One of the comments I have heard most frequently from wind power users is that they wished they had installed the system on a taller tower. When the tower is too short, the effect of nearby obstacles is evident as soon as the wind system is installed. During periods of apparently "steady" winds, the rotor will suddenly slow down or even stop as the wind shifts to a direction in which the obstacle is causing turbulence. The wind turbine appears to "hunt" for the wind for a few seconds and then will speed up again as the wind shifts to a direction that is clear of obstacles.

My own wind system is located in an open field with trees over 400 feet away to the north and south, and with open fields for a mile to the west and a quarter of a mile to the east. Even with this favorable location, I have observed the familiar "hunting" of the wind system when the winds are from the northeast. Yet the trees

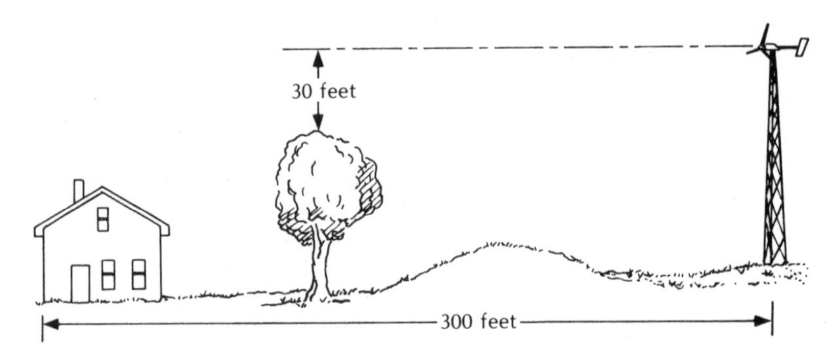

Illus. 1-4 Tower Height Rule of Thumb—Install your wind system on a tower that is at least 30 feet higher than any obstacle within 300 feet of the tower.

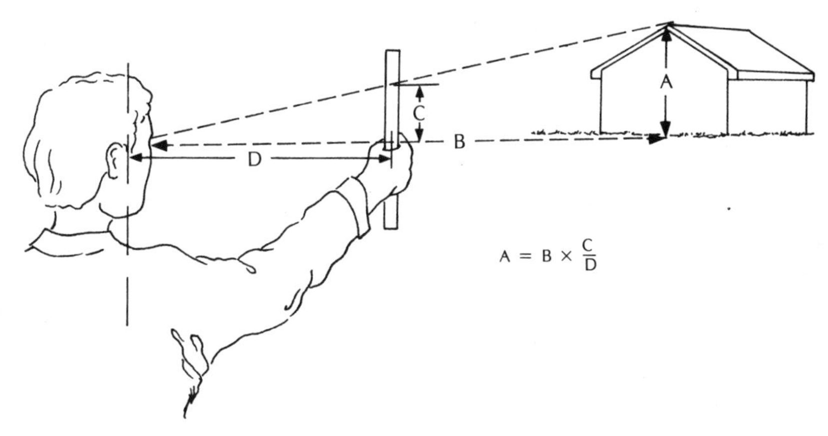

Illus. 1-5 Estimating Obstacle Height Using the Triangulation Method—The height of surrounding objects can be estimated by measuring the distance from the object and siting to the top of the object along a ruler.

are over 400 feet away in that direction! Fortunately, northeast winds are rare in the area so that the effect of the trees is not significant.

In cases where the height of the trees or other obstacles is not known, their height can be estimated by triangulation as described in Illustration 1-5. The height can be found by siting along a ruler to the top of the object and then measuring the distance from where you are standing to the object.

Even on perfectly level land with no obstacles, the wind speeds increase with the height above the ground. Tests with wind instruments at several different heights at the same location show the "wind profile," or variations in wind speed according to the height above the site. Winds very near the ground are greatly reduced due to the drag caused by the roughness of the terrain, including grass, crops, brush, trees, and the like. Wind speeds continue to increase up to several hundred feet above the ground (see Illus. 1-6).

The differences in wind speeds at various heights above the ground can be estimated by a simple equation called the "$1/7$ power rule." According to this rule, the wind speed increases as the $1/7$ power of the height above the ground. Calculating by the "$1/7$ power rule" can be tedious and difficult, so a summary of this rule for various heights above the ground is shown in Appendix C. If an average wind speed of 10 mph at a height of 30 feet above the

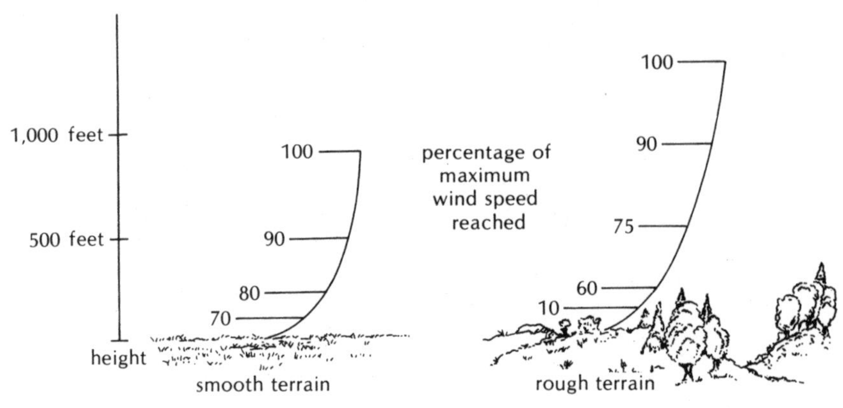

Illus. 1-6 Wind Speed Profiles for Different Terrain—Wind speed profiles vary in shape and height according to the differences in surface features at their sites.

ground has been measured at a site, the average wind speed at a height of 60 feet will be 1.10 × 10 mph or 11 mph. This rule is only an estimate, and the actual wind speed at 60 feet may be modified by the surrounding topography. The example does illustrate the importance of proper tower height because even though a change in average speed from 10 to 11 mph sounds small, it represents a 33 percent increase in the available power.

On level land with no nearby obstacles, a 40-foot tower should be about the minimum height for a wind system. Most towers come in increments of 10 or 20 feet in height. Adding 10 feet to the height of the tower increases the cost of the whole wind system by only about 5 percent but may increase the output of energy up to 10 percent.

As an example, consider a site where it is estimated that a certain wind system will produce 300 kwh. of electricity per month if installed on a 50-foot tower. Would it be worth making the tower 60 feet tall instead? If adding the 10 feet increases the power output of the system by 10 percent, an additional 30 kwh. per month would be generated. The additional tower height would be worth $3.00 per month or $36.00 per year if the electricity is valued at $.10 per Kwh. Next, if the additional height increases the initial cost of the system by $200.00, the equivalent amount of electricity would be generated in 5½ years (200 ÷ 36). Finally, if the wind system is used for 20 years, the extra cost for the 60-foot tower will

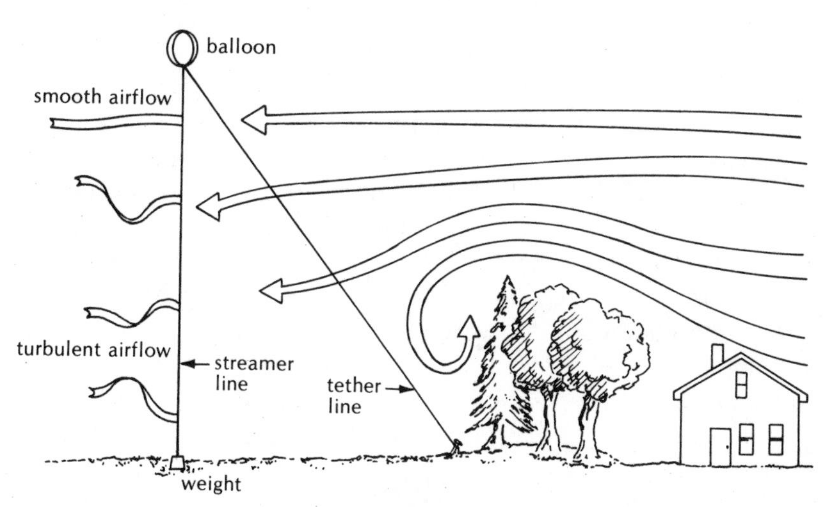

Illus. 1-7 A Simple Method for Detecting Turbulence—Streamers attached to a pole, kite, or balloon at 4-foot intervals indicate how high you'll have to go with your tower to rise above turbulence caused by obstacles.

be paid for almost 3 times over by the value of the added electricity.

Generally speaking, towers of from 60 to 80 feet in height are used for home-size systems. However, towers of 80 to 100 feet are not uncommon when it is necessary to overcome the effects of nearby obstacles, or where the wind power user feels that the additional energy obtained is desirable. In some cases, though, the tower height may be restricted by zoning codes, as we will discuss in Chapter 3.

If there is still a question about turbulence after selecting a wind site, it is possible to perform another check by attaching streamers (about 4 feet long) to a tall pole and observing them. Or, you can use a large helium-filled balloon with streamers attached to a string for the same purpose.

Topography

On the global scale, winds have a prevailing direction over different parts of the earth. On the North American continent, winds are most frequently from the west. These prevailing westerlies are very evident in the flat Plains states. But on a regional or local scale the prevailing winds are modified by the topography and by thermal conditions.

Thermal winds are created by temperature differences between the air and land, or the air and a body of water. In mountainous areas, air is heated throughout the day and rises up the mountain as a thermal wind. At night, the opposite effect takes place and the cooled air moves down the mountainside. Similarly, sea breezes move toward the warmer coastal area during the daytime and back toward the sea at night as the land cools down. In mountainous areas wind conditions can vary considerably over short distances, so comparisons of wind speed data with local airport data are very difficult to make. Furthermore, in regions such as some areas of the Rocky Mountains, very high wind speeds occur with great turbulence, causing such places to be poor sites for wind power.

On the other hand, hills and ridges can be assets as wind sites if the tower can be located properly. Any hill acts as a natural tower, putting the wind system into higher wind speeds than those of the

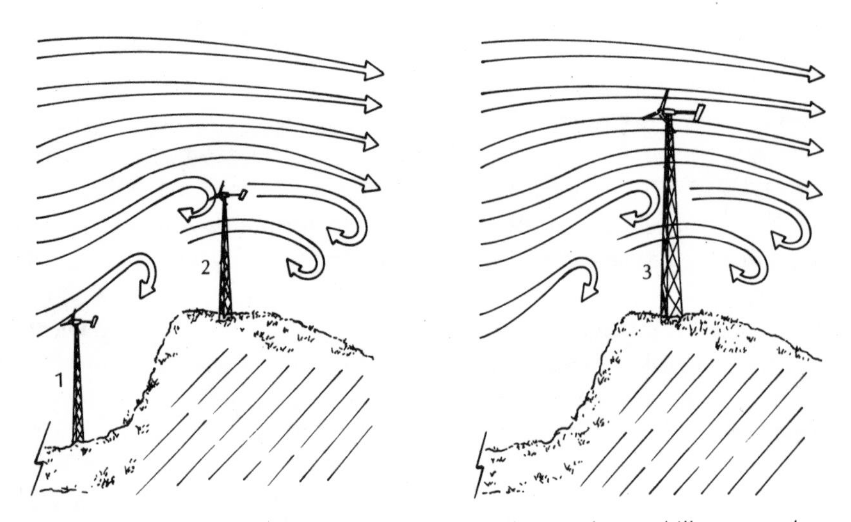

Illus. 1-8 Placing a Wind System Tower on a Hill—Winds over hills can produce significant turbulence but can also increase wind speed. Tower 1 is too near the foot of the hill; wind coming toward it can be reduced in speed by 50 percent, and wind coming from the hilltop direction can be reduced in speed to virtually zero. Tower 2 is located too low on the top of the hill in the midst of turbulence. Tower 3 is correctly positioned at the right height on top of the hill, and the speed of the wind can increase up to 200 percent.

lower surrounding area. Also, at the top of a hill or ridge the wind actually speeds up as much as twice as fast before approaching the hill. Remember Betz's rule again: the power in the wind cubes as the wind speed increases. Such acceleration of the wind over the top of a hill is due to the fact that the hill creates a low pressure area on the down-wind side. This low pressure area causes the wind to speed up near the top of the hill.

Although the wind speeds up as it goes over a hill, there is a certain amount of turbulence created at the crest, and especially on the down-wind or lee side of the hill. For this reason, the tower should be high enough to be out of the turbulent area. The least desirable location is at or near the foot of a hill since wind speeds at such sites can be reduced by as much as 50 percent when the wind is blowing toward the hill, and the wind speed can be reduced to almost nothing when the wind is blowing from the opposite side, down over the hill (see Illus. 1-8). Sites near lakes can also be assets since breezes are created by differences in water and land temperatures, and also because the wind is unobstructed as it passes over the lake.

Wind Speed Measurement

After selecting the site, wind speeds should be measured for at least three months—preferably in the spring or winter. Several types of wind measuring instruments can be used for these surveys.

Anemometers

Wind speeds can be measured with instruments called anemometers, most of which are basically miniature wind systems. They're usually constructed as a small turbine consisting of either funnel-shaped cups or a small rotor that turns a small generator. The electrical signal from the generator varies along with the speed of the wind (see Photo. 1-1). Many anemometers are also equipped with a wind direction indicator, which may come in handy if you suspect that wind direction could present a problem as far as obstructions are concerned. The anemometer should be mounted on a pole, a mast, or a tower for a TV antenna that is about as high as the wind system tower will be.

Photo. 1-1 A Wind Cup Anemometer with a Wind Direction Indicator—The anemometer acts as a small wind system, spinning in the wind to record the wind speeds below. The wind direction indicator can be valuable by showing where the highest-speed winds come from, and if these winds are impeded by obstacles.

Some low-cost wind measuring devices consist of an anemometer and a meter that indicates instantaneous wind speed. While it is interesting to watch the wind speeds change, such a device is of little use in gathering wind data since no average reading is recorded. Likewise, hand-held wind measuring devices are not very useful for collecting information since no data is recorded and the wind speeds near the ground differ greatly from the wind speeds 40 feet or more in the air.

Wind Odometers

A simple type of recording anemometer is called a wind odometer. As the anemometer turns, an electronic or mechanical odometer connected to it records the number of revolutions that the anemometer makes (see Photo. 1-2). The number of revolutions is calibrated in miles of wind and the reading on the odometer is called the *wind run.* The average wind speed is found by dividing the wind run by the period of time over which the wind speeds were recorded. As an example, if over a period of one day the odometer reading was 7,200 and each count represents 1/60th of a mile, the wind run would be 7,200 ÷ 60, or 120 miles. The reading was taken over a 24-hour period, so the average wind speed for that day is 120 miles ÷ 24 hours, or 5 mph. In a similar manner, the average wind speed for a week (168 hours) or month (720 hours) can be measured. Daily readings of the odometer are most valuable since you can get an account of the daily variations in the average wind speed. Weekly or monthly averages may not account for a brief period of intense winds, thereby misleading you on what you can regularly expect as the daily average wind speed.

Priced from $125 to $175, wind odometers are the lowest in cost of the various wind-recording instruments. Also, they can often be rented from distributors of wind equipment for a nominal fee. In fact, some state energy offices lend anemometers to the public as part of their wind-testing programs.

After the average wind speed has been measured at a site, the energy output of a particular wind system can be predicted based

Photo. 1-2 A Wind Odometer—A wind anemometer is connected to this device which records the number of revolutions it makes, calibrated in mph.

upon the estimates of the wind system's manufacturer. Some sample ranges of monthly outputs in kwh. are shown in Table 1-3 for four types of wind systems. The energy output at different average wind speeds for many of the available wind systems is presented in greater detail in Chapter 6.

The numbers shown in Table 1-3 for the Aeropower system indicate the typical variations of energy output with average wind speeds. At an annual average wind speed of 10 mph, the Aeropower system is predicted to generate 125 kwh. per month. At 12 mph, the output is 258 kwh.—twice the energy with a difference of only 2 mph in the average wind speed.

Wind Energy Monitors

Another simple way to estimate the possible energy output of a wind system is to use a wind energy monitor. A monitor is similar to a wind odometer except that its reading directly indicates the potential kwh. of energy attainable at the site instead of indicating the wind run. The unit is programmed before installation to simulate the performance of a particular wind system. At the end of the measurement period, the reading on the monitor gives the es-

TABLE 1-3
Estimated Monthly Output in Kilowatt Hours
(Manufacturer's Estimate)

Wind System	Rotor Diameter, Feet	Average Wind Speed, Mph			
		10	12	14	15
Winco/"1222H"	6	20	26	30	. . .
Aero Power/ "SL 1500"	12	125	258	. . .	406
Pinson/ "Cycloturbine C2E3"	15	250	458	. . .	800
Jay Carter/ "Mod 25"	32	2,160	3,528	. . .	5,400

timated amount of energy that the wind system would have generated during that time period. Since a wind energy monitor costs about twice as much as a wind odometer, you may want to rent one rather than buy one.

Wind Frequency Charts

Another valuable tool for estimating wind energy is a wind frequency chart, which shows the number of hours per month that the wind blows at different speeds. Used in conjunction with the power output curve of a particular wind system, a wind frequency chart can help to predict the amount of energy that can be generated at the site. A power output curve shows the power that a particular wind-powered generator puts out at various wind speeds, and specific model's curves should be available from the manufacturer. The estimated energy that can be produced by the generator at your site is found by multiplying the number of hours that each wind speed occurs multiplied by the power output of the generator at that speed (power output curves are discussed in greater detail in Ch. 6).

Wind frequency charts can also help you obtain a good long-term overview of how your prospective system will perform by comparing them to charts available for your local airport. The National Climatic Center maintains wind frequency charts for many airport locations. These are called either *Percentage Frequency of Wind Direction and Speed Charts* or *Frequency of Occurrence Charts,* and they can be obtained for a nominal fee from the center. The percentage frequency charts compiled for many airports give the percentage of time that the wind occurs in 11 different speed ranges or classes. The percentages are broken down for each speed class into the 16 directions of the compass.

The Frequency of Occurrence Charts available from other airports give the number of times that each wind speed class was observed to occur over a period of time. The percentages of frequency can be calculated using Frequency of Occurrence Charts by dividing the total number of observations indicated on the chart into the number of observations for each speed class (for an example of a Frequency of Occurrence Chart, see Table 6-2 in Ch. 6).

Other Wind Measuring Devices

More sophisticated and expensive wind measuring devices such as strip chart wind recorders and wind frequency analyzers can be used for collecting wind data. In general, these are not recommended for use in siting home-size systems. They are much more expensive than wind odometers and monitors, and do not necessarily produce any more useful data because of the wide variations in wind speeds from month to month. They are, however, most useful for research projects or for special situations.

Strip Chart Wind Recorders

The most common type of strip chart wind recorder models use an anemometer connected to a machine that records the readings in detail on a long strip of paper. A mark is made on the chart paper every two seconds to indicate the wind speed at that time. The paper is marked off in grids so that one square, or grid, represents 15 minutes, and one inch represents one hour. A 60-foot roll of paper is sufficient to measure wind speeds for one month. The average wind speed for each 15-minute period is found visually by finding the center of the mass of dots on the paper for that period.

Understandably, it is very time-consuming to analyze the wind data from a strip chart recorder, and it takes many hours to derive a wind frequency chart or average wind speed figure from such equipment. They are useful, however, in showing patterns of wind speeds, since the wind speeds are recorded in the time sequence in which they occurred. I have used a strip chart recorder for about a year and found it painstaking to read the data, but also found it very interesting. I was able to observe, for example, the very definite day-night pattern of winds so typical in many parts of the country. Usually, "thermal" winds start up in the early morning and drop off as soon as the sun sets. This pattern is disrupted whenever a cold or warm "front" moves through the area, in which case an irregular pattern of wind can continue overnight and even over a period of days.

An often asked question is, "If your system is connected to storage batteries, how long can you go without wind before the batteries are discharged?" The answer depends on how many kwh. of electricity are used from the storage batteries each day. However, knowing the return time, that is, the time between winds of high enough speed to allow the wind system to "charge" the storage batteries, is also necessary to answer the question of "How long?" Using a strip chart recorder at my own site during May of 1977, I found that the longest period of "charging" winds (winds exceeding 8 mph) was 44 hours, with the average for the month being 5.8 hours. The longest return time or period of "noncharging" winds was 51 hours, with the average period of 6.3 hours. Such data is helpful in estimating how large a storage system you'll need for storing energy during nonwindy periods (discussed further in Chs. 5 and 9). At a location with a very low average wind speed, the return time can be as long as seven or more days several times throughout the year. With this example in mind, it could prove to be useful to take the time to examine a chart recording of wind speeds at a local university or research center, if such recordings are available.

Wind Frequency Analyzers

An instrument that automatically compiles a wind frequency chart for a wind site is called a wind frequency analyzer. Like strip chart recorders, analyzers are much more expensive than odometers or monitors and are designed more for research than for home-size wind power use. Nevertheless, they may be useful in some cases, and often they can be rented.

An analyzer separates the wind speed data from the anemometer into different speed ranges, or increments also called bins. Typically, it can measure the frequency of each wind speed from 0 to 64 mph in increments of, say, 2 mph. The wind data from an analyzer is essentially the same as provided in the Percentage Frequency of Wind Direction and Speed Charts.

Both strip chart recorders and wind frequency analyzers can also record wind direction. By plotting the percentage of time that the wind blows from each direction on a chart, the resulting wind rose chart indicates the direction from which prevalent winds arrive (see Table 1-4).

Of all the methods of measurement considered here, the use of a wind odometer remains the simplest and least expensive one for siting a wind system for home use. Finding the average wind speed at the site and relating it to the proposed system's estimated

TABLE 1-4 Wind Rose for St. Cloud, MN*

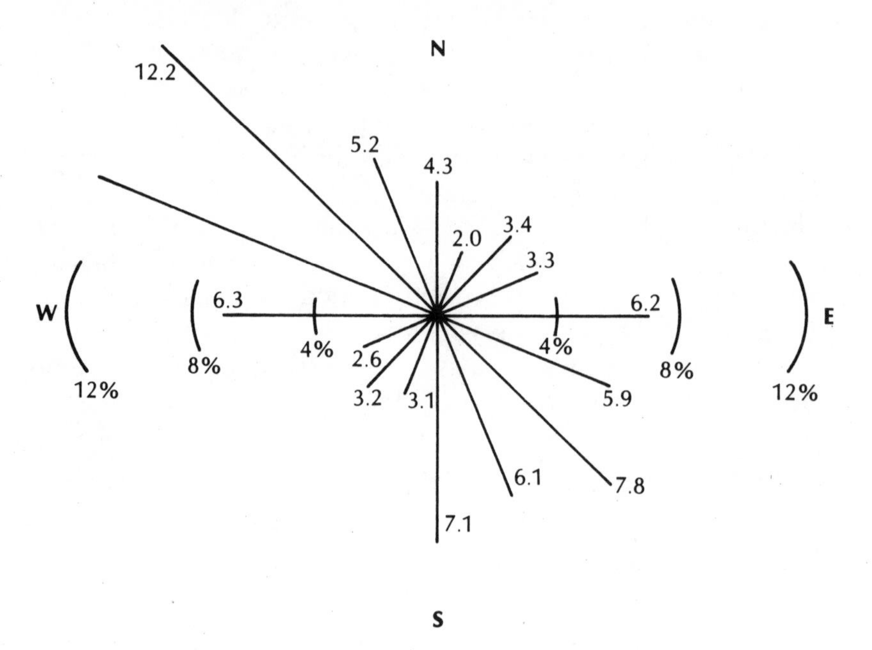

*Each direction indicates percentage of 8,760 hours (1 year).

energy output at that speed is sufficient for most sites. Of course, any other data such as that available from your local airport should be used for confirmation of your own findings. The larger the system, the larger is the investment and the more important siting and measurement become. Whereas three months of testing is sufficient in most cases for the smaller wind systems, six months to a year is a better range for systems rated above 20 kw. As a rule, figure on spending about 1 percent of the installed system cost on siting and testing.

Chapter 2

Wind System Economics

Upon completion of a thorough analysis of your proposed wind site, you probably have determined that the average wind speeds in your area can support most available wind power systems. However, as shown in Appendix F, the energy output of these wind systems is limited according to your average wind speeds. What you want to learn at this point is if the energy output of these systems can supply you with the electricity that you need to power your house.

In order to find out this information, you will have to do a simple survey of the electricity you use at present, and then compare it to the output of the various systems in Appendix F. This comparison will help you to figure out what percentage of your electrical requirements can be supplied by a particular wind system, or to what extent you may have to modify your use of electricity in order to maintain all of your use of power within the limits of your wind system's capacity to produce it.

The comparison can also give you an idea of the types of systems that you should consider to fulfill your electricity requirements. Later on you can pick one system as an example, and figure out its economical payback period by comparing the value of the electricity generated by the wind system to the cost of utility power. But, first, you have to estimate your own electrical needs.

Counting Kilowatts

To find out how much electricity you use, take a look at your present electricity bill. The amount you pay is based upon how many kilowatt hours of electricity you use during a month. The number normally appears on the bill and is usually abbreviated as KwH or KwHr (we will use kwh. in this book). If you are a "typical" consumer, this number is somewhere between 500 to 1,500 or even more kwh. per month. If you are an energy-conscious consumer, you may be using less than 500 kwh.

If the number of kwh. does not appear on your bill, or if you are not sure, take a look at your monthly charge. Next, call your utility company and find the average rate per kwh. in your area. Then, knowing your monthly bill and the rate at which the utility company charges, you can estimate the monthly electrical usage. For example, if your bill was $30 for a given month and the rate charged was $.05 per kwh., your estimated usage would be 600 kwh. per month ($30 ÷ $.05 per kwh.). Utility rates vary with the amount of electricity used and there are often various surcharges added to the bill, but we are concerned only with average rates.

You can also read your kilowatt-hour meter directly to find your use of electricity. Many new meters have easy-to-read dials just as on a car odometer, so that reading the meter at the beginning and end of the month is simply a matter of writing down the numbers indicated on the dial. Older meters have pointer dials for each digit and reading these meters takes a little practice. For the beginning and ending of the monthly readings shown in the example in Illustration 2-1, read the meter from left to right and note down the number that each dial pointer is on or has just passed. In the example shown, the beginning reading is 1,575 kwh. and the end of the month reading is 2,159 kwh., for a difference of 584 kwh.

Typical Electrical Usage

Besides knowing the total amount of electricity you use per month, it's helpful to know how much each appliance contributes to the total. This can aid you in deciding whether or not to cut back on the use of different appliances, to eliminate their use, or to modify their use in some way.

Most appliances have their power rating stamped on the nameplate. This power rating multiplied by the number of hours of opera-

(*continued on page 32*)

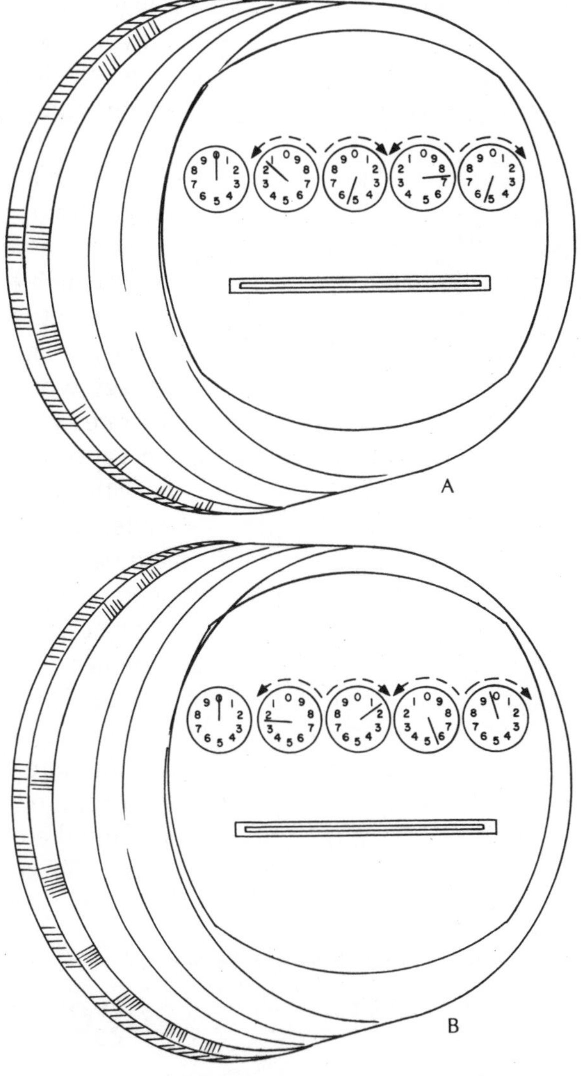

Illus. 2-1 Reading an Electric Meter—The reading on meter A is 1,575 kwh., and the reading on meter B is 2,159 kwh. Always read the number that the pointer has just passed.

TABLE 2-1
Typical Monthly Use of Energy by a Family of Four

Appliance	Approximate Wattage	Average Kwh. per Month	Monthly Cost in Dollars at $.07 per Kwh.
Air conditioner (window)	½–2½ tons	800–2,500 per season	56.00–175.00 per season
Blanket, electric	100–200	25–50	1.75–3.50
Blender, food	200–300	1	.07
Broiler	1,300–1,600	15	1.05
Can opener	175	1	.07
Casserole	1,500	20	1.40
Cleaner, vacuum	400–1,200	5–10	.35–.70
Clock	3–10	2–8	.14–.56
Clothes dryer (electric element)	4,500–6,000	80–400	2.10–7.00
Coffee maker	600–1,000	3–10	.21–.70
Cooling, attic fan	⅙–¾ hp.	60–90	4.20–6.30
Corn popper	450–1,300	1	.07
Dehumidifier	300–650	100	7.00
Dishwasher (with electric heating element)	1,000–1,600	30–45	2.10–3.15
Dishwasher (without heating element)	250	10	.70
Fan, portable	35–210	4–10	.28–.70
Floor polisher	200–400	1	.07
Food grinder	500	2	.14
Freezer, food (5–30 cu. ft.)	300–800	75–250	5.25–17.50
Fry pan, electric	1,000–1,500	5–10	.35–.70

SOURCES: Donald Bates and Harold Cloud, "Energy Requirements of Electrical Equipment," *Agricultural Engineering Fact Sheet No. 1* (St. Paul, Minn.: University of Minnesota, 1969).

Edison Electric Institute, *Annual Energy Requirements of Electric Household Appliances,* EEI-PUB #75-61 (New York: Edison Electric Institute).

TABLE 2-1—Continued

Appliance	Approximate Wattage	Average Kwh. per Month	Monthly Cost in Dollars at $.07 per Kwh.
Furnace, blower	300–600	200–450	14.00–31.50
Furnace, gas (electric control)	25	12–25	.84–1.75
Furnace, oil	300	40–80	2.80–5.60
Garbage disposal	400	1–2	.07–.14
Griddle, automatic	1,000–1,500	10	.70
Grill	650–1,200	5	.35
Hair dryer	200–1,200	½–6	.04–.42
Heater, portable	600–2,000	15–30	1.05–2.10
Heating pad	50–150	1	.07
Hot plate	500–1,650	10–30	.70–2.10
House-heating	10 per sq. ft. floor area heated	12–15 per sq. ft. floor heated	.84–1.05 per sq. ft. per season
Iron, hand (steam or dry)	1,000–1,500	12–18	.84–1.26
Knife, electric carving	100	6	.42
Knife sharpener	125	¼	.03
Lighting, home	7½–300 (bulb size)	50–100	3.50–7.00
Mangle	1,000–1,500	8–15	.56–1.05
Microwave oven	14½	16	1.12
Mixer, food	100–350	1	.07
Radio, console	100–350	5–15	.35–1.05
Radio, table	40–100	5–10	.35–.70
Range	8,500–16,000	100–150	7.00–10.50
Record player, hi-fi, stereo	75–100	1–5	.07–.35
Refrigerator, freezer combination	300–750	100–150	7.00–10.50

TABLE 2-1—Continued

Appliance	Approximate Wattage	Average Kwh. per Month	Monthly Cost in Dollars at $.07 per Kwh.
Sewing machine	75–100	1–2	.07–.14
Shaver	12	1/10	.01
Sunlamp	250–500	3	.21
Television, black and white	100–450	10–120	.70–8.40
Television, color	500–600	45–150	3.15–10.50
Toaster, two-slice	1,000–1,200	4–6	.28–.42
Toothbrush (electric)	1 1/10	1/10	.01
Waffle iron	550–1,300	1–2	.07–.14
Washer, automatic	300–700	3–8	.21–.56
Water heater	1,200–5,000	200–400	14.00–28.00
Water pump (deep)	1/3–1 hp.	10–60	.70–4.20
Water pump (shallow)	1/4 hp.	5–20	.35–1.40

tion per month gives the energy use per month for that appliance. For example, a 100-watt television set operated for 60 hours per month uses 60 × 100, or 6,000 wh. (6 kwh.) per month. If necessary, a separate kilowatt-hour meter can be connected to an appliance to measure its electrical use directly. Sometimes these meters can be obtained on loan from utility companies or electricians.

For most situations, these measurements are not necessary. A number of studies have been done on "typical" homes and farms showing the average monthly electrical use for different appliances. One such study, based on a "typical" family of four, is summarized in Table 2-1. The table gives a list of often-used appliances, their wattage rating, the kwh. consumed, and the cost at a rate of $.07 per kwh.

Make a list of the appliances you use at present, along with their kwh. consumed. (An example is shown in Table 2-2 for which the range is 773 to 1,683 kwh. per month at a cost of $54.11 to $117.81.) Then, by comparing the number of kwh. used with the estimated kwh. available from a wind system you are considering, you can tell whether or not the system will furnish all or part of your electrical needs. Taking an example from Appendix F, you can see that the estimated output of an Enertech model "1800" at an average wind speed of 12 mph is 370 kwh. per month. For the example shown in Table 2-2, this system would supply 20 to 43 percent of the electricity used. In remote areas where utility power is not available, or where a homeowner intends to provide all or nearly all of his or her electricity using a wind system, some choices will have to be made on what appliances to operate with the wind system as the energy source.

One fact you will certainly notice from the example in Table 2-2 is that your electricity bill is not necessarily a "light bill," as it is so often called. In the example given, lighting amounts to only about 6 percent of the total usage. The much-maligned electric toothbrush is not really the villain either. Small appliances generally do not use much power and are only operated for brief periods of time. Major appliances, however, especially electric hot water heaters, ranges, and furnace blowers, consume the greatest amount of

TABLE 2-2
A Sample Estimate of Electrical Use for a Month

Appliance	Kwh. per month	Cost at $.07 per Kwh.
Clothes dryer	75–150	$5.25–10.50
Freezer, food	75–250	5.25–17.50
Furnace, blower	200–450	14.00–31.50
Iron, hand	12–18	.84–1.26
Lighting, home	50–100	3.50–7.00
Range	100–150	7.00–10.50
Record player	1–5	.07–.35
Refrigerator	50–100	3.50–7.00
Water heater	200–400	14.00–28.00
Water pump	10–60	.70–4.20
Total	773–1,683	$54.11–117.81

electricity. Many major appliances also contain heating elements that use great quantities of power.

As an example, compare the clothes washer to the dryer. The washer has a motor for turning the tub and pumping water. The dryer has a heating element to dry the clothes and a motor to turn the drum. The dryer takes 30 kwh. per month to operate, while the washer takes only 3. The difference is in the amount of electricity it takes to operate the heating element. Similarly, an electric range uses about 100 kwh. per month and an electric hot water heater from 200 to 400 kwh. per month. These considerations often prompt the energy-conscious wind power user to consider nonelectrical sources of energy, such as solar energy, natural gas, or wood heat, for all or part of household heating applications, including heating water, cooking, and space heating. Since most home-size wind systems generate in the range of 200 to 600 kwh. per month, powering the heat-using appliances from nonelectrical sources makes it possible to generate all or most of the remaining electrical needs from wind power.

In general, it is cheaper to provide heat from a nonelectrical source if possible. Even though electric heating is supposed to be 100 percent efficient, the total process is only about 30 percent efficient. This is because it takes three pounds (lbs.) of coal back at the coal-fired utility plant to generate enough electricity to be equivalent to the heat available from burning one lb. of coal directly. However, where heating element appliances are being used, it makes sense to use any excess wind power to operate these appliances, as you'll discover in detail in Chapter 7.

Load Demand

The daily variation in home electrical use, known as the load demand, should also be assessed in relation to using wind power. This is especially important if you're thinking of installing a utility-connected wind system and you live in an area where the utility company pays a low rate for any excess electricity that you generate and return to their lines (see Ch. 3 for further information). Excess electricity results from your wind system providing more energy than necessary to run the appliances operating at that time. The amount of excess energy generated can be measured as the percentage of time when extra energy is produced compared to the regular functioning of the wind system, when only enough energy is produced to power just those appliances in use at a given time.

If your electrical use is 500 kwh. per month (or 720 hr.), then the average power use would be 0.7 kw. per hr. This does not

TABLE 2-3
A Typical Daily Load Demand for a Family of Four

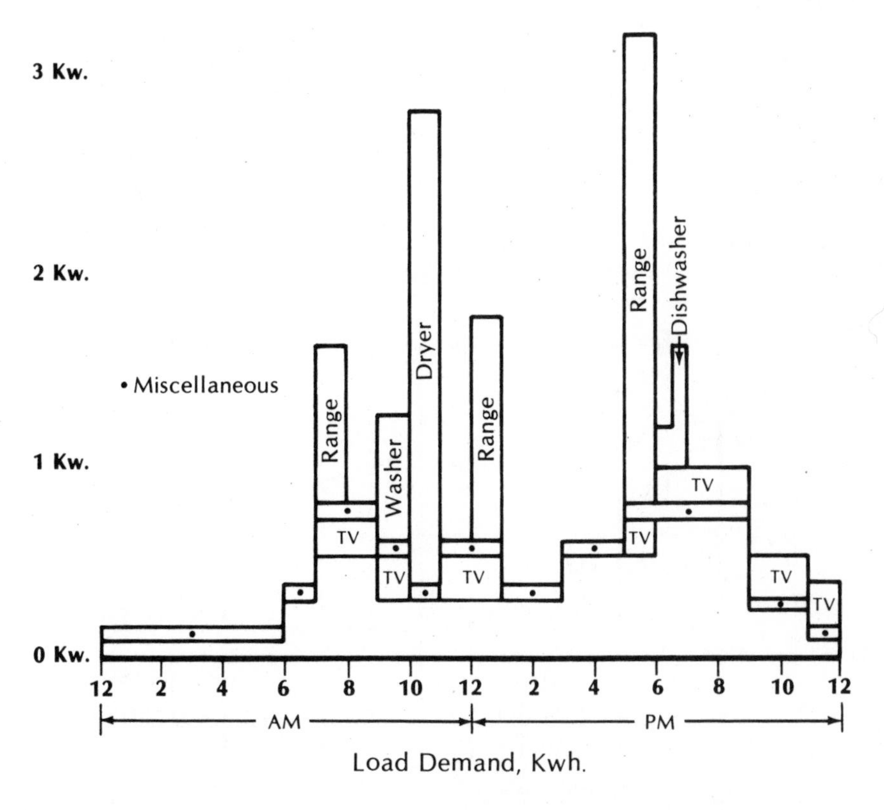

SOURCE: Dr. Henry Kelly, "Application of Solar Technology to Today's Energy Needs" (Washington, D.C.: Office of Technology Assessment, Congress of the United States).

present a completely accurate picture, however, because most of the electrical use occurs at specific times when various cooking appliances or a great deal of hot water is being used. This creates a peak demand for electricity which is several times greater than the average demand. On the other hand, certain appliances are apt to be running throughout the day, thus providing a base load. Such appliances include the refrigerator, freezer, air conditioner, water heater, furnace blower, lights, and any other device that operates fairly continuously.

A refrigerator, for example, operates from 20 to 50 percent of the time. Typically, it will be running from 6 to 15 minutes every half hour. A hot water heater may operate about 100 hours per month, or 3.3 hours per day. Most of this time will occur during the dishwashing, bathing, and clothes-washing activities of the family. Also, in the evening hours lighting provides an almost constant load for a utility-connected wind system. For example, five 75-watt bulbs burning for 5 hours each evening would require a steady load of 375 w.

TABLE 2-4 A Typical Household's Monthly Load Demand

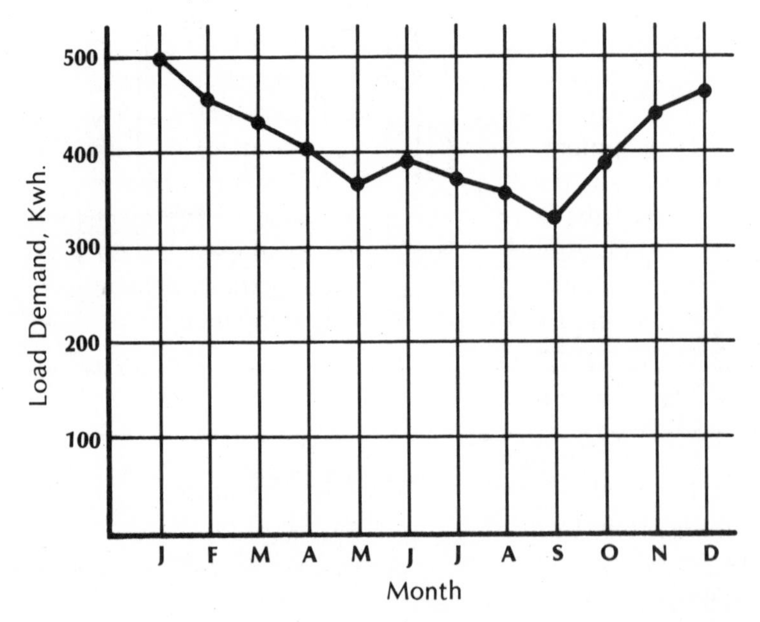

Although normally only a small percentage of the power generated from a utility-connected wind system is fed back to the utility grid, if you intend to choose a system of sufficient size to provide 70 percent or more of your home's electricity, it is advisable that you assess the daily variation in load demand. You can do this by using a strip chart recording wattmeter, available from electrical supply houses and some electricians. The chart recording will show a daily picture of how much power is used at all times.

The household load demand also varies throughout the year. Generally speaking, the greatest demand occurs during the winter months because of increased lighting and heating needs. The furnace blower on a forced-air furnace operates about 30 percent of the time during the cold weather seasons, but not at all during the summer. Fortunately this demand pattern fits the wind patterns in most parts of the country, in which the winter wind speeds are higher than the summer speeds. Homes with large air conditioning loads, however, do not usually match well with the lower summer wind patterns. Likewise, summer vacation homes in areas with low summer winds may not be suitable for a wind system, either.

Wind Power Finances

After determining the energy available in the wind and what your energy requirements are, the next question to answer is, "What will it cost?" Like any other renewable energy device, the cost of the equipment, the *first cost,* is the major cost. After the system is installed, the "fuel"—the wind—is free. To assess the relative value of the wind system, the value of the electricity it generates must be estimated. The "return on the investment" from the wind system includes the value of the electricity generated, the benefits of state and federal tax credits, and the intangible value of the satisfaction of using a nonpolluting energy source.

Payback

After the wind system is installed, the accumulated value of the electricity generated will equal the original cost of the unit at some point in the future. Called the payback time, this period typically varies anywhere from 6 to 12 years.

The payback period is determined by many factors and, in choosing a wind system, all of the possible cost factors should be listed. The cost factors include:

- The initial cost of the system.
- Loan interest rates and taxes to be paid.
- Insurance
- Yearly maintenance and upkeep costs.
- The yearly kilowatt-hour output of the unit at the average annual wind speed of the local area.
- The cost of utility power in the local area, and the expected rate of price increases.
- The expected useful lifetime of the system.
- The expected resale value of the system.

This list may seem overwhelming, but just keep in mind that the effective cost of the wind system can be reduced significantly by taking advantage of available state and federal tax credits, which are discussed in the following chapter.

First Cost

The initial cost of the wind system includes all of the expenses, not just the price of the wind turbine itself. The total system cost includes the cost of the wind turbine and generator; the tower and its accessories and footings; the cost of batteries for storage, inverters for changing direct current (d.c.) electricity to alternating current (a.c.) electricity, and wiring; and the charges for installation. Although you can easily obtain the list prices of equipment from the suppliers, factors such as the cost of wiring are often overlooked. As mentioned before, wiring from the generator to the house can be an important expense, especially for distances over 1,000 feet. Similarly, towers sometimes require up to several yards of concrete in the footings so that the cost of concrete can be significant. This is particularly true if premixed concrete is used. Installation charges may be minimal if you do the work yourself and do not "value" the time. Professional installation charges vary considerably, but they can account for 10 to 30 percent of the cost of the system.

Interest Rates, Taxes, and Insurance

If you borrow money to purchase a wind system, the interest rate paid on the loan is a significant factor in comparing the cost of the wind-generated electricity to other alternatives. The cost of a high interest loan can offset the value of the electricity generated.

In localities where wind systems are taxed as part of your real estate, the yearly addition to your tax bill should be accounted for. Some states have incentive laws that exempt wind systems and other renewable energy equipment from property taxes. Similarly, some insurance companies may require a separate clause for your wind system, and thus an additional fee per year.

Maintenance Costs

Yearly maintenance of a wind system typically involves a once- or twice-a-year inspection and lubrication. This can be done by the owner or, under a service contract agreement, by the installer of the system. In the latter case the yearly service agreement cost can be used as the maintenance cost figure. In general, it is reasonable to assume that there will be some major component repair or replacement expenses during the life of the system such as blade replacement, or a generator overhaul. To allow for this probability, assume a value of about 1 percent of the initial system cost as the average yearly maintenance and repair bill. For example, for a $5,000 system, the annual upkeep would be $50.

Average Output

The average yearly energy output of a wind system can be obtained from the manufacturer's data, if available, or from estimated outputs listed in Appendix F.

Utility Costs

Where utility power is available, the value of wind-generated electricity is closely tied to the cost of power from the utility company. In remote areas, it is generally compared to the cost of diesel-generated electricity.

It is important to estimate the future cost of electricity as well as the present cost since nearly all indicators show a steady increase in the cost of utility power and fossil fuels far into the future. Historically, the average cost of electricity for homes in the United States declined from $.175 per kwh. in 1900 to a low of $.021 in 1970. One industry forecast indicates that electricity costs will rise at a rate of 1 percent less than the Gross National Product (GNP) through the year 2000. This means that prices would triple from the year 1980 to the year 2000, increasing at a rate of 6 to 7 percent per year. This is a minimum estimate, however, since it assumes an uninterrupted supply of fossil fuels during that period. A more pessimistic estimate is that electricity costs will increase at an average yearly rate of 10 to 15 percent for the next 20 years or more.

The future is unpredictable, of course, but your personal best estimate of future energy prices is important in comparing the cost of wind-generated electricity to other sources. If utility power costs $.05 per kwh. at present it will cost $.152 in 20 years if prices increased at 6 percent per year. At a rate of 10 percent per year, the cost in 20 years would be $.306. Clearly, the rate at which prices do increase will have a large effect upon the relative value of wind-generated electricity.

In semiremote areas the cost of bringing in utility lines for distances of a mile and over are considerable, up to $10,000 per mile. In such cases, the initial cost of the wind system is sometimes equal to or less than the cost of the utility connection, making wind energy a very attractive alternative.

Expected Lifetime of the Wind System

Designers of wind equipment in general aim to make the equipment durable enough to last at least 15 to 20 years or more. Although a lifetime of 15 to 20 years is assumed in estimating the value of a wind system, there are no long-term statistics yet available to corroborate such estimates. Stories of pre-Rural Electrification Administration (R.E.A.) wind systems and their durability abound but no "frequency of repair" records were kept or made public. Obviously, it is important that the lifetime of your wind system should at least equal the payback period. Long-term testing being conducted at the Rocky Flats site in Golden, Colorado will help to determine projected lifetimes more accurately. In the meantime, the more conservatively designed a wind system is, the more likely the 20-years-plus lifetime figure will hold true. With regular maintenance and perhaps an overhaul every 10 to 15 years, there is no reason why a well-designed system cannot last much longer.

The lifetime of a battery storage system should be estimated separately from that of the wind turbine. A set of new deep-cycle batteries (such as those used in golf carts) may last up to 10 years in a wind system even though they are generally guaranteed for only

5 years in electric vehicle use. A set of industrial, lead-acid batteries can last from 15 to 25 years (see Ch. 5 for further information).

Resale Value

For purposes of computing the payback period of the wind system, the unit is usually assumed to have no value at the end of its lifetime. In reality, though, it is reasonable to assume that the unit will have a resale value even after 15 or 20 years. During the period of cheap fossil fuels in the 1950's, the resale value of wind systems fell to scrap price levels. Many wind generators in perfect working order were scrapped for the value of the copper wire in the generator. In other cases, the units became targets for rifles or were pushed off the top of the tower (a dangerous practice at best) to make way for a TV or CB antenna. At the time, utility power was averaging $.02 to $.03 per kwh. with $.01 and less being forecast.

In the 1970's, however, the value of pre-R.E.A. wind systems increased almost directly in proportion to the cost of utility power. In many cases, their value rose to their original prices, including an allowance for inflation (consumer prices increased roughly by a factor of 3 from the 1950's to the 1970's). In part this increased value was due to the relative scarcity of available machines at the time. However, inflation and the rising costs of utility power were also considerable factors.

Any well-built machine or durable product becomes an investment with a sustainable resale value during periods of inflation. To the extent, then, that inflation and increased energy costs are a part of our future, the value of a well-designed wind system should remain stable for many years. Also, the tower is likely to last longer than the generator, and towers represent a considerable part of the cost of wind systems. But since this resale or end-of-life value is not a "sure thing," it is best to assume that the system will have no value at the end of its expected useful lifetime. Any value at that time will simply be a bonus.

Payback Analysis

Having noted all of the costs involved in purchasing and maintaining a wind system, you can now calculate the payback period, which gives an idea of the relative value of the wind system compared to utility power or other power sources. Although there are many possible methods of estimating the economic value of wind power, the two methods considered here are the *average system cost* method and the *cumulative energy value* method. The former method assigns value to the wind-generated electricity based upon the cost of the wind equipment. The latter method assigns the value based upon money saved in the utility power not purchased.

Either way of assigning a value to the wind-generated electricity is useful to compare wind systems as well as to estimate payback periods. Keep in mind that a system that is initially more expensive may prove to be more economical if it has greater output at the average wind speed for your local area.

The payback periods themselves should not be taken as absolute numbers. Since the calculations are based on many assumptions, the method is most useful in comparing one wind system to another, or wind power to other energy sources.

Interest Tables

To make allowances for inflation rates and the future value of electricity, the Future Value of Electricity Table in Appendix D can be used. For each rate of interest or inflation (as the case may be) from 1 to 14 percent, there are three columns in the Future Value Table. Column A is a compound interest table and shows the future value of electricity if electrical costs increase at the given percentage. If electricity costs $.05 per kwh. now, and it increases at a rate of 8 percent for the next 15 years, what will it cost in the 15th year? From the table (8 percent, Col. A, Year 15), the cost would be 2.94 × $.05 per kwh. or $.147 per kwh.

Column B is a cumulative value table and can be used to find the cumulative value of the electricity generated by a wind system over a period of years. For example, if utility power prices increase at a rate of 10 percent per year, then each dollar saved in the first year will amount to a total savings of $57.28 over 20 years (10 percent, Col. B, Year 20).

Column C is an average value table and simply shows the average value of the figures in Column B. The average value of the electricity saved in the above example would be $2.86 over 20 years for each dollar saved in the first year.

Average System Cost Method

The most straightforward way of evaluating the cost of wind power is to average the cost of the wind system over its expected lifetime. The total cost of the unit divided by its output over its expected lifetime gives an average cost per kwh. This method is useful in comparing different models of wind systems since the lowest cost wind system may not necessarily produce the lowest cost per kwh. produced. This method is also useful in evaluating the cost of wind power in areas where utility power is not available and comparisons to the cost of utility power are not meaningful.

Consider a wind system that costs a total of $5,000 after tax credits are accounted for (see Ch. 3), and that generates 220 kwh. per month (2,640 kwh. per year) in a 12-mph average wind speed area. Assuming a 15-year lifetime for the system, 39,600 kwh. of electricity would be generated over that period. The average cost of the wind-generated electricity would then be $5,000 ÷ 39,600 kwh. or $.126 per kwh.

Another wind system may cost $4,000 after tax credits but generates only 150 kwh. per month in a 12-mph average wind speed area. At that rate the average cost of the second wind system would be $4,000 ÷ 27,000 kwh. (15-year output), or $.148 per kwh. In this case, the second wind system is not the better buy even though its initial cost is lower, because its average power cost is $.148 as opposed to $.126 per kwh.

Actually, if the wind power is stored in batteries the lifetime of the batteries should be considered separately from the lifetime of the rest of the system. For example, if the first wind system uses a battery bank costing $1,500 (before tax deductions) and with an expected lifetime of ten years, the cost of purchasing a new set of batteries in the tenth year must be considered.

If battery prices increase at 10 percent per year, the battery bank would cost 2.36 × $1,500, or $3,540 to replace in the tenth year (App. D, Future Value Table, 10 percent, Col. A, Year 10). Only one-half of that cost will be used up in the last five years, so an additional $1,770 is added to the cost of the system to account for battery replacement in ten years. The power cost then becomes $6,770 ÷ 39,600 kwh., or $.171 per kwh.

This cost can also be compared to the average cost of utility power over the same period. If utility power costs $.08 per kwh. and increases by 10 percent each year, then the average cost of utility power over the 15-year period would be 2.12 × $.08, or $.17 per kwh. (App. D, Future Value Table, 10 percent, Col. C, Year 15). Of course, this cost of utility power does not take into account the expense of running power lines to a remote site, which would raise the cost per kwh. for utility power substantially.

Cumulative Energy Value Method

Each kwh. generated by the wind system represents electricity that does not have to be purchased from the local utility company. After several years—the payback period—the *cumulative value* of the electricity saved equals the amount originally paid for the wind system.

Take the example of the first wind system discussed above. At $.08 per kwh., the 2,640 kwh. generated by the wind system replaces $211.20 worth of utility power in the first year. If utility power prices increase by 10 percent each year, the savings are 1.1 × $211.20, or $232.32 in the second year and the accumulated value is $443.52. In the 12th year the accumulated savings are 21.38 × $211.20, or $4,515.46, and in the 13th year, the savings are $5,178.62, slightly greater than the original systems cost of $5,000. Thus, the payback period in this example is 13 years. If the lifetime of the system is 20 years, the total savings in electricity cost would be 57.28 × $211.20, or $12,097.

It is not necessary to find the accumulated savings for each year as in the above example, however. In general, the payback period can be found by dividing the system cost by the first year's savings and looking up the equivalent year in the Future Value Table Column B. For the example, $5,000 ÷ $211.20 is 23.67. From the Future Value Table (10 percent, Col. B.) 23.67 is between 21.38 and 24.52 or between the 12th and 13th years.

The payback period varies considerably with the assumed rate of inflation. In the example, the payback period varies from 11 years at 14 percent inflation rate to 15 years at a 5 percent inflation rate.

The payback period also varies considerably with the average wind speed. If the output of the hypothetical wind system being discussed is 120 kwh. per month at an 8 mph average wind speed and 350 kwh. at 14 mph, the payback period would vary from 18 to down to 10 years.

If the money for the original purchase of the wind system is borrowed, the cost of borrowing should also be calculated. The annual repayment rate for each dollar borrowed at different interest rates is shown in Appendix D in the Loan Repayment Table. If the $5,000 in the above example is borrowed at a rate of 12 percent annual interest for a period of five years, then the annual loan payment would be 0.277 × $5,000, or $1,385. The real cost of the wind system would then be $6,925 (5 × 1,385) and the payback period would be 15 years instead of 13 years at a 10 percent rate of inflation. An example is shown in Table 2-5 of a complete payback analysis, including the impact of loan rates, maintenance costs, and state and federal tax credits.

TABLE 2-5
A Sample Payback Calculation Using the Cumulative Value Method

1. Cost factors	
System cost	
Wind turbine generator	$4,000
Tower	1,500
Installation/miscellaneous	1,000
	$6,500
Expected lifetime	20 years
Output	400 kwh. per month at 12 mph average, 4,800 kwh. per year
Utility power	$.09 per kwh., increasing at 9% per year
Interest rates	6-year loan at 11%
Maintenance, insurance, property taxes	$75 per year
Available federal and state tax credits	
Federal: 40% of first $10,000	
State: 20% of first $10,000	

TABLE 2-5—Continued

2. Net cost calculation

System cost	$6,500.00
Less federal tax credit	2,600.00 (.4 × 6,500)
Less state tax credit	1,300.00 (.2 × 6,500)
Subtotal	$2,600.00
Plus loan cost*	1,081.60
Net system cost	$3,681.60

3. First year's savings

Electricity saved	$432 ($.09 × 4,800 kwh.)
Less maintenance and related costs	75
Net savings	$357

4. Payback analysis

$3,681.60 ÷ $357.00 = 10.31

From Appendix D, Future Value Table (9%, Col. B), the payback period is thus 7½ years.

5. Lifetime system savings

From Appendix D, Future Value Table, (9%, Col. B), the value of the electricity generated by the wind system over 20 years is 51.17 × $357 = $18,267.

*Loan cost (From Loan Repayment Table, Appendix D)

.236 × $2,600 = $613.60—the yearly loan payment on the 6-year, 11% loan.

6 × $613.60 = $3,681.60—the total principal plus interest cost.

$3,681.60 − $2,600 = $1,081.60—the total cost of the loan (interest charges).

Although there are many things to consider in assessing the economical feasibility of wind power at your site, you can assume in general that any site with an average wind speed of 12 mph or greater with utility power costing over $.05 per kwh., or in an area where utility power costs $.10 per kwh. or more, and wind speeds average over 10 mph, will prove to be "economical," meaning that payback periods will not be more than six or seven years. Furthermore, at remote sites where utility power is not available, wind power will usually prove to be lower in cost than a diesel generator system, due to the high operating costs of diesels.

In other cases, where utility rates or wind speeds are relatively low, the payback period may be as long as 10 to 15 years. For many people, however, strict economics is not the only consideration. If energy independence is your goal, or if you gain satisfaction from using a renewable energy source, or if your estimate of the future cost of utility power is much higher than that used in the examples here, then wind power still makes sense for you. In effect, you are either assigning a higher intrinsic value to wind-generated power or projecting a much higher rate for the price of future utility power.

Chapter 3

Wind, the Law, and Utility Regulations

Installing a wind system is not solely a personal decision for you to make. Before going ahead with the purchase of equipment, it is best to find out whether any zoning ordinances, building codes, or other legal requirements exist that might be a barrier to your installation. There are also electrical codes to follow and, for utility-connected systems, utility company approval will be required. Your insurance policy may have to be modified and you will want to make inquiries to see if the wind system will increase your property tax evaluation. In most cases, legal considerations should not prevent you from installing a wind system. It is true, however, that the more densely populated an area is, the greater the possibility that a legal barrier will exist.

On the plus side, it is worth learning about the federal and state tax incentive programs that can save you money. In addition, you may be able to sell excess electricity from your utility-connected system back to the utility company under guidelines set by the federal government.

Zoning Ordinances

Of all the possible legal considerations, the most potentially restrictive ones are zoning ordinances and building codes. In

TABLE 3-1
Checklist of Institutional and Legal Concerns for the Home Wind Power User

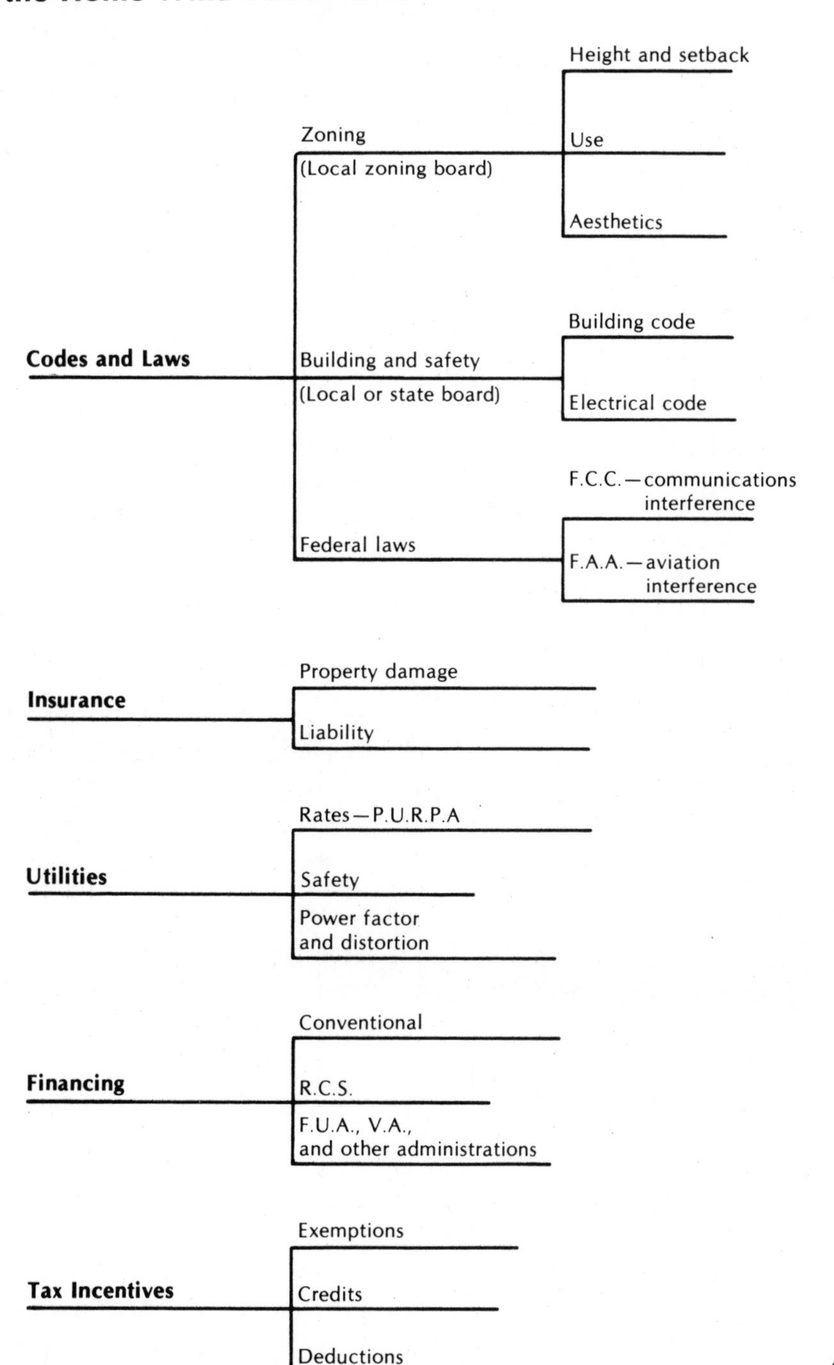

densely populated urban areas, the chances of being able to install a wind system are not very good, unless the installation is considered a "demonstration" project. In suburban areas and towns, zoning and building restrictions vary greatly with each locality. Zoning laws define how a piece of property can or cannot be used. As applied to wind systems, these restrictions generally concern *height, setback,* and *aesthetics.* The existence of a restriction does not necessarily prevent the installation of a wind system, however. A hearing can be scheduled with the local zoning board to obtain a variance to allow the installation. Variance hearings call for pro and con testimony by all parties concerned.

Regardless of what the local ordinance permits, the testimony of neighbors usually is weighed heavily by the board. For this reason, many "pioneers" who are the first to install a wind system in their community have found it advantageous to spend some time educating their neighbors and the community prior to any hearings. In some cases this groundwork has led to wind systems being specifically mentioned in the local zoning laws.

Height Restrictions

Zoning ordinances often contain restrictions on the height of structures, particularly in residential areas. Typically, communications towers for TV, CB, or ham radios are mentioned with an attendant maximum allowable height, and their towers are defined as permissible accessory structures. In such cases, it may be desirable to have the wind system classified under the same permitted use as communications towers. A possible point of contention is whether the height ruling applies to the top of the tower, to the highest point swept by the rotor, or to the center of the rotor. Height restrictions are often in the 15- to 45-foot range, so it may be necessary to obtain a variance in order to get the rotor high enough to be above turbulence from nearby buildings and trees.

In addition to local zoning ordinances, the Federal Aviation Administration (FAA) requires notification before building structures over 200 feet in height. While this law would affect few, if any, home-size wind systems, the FAA also limits the height of structures within an area of about four miles of any airport runway. In such a case, it would be necessary to contact the FAA to find out the allowable height for that particular location.

Setback Restrictions

Setback requirements state the minimum distance that a structure can be from the property line, and these requirements are closely related to height restrictions. A particular zoning board may have to decide if the setback requirement applies to the location of the nearest leg of the tower or to the nearest guy wire on a guyed

Photo. 3-1 Douglas Brooks's Home Wind Power System in Central Michigan—Brooks overcame setback restrictions for his 100-foot tower and system by citing that hazards that his system might present already existed in his residential/commercial area.

tower (see Ch. 4). If there is no height restriction in the local ordinance but there is a setback requirement, it may simply be a matter of locating the tower so that if it falls over, it won't touch any neighboring property line.

This was the case with Douglas Brooks, who installed a wind system on a 93-foot tower in the middle of a medium-size, central Michigan town. When Brooks applied for a permit to put up the tower, he met resistance from the town zoning board, which ruled that the tower could not be raised because it could fall on adjoining property. Brooks appealed and won his case by arguing that advertising signs on adjoining property could fall on *his* property. He was probably aided by the fact that his house was in a residential/commercial area rather than a purely residential area.

In another case involving setback requirements, James Campion IV, of Hanover, New Hampshire, was denied a permit until an ordinance could be drawn up that would eliminate the height restrictions on the wind system but would base the setback on the expected maximum distance that a blade could be thrown from the rotor. Campion had at first obtained a zoning variance to install a tower. However, the permit was then denied at a public hearing when a group of neighbors expressed concern about the aesthetics of the tower, possible TV reception interference, and the possible hazard of a blade being thrown from the rotor during a storm.

From the safety point of view, zoning requirements that single out wind systems for more restrictions than for any similar structures, such as in the Campion case, are unnecessary. A wind system that has been properly engineered, tested, and installed is in much less danger of falling during a storm than are trees or power lines. An extremely strong wind, of tornado or hurricane force, can damage towers just as it could damage any other structure. However, the most likely damage from such forces is that the tower would bend over near the top or middle rather than actually topple like a tree. This would occur in winds exceeding 100 mph, since most towers are designed to withstand winds below these speeds. In contrast, trees can blow over in even a 60-mph wind.

The chances of a blade being thrown from a rotor during a wind storm are also minimal (see Ch. 4) By far, the much more likely occurrence is that a tree limb will fall on a structure during a storm and damage it. Despite this possibility zoning boards have not seen fit to restrict the planting of trees or require them to be set back from property lines based upon the distance that a branch will fall during a storm.

Related to setback requirements is the question of *wind rights,* or access to the wind. What can be done if a tall building that interferes with the operation of the wind turbine is built on the adjoining property? At present, there is no body of law which addresses the question of wind rights. The only possible remedy is to obtain a negative easement from the adjoining property owner. The easement would be an agreement to restrict the height of structures on the adjoining property that are within a certain distance of the proposed wind system. For example, using the rule of thumb that the wind system should have unobstructed wind for a distance of 10 times the rotor diameter, a distance of 150 feet would be required for a system with a 15-foot-diameter rotor. This distance would present no problem on a large suburban lot but would clearly require an easement on a small lot, say, in a subdivision of 75- to 100-foot-wide lots. Such an easement would be expensive, since it would limit the use and thus the possible resale value of the property.

Actually, in purely residential areas where multistory apartment buildings are not allowed, the question of wind rights should be of minimal concern. Once a permit has been obtained for a tower sufficiently tall to be above the nearby buildings and trees, the only consideration might be that a neighboring tower for communications or wind power use would interfere with yours. Again, lot size would be the determining factor in such a case. Possible blockage of access to the wind would be a concern in any area where multistory buildings are allowed and the lot size is relatively small.

Aesthetics

Your wind system may be a beautiful sight to behold for you but it may be considered an eyesore by a neighbor. Ordinances which require that structures be aesthetically compatible with local architecture can be stumbling blocks for the potential wind power user. In such cases it may again be helpful to have the wind tower classified under the same category as a communications tower. If

TV and CB towers are allowed, it can be effectively argued that there should be no objections to a wind system tower. This is particularly true considering the fact that the towers for use with wind systems made by at least one tower company are almost identical in appearance to towers for communications use.

In other cases it may be best to deliberately choose an "aesthetically pleasing" tower. Although this is obviously a matter of taste, many people consider a self-supporting steel pole or a similar type of tower to be more appealing than a lattice tower with guy wires. This is similar to instances in which painted steel utility poles are used instead of lattice towers in some urban areas.

Since defining what is pleasing or compatible is highly subjective, the question of aesthetics is one easily used by someone wishing to oppose the installation of a wind system in the neighborhood. Such opposition was experienced by Joe Laleman of East Moline, Illinois. Laleman lives in a residential area where auxiliary structures such as towers are limited to a height of 15 feet. Laleman applied for a variance to install a 60-foot tower for his wind system. He pointed out that the tower would be positioned so that it was more than 60 feet from any adjoining property line.

Not taking any chances, Laleman submitted a 20-page report to educate the zoning board, obtained support from several groups, and solicited a letter from the local power company stating that it had no objections to the project. The local newspaper wrote favorable reports on the proposed wind system and what Laleman was doing. Despite the favorable publicity, some of Laleman's neighbors attended the public hearing and spoke against the granting of a variance, calling the wind generator an eyesore or saying that it would look like "an oil field." With this opposition, the board voted against the variance. In effect, the board ruled that the very fact of opposition on the basis of aesthetic appearance was sufficient cause to deny the permit.

William Stuart of upstate New York won his case on the basis of aesthetics. Stuart lives in what is called in his locality a "Class I" residential area where hearings are required before both the zoning board as well as an architectural review board. Stuart contacted the local newspaper after applying for his permit and told his side of the story. An interested reporter wrote a favorable story which was helpful in publicizing the facts about wind power.

About the hearing, Stuart says, "I had one neighbor who was a nuclear power design expert who objected, saying he was worried about his kids climbing the tower! But overall, there was hardly a dissenting vote."

To make sure that the wind system met the aesthetic requirements, Stuart painted everything white—the tower, generator, rotor, and even the tail vane, since a logo on the tail vane might have been considered advertising. He also left off the lower ladder section on the tower so that children could not climb it. Some neighbors expressed concern about noise and electrical interference with TV reception, but neither concern has been borne out, Stuart reports.

Noise Pollution

The question of whether a wind system will create a noise problem may also be brought up at zoning hearings. While the rotor does create a slight "whooshing" sound when operating, it is not a loud or objectionable noise and it is generally masked by the sound of the wind itself. A rotor that does create considerable noise is either not designed properly or is out of balance.

Building and Safety Codes

In many areas, a building permit must be obtained after the zoning permit has been approved before erecting a tower. Also, electrical codes apply to wind systems just as to other electrical devices, and a permit or inspection may be required for the electrical system.

Building Codes

Building codes address themselves to the safety of a structure and its construction. Codes are generally in force in any incorporated area, and some states have building codes that apply to all areas.

As with zoning ordinances, building codes vary greatly from locality to locality. They are generally derived in part or fully from one of the model codes, such as the Uniform Building Code of the

International Conference of Building Officials, also known as ICBO; the National Building Code of the American Insurance Association; the Basic Building Code of the Building Officials Conference of America, also known as BOCA; and the Southern Standard Building Code of the Southern Building Code Conference.

Your local building code may be the older "specification" type, in which materials to be used or not used are listed, but it is more likely one of the "performance" codes listed above. The performance codes may specify the structural and foundation loads that the tower must be capable of withstanding. This includes snow loading, which is generally minimal on a tower, and wind loading, which is the force of the wind that the tower must be capable of withstanding. Some state or local building codes may require that a registered professional engineer (P.E.) prepare the structural design plans.

In those cases in which a tower from an established manufacturer is being installed, the various loads and stresses that the tower is capable of withstanding are usually spelled out in drawings supplied by the manufacturer. Copies of these drawings can be supplied to the building code office when applying for a permit. If the code requires the signature of a registered engineer, the manufacturer's data can be supplied to the engineer for approval. Normally, the cost should be nominal for such a service. In most states, however, once a model of a tower design is approved in a particular area it does not have to be certified each ensuing time another individual tower of its model is installed. Also, many states have reciprocal certification agreements so that approval in one state is valid in another. If a building code official requires proof of this certification, contact the manufacturer of the tower.

In places where construction of a tower by the owner requires a building permit, the extra cost of having the tower plans approved or designed by a registered engineer should be considered. These services are often available through architectural or contracting firms, or simply through an engineering company.

Another safety provision that may be included in a local building code is the "attractive nuisance" provision that requires a protective enclosure to be built around the base of the tower to prevent children from climbing it. This is similar to requirements for fencing in swimming pools to prevent children from falling into them.

Electrical Codes

The building code will also contain an electrical code section that lists wiring specifications and inspection requirements. Electrical codes are based on the National Electrical Code with state or local provisions added (see Ch. 9 on wiring for wind systems). Some codes exempt from inspection low voltage wiring, such as for doorbell or burglar alarm systems. If that is the case, a low-voltage wind system could be exempt. Similarly, devices that plug into an outlet are often considered appliances and thus not subject to inspection.

The decision as to what devices are exempt from permits or inspections is not the individual's to make, however. It is best to check with the local electrical inspector or state electrical board to find out what requirements will have to be followed before any work is done. Also, even if the wind system is not connected to utility power, the electrical code still applies.

The Federal Communications Commission

While electrical safety is regulated through the state or local building code, electrical (or electromagnetic) interference is regulated by the Federal Communications Commission (FCC). An electrical generator that causes "harmful interference" to nearby communications reception—radio, TV, CB radio, and the like—is subject to regulation by the FCC. The harmful interference results from radio frequency energy (radio waves) "emitted" or radiated through space by the generator. In general, any electrical generator used in a wind system should be constructed so that the electrical interference or "noise" that it produces in communication equipment is no greater than that allowed for any other type of generator. If it does cause unacceptable interference, the manufacturer would also be subject to FCC regulation, just like manufacturers of standby generating plants.

Electrical interference from a wind system generator is similar to the noise heard on a radio or television due to a passing car or motorcycle. To date, I have not heard of any electrical interference problems with wind systems, but radio frequency energy is sent out at some level by any electrical generator. However, it is not commonly noticeable if a generator is properly designed, especially with the necessary filtering. Also, modern electronic radio and television receivers are designed to eliminate most interference. For example, I have noticed that slight arcing at the carbon brushes on my pre-R.E.A. Jacobs d.c. generator causes some interference on my "antique" vacuum-tube radio, but not on my modern transistor radio.

Another form of communications interference that a wind system may produce is the scattering of television signals by rotors with metal blades. Television signals travel in a direct "line of sight" from the TV station's antenna to your TV receiving antenna. A rotating metal blade located in that line of sight could scatter some of the signals and interfere with reception. This would be most likely in a fringe reception area in which your primary channels are channels 2 through 6. If you live in such an area you may want to use wood or fiberglass blades with your wind system. The wind system manufacturer or dealer can also advise you on whether or not any such problems have been reported with that particular system.

The possibility of a wind system scattering TV signals is not usually a concern for people with home-size systems. However, very large turbines with metal blades can cause problems in some cases. It was for this reason that cable TV was installed for the residents of Block Island, just off the coast of Rhode Island. The United States Department of Energy (DOE) built a large, 200-kw. wind turbine with 125-foot-long blades that scattered the island resident's TV signals until the TV cable was finally installed.

Insurance and Liability

You can obtain property damage insurance that covers any losses to the wind system due to fire, vandalism, lightning, and other catastrophes. But not every insurance company will cover a wind system, while some will require special clauses or "riders." So, it is best to check with your insurance company beforehand. Generally, the wind system will be considered an "adjacent" or "auxiliary" structure, like a garage, as opposed to being considered personal property.

There are some exceptions to watch for when inquiring about insurance. Most homeowner policies exclude auxiliary structures that are used for commercial purposes. A utility-connected wind system in which power is sold back to the utility could conceivably be construed as a commercial activity. Also, some policies exclude antennas and masts from coverage for damage due to wind and weather. The wind system tower could fall into this category, so this would have to be investigated in considering a particular policy. Another possible exception are "acts of God" or "infrequent risk" clauses which may exclude coverage of the wind system. In any case of exclusion or doubt, it is possible to add a specific rider to the policy, or, if the company objects, to seek another company. The cost for this insurance is usually nominal, only a few dollars per $1,000 of value.

Besides property damage insurance, you should have liability protection to cover any personal injury or damage to another person's property due to the operation of the wind system. Again, the wind system may be covered under your homeowner's policy, but it is best to check with the company to make sure that it is, so that you can add a rider clause if necessary.

A notable exception of liability insurance is that in most policies, helpers engaged in erecting the tower and installing the wind system are not covered. A paid contractor would be covered under the contractor's own workmen's compensation or liability insurance, but other helpers, such as interested friends, are not covered by most homeowner's policies. If personal injury or property damage results from an inadequate design or from negligence, the manufacturer or supplier may be liable. Besides being liable due to negligence, a manufacturer or supplier may be liable if the wind system does not live up to its warranty, and damage or loss results. The warranty may be a written warranty or it may be an expressed warranty such as advertising or a sales brochure. This could also apply to claims about the lifetime or energy output of the wind system.

Tax Incentives and Financing

What financial aids are available to the potential wind power user? Generally, loans for the purchase of wind equipment are available through normal channels of credit. Also, various state and federal tax incentive programs are available that in effect reduce the amount that you, as a consumer, pay for the wind system.

Loans

Many wind systems are purchased through the use of personal savings. However, it is possible to obtain financing through a regular bank loan or through special loan programs such as from the Farmers Home Administration (FmHA) or the Veterans Administration (VA). Although these institutions may not have funds set aside specifically for wind systems, it is possible to purchase the equipment as part of a home construction or improvement loan. Of course, approval varies with local conditions and the buyer's qualifications. As discussed in Chapter 2, special attention should be paid to the effect of the interest rate and the number of years of the loan on the payback period of the wind system.

Residential Conservation Service

Another avenue for help in financing a wind system is through the Residential Conservation Service (R.C.S.) program which is available through many utility companies. The R.C.S. program was mandated as part of the National Energy Conservation Act of 1978. The legislation directs utilities to offer help to customers who wish to install energy conservation or renewable energy devices, including wind systems.

As part of the program, utilities are required to make available the services of an auditor who, in the matter of wind systems, will offer help in siting and in evaluating the economic benefits of using wind power. An additional feature of the R.C.S. program is that utilities are required either to offer you help in obtaining financing for a wind system through regular channels or to offer financing themselves with payments being made as part of your monthly utility bill. For further details on the R.C.S. program check with your state energy office or your local utility company.

Tax Incentives

Available tax incentives are well worth considering when purchasing wind equipment. Incentives may come in the form of a tax credit, a deduction, or an exemption. A tax credit allows you to deduct a percentage of the cost of the wind equipment from the amount of income tax you owe. This is different from a tax deduction that allows you to deduct all or part of the cost of the system from your gross income figure before you compute your taxes. Finally, some states have tax exemptions available which exempt the wind equipment from either sales and/or use taxes, or from property taxes.

Tax Credits

The National Energy Act of 1978 allows you to deduct 40 percent of the first $10,000 spent on your wind system from your federal taxes. For example, if the wind system costs $8,000, the tax credit would be $3,200 (40 percent of $8,000). This $3,200 can be deducted from the amount of federal taxes owed for that year. If the credit is more than the amount of taxes owed, the remainder of the credit can be carried over to the next year.

To claim the tax credit, IRS Form 5695 must be filled out. The claim can be made only if the equipment is new, for use in your primary residence, and is expected to last at least five years. The federal tax credit is available only for equipment purchased from April 20, 1977 through January 1, 1986.

Several states have tax incentive programs that allow credits or deductions for the installation of renewable energy equipment or energy conservation devices. However, not all of these states include wind energy as one of the allowable deductions. As with the federal tax credit, these programs generally have termination dates after which the program will be reviewed for renewal (see Table 3-2). Check with your state's department of revenue for updates and changes in your state's tax program.

In Minnesota, the tax credit is 20 percent of the total cost of the wind system, with a maximum credit of $2,000 allowed. Here, the state tax credit of an $8,000 system would be $1,600 (20 percent of $8,000) and the wind system would cost only $3,200 after state and federal credits were claimed. This is only 40 percent of the initial cost for the given example.

Tax credits vary widely from state to state, however. Again, you should contact your state's department of revenue for the correct guidelines on tax credits.

(continued on page 64)

TABLE 3-2 State Wind Energy Tax Incentives

State	Program Termination Date	Income Tax*	Property Tax*	Tax Exemptions Sales	or Use
Alaska	12/31/82	C 10% of cost, maximum of $200	...	...	...
Arizona	12/31/83	C 35% of cost, maximum of $1,000	D	...	X
California	12/31/83	C 55% of cost, maximum of $3,000 deducted first	...	...	...
Colorado	...	C 30% of cost, maximum of $3,000	...	...	...
Connecticut	...	...	D, for 15 yrs.	X	X
Hawaii	12/31/81	C 10% of cost	D, 5% of value	...	...
Idaho	...	D 40% of cost, 1st yr. 20% for next 3 yrs. maximum of $5,000 in 1 yr.	...	...	...
Illinois	...	...	D	...	...
Indiana	12/31/82	C 25% of cost, maximum of $3,000	...	...	...

SOURCES: Stephen B. Johnson, "State Approaches to Solar Legislation: A Survey," *Solar Law Reporter,* vol. 1, no. 1, May/June 1979, pp. 55–127.

National Heating and Cooling Center, *Solar Legislation* (Rockville, Md.: National Heating and Cooling Center, 1979).

NOTE: Contact your state Department of Revenue for further details. (Not all states have programs as of this writing.)

*C = Credit
D = Deduction

TABLE 3-2—Continued

State	Program Termination Date	Income Tax*	Property Tax*	Tax Exemptions Sales or Use	
Kansas	7/1/83	C 25% of cost	C, 35% refund up to 5 yrs.	. . .	. . .
Maine	. . .	C 20% of cost or $100, whichever is less	. . .	. . .	. . .
Massachusetts	12/31/83	C 35% of cost, maximum of $1,000, federal tax credit deducted first	D, 5% of actual value for 20 yrs.	. . .	. . .
Michigan	12/31/83	C Variable	D	X	X
Minnesota	12/31/82	C 20% of cost, maximum of $2,000	D	. . .	. . .
Montana	. . .	C Variable	. . .	. . .	. . .
Nevada	. . .	. . .	C	. . .	. . .
New Hampshire	. . .	. . .	D	. . .	. . .
New Jersey	. . .	. . .	D, 5% of actual value	X	X
New York	. . .	. . .	D	. . .	. . .
North Dakota	. . .	D 5% of cost for 2 yrs.	. . .	. . .	. . .
Ohio	12/31/85	C 10% of cost, maximum of $1,000	D	. . .	. . .
Oklahoma	1/1/88	C 25% of cost, maximum of $2,000	. . .	. . .	. . .

*C = Credit
D = Deduction

TABLE 3-2—Continued

State	Program Termination Date	Income Tax*	Property Tax*	Tax Exemptions Sales or Use
Oregon	1/1/85	C 25% of cost	D, 5% of actual value	... ...
South Dakota	...	...	C	... ...
Tennessee	...	...	D	... ...
Texas	...	...	D	X ...
Vermont	...	...	D	... ...
Wisconsin	12/31/84	D 24% 1979–80 18% 1981–82 12% 1983–84	...	... ...

*C = Credit
D = Deduction

Tax Deductions

Some states, such as Idaho, permit tax deductions rather than credits. In Idaho, you can deduct 40 percent of the system's cost from your gross income tax, and 20 percent during the next three years, with a $5,000 maximum deduction allowed for each year. If your gross income is $17,000 and you purchase an $8,000 wind system, you would pay state income taxes on an adjusted gross income of only $13,800 instead of $17,000 in the first year. The second through the fourth year your adjusted gross income would be $15,400. This not only reduces the amount of taxes you owe, it also changes the percentage rate (tax bracket) at which the taxes are figured.

Property Tax Exemptions

Besides income tax credits, many states allow credits or deductions to be taken from property taxes. A property tax deduction exempts a percentage of the cost of the wind system from being assessed on your property taxes. A few states, such as South Dakota, Kansas, and Nevada, have property tax credits that are actually refunds on your property tax bill based upon a percentage of the assessed value of the wind system. Similarly, some states such as Massachusetts and Michigan exempt wind systems from either sales taxes, use taxes, or both.

Although income tax credits and deductions are helpful, they do not help everyone. Obviously, if you pay little in taxes or live on a retirement income, the credits are not useful to you. For an incentive program to be available to the most people, a direct rebate of the purchase price would have to be implemented. Another option is to make low-interest loans available to purchasers of wind systems and other renewable energy equipment.

If wind power is so desirable, why does it require a government subsidy? Proponents of tax incentive plans argue that wind and other renewable technologies need widespread marketing to lower the cost and that the incentives help this process. Also, it is argued that energy from fossil fuel sources are subsidized in various ways and that incentives help equalize the difference between the artificially low utility power costs and the unsubsidized cost of energy from wind and solar energy.

Opponents of tax incentives argue that they will not have any impact in the long run and may have negative effects on prices and competition. Whatever the reasons are, as long as tax incentives exist, it is to your benefit to check into them before purchasing a wind system to see which ones are available. When you do install a system, be sure to keep all receipts for financing your wind system and its installation in order to document for tax purposes your purchase of wind equipment.

Connecting to Utilities

You can install a utility-connected wind system and sell power to the utility company. Most utility companies are quite cooperative in allowing the installation of utility-connected systems as long as the required procedures are followed.

The primary concerns of the power company are that the utility-connected wind system will be electrically safe, that the wind system does not adversely affect the power wave form (see Ch. 5), and that a rate for power sent back to the utility be specified. In cases where the wind system is one of the first to be installed in the local area, utility companies have been known to install extra meters for monitoring the system, to help install towers, and to provide other services as well.

Such was the case for Thomas Zaborski of Calumet City, Indiana. Zaborski was not only the first to install a synchronous inverter system in his area, he was the first to do so in Indiana. The local utility, the White County Rural Electric Membership Corporation (R.E.M.C.), was very cooperative and installed three ratcheted (one-way) meters to record power usage on a magnetic tape recording system. One meter recorded the power used from the utility, one measured the power sent back to the utility, and the third measured the power produced by the wind system.

Derek Schruers of Edinboro, Pennsylvania received help from

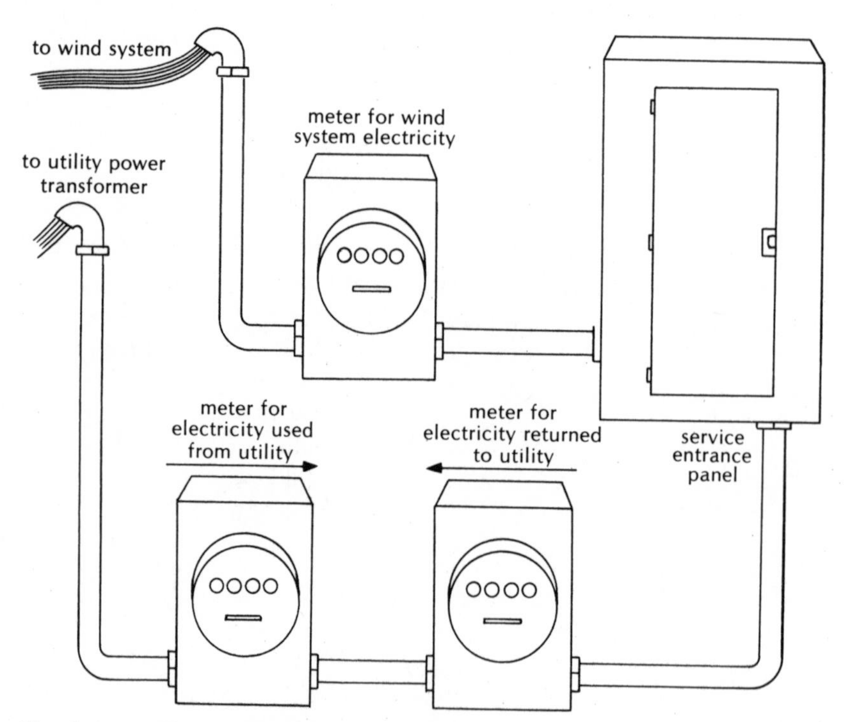

Illus. 3-1 Installation of Kilowatt-Hour Meters—Ratcheted meters installed with a utility-connected wind power system record for credit the wind power electricity that is fed back onto the utility's grid.

the Northwestern Rural Electric Co-op in installing his telephone pole-type tower. The Co-op sent two trucks and three men to plant the pole as well as to set up the guy wires. The Northwestern Co-op also installed ratcheted meters to monitor the wind system's output.

Safety

If the wind system generates power on the utility lines when those lines are down, a lineman making repairs could be injured or electrocuted. It is for this reason that the commercially available synchronous inverter systems and the induction generator class of wind systems (see Ch. 4) have all been designed to automatically stop sending power over the utility lines whenever utility power is not present. If a local utility has questions concerning safety, the manufacturer can refer it to other utilities that have verified their safety characteristics. In most customer agreements, however, the utility reserves the right to inspect the equipment for safety.

Utility Rates

The rates that utility companies will pay you for any power backfed over utility company lines vary considerably from area to area. In many cases, the rates will be set under guidelines specified by the Federal Energy Regulatory Commission (FERC). The Commission developed these guidelines as a result of the passage of the Public Utility Regulatory Policies Act of 1978 (PURPA). The PURPA legislation is aimed at encouraging energy conservation and the use of renewable energy sources such as wind power. As a result of PURPA, utilities regulated by a public utilities commission are required to sell electricity to "qualified" small power producers and to buy electricity back at reasonable rates. Rural electric cooperatives also may be subject to the FERC guidelines, but only if your state public utilities commission (P.U.C.) decides so.

The FERC guidelines define a qualified small power producer as one in which the system has a rated power of 30 megawatts (30 million w.) or less. This, of course, includes any home-size wind system discussed in this book. The qualified small power producer is exempted from many regulations that would otherwise apply to a regular co-generator (a facility that generates electricity at the site while being connected to utility power for occasional purchase of electricity from the utility grid). Also, the small power producer is protected from being charged unreasonable or discriminatory rates for such services as meter hookups and meter reading charges, or from being put on a different rate schedule simply because they are using wind power or some other renewable energy source.

Demand Charges

The FERC guidelines also protect the wind system user from the imposition of *demand charges.* Traditionally, utilities have added an extra charge, called the demand charge, on co-generators of electricity. The reason for the charge is that the utility must still maintain the distribution lines and its own standby generating capacity to supply power to the part-time customer whenever it is demanded. In addition to paying a demand charge, customers classified in this category are often put on an interruptible service basis, which means that the utility can disconnect power whenever it needs to.

In a much publicized test case that predated the passage of the PURPA legislation, the Energy Task Force group of New York City challenged the Consolidated Edison Company (Con Ed) in court on the matter of rates for backfed power. The Energy Task Force installed a rebuilt pre-R.E.A. Jacobs wind system on top of a five-story building on Manhattan's Lower East Side and requested reimbursement from Con Ed for backfed power. The case went to court and the New York State Public Service Commission (P.S.C.) ruled that Con Ed had to pay for the power. The rate specified was to be equivalent to Con Ed's average fuel cost or approximately $.025 per kwh. At that time, Con Ed did impose a demand charge for the 2-kilowatt Jacobs wind system, which came to a total charge of $13.60 per month. This charge was waived if the monthly bill was greater than the minimum charge of $4.96 plus the demand charge of $13.60. This precedent-setting case has since been superseded by the PURPA legislation.

Avoided Cost

The FERC guidelines also specify that the rate paid for backfed power should be based on the utility's *avoided costs* instead of its *imbedded costs.* The avoided cost is that which reflects the expense that the utility avoids by not having to produce the power itself or purchase elsewhere. It is based on the most expensive fuel that the utility uses. This differs from the imbedded cost which represents the average cost of all of the fuels that the utility uses such as coal, nuclear fuel, or oil.

Utilities subject to the FERC guidelines must publish their avoided costs with the state P.U.C. and these costs are to be reevaluated every two years. In some states the P.U.C. establishes standard rates for all qualifying power producers whose systems have a power capacity of 100 kw. or less. In other states, it may be necessary to petition the P.U.C. in order to establish the rate.

The P.U.C. can base the avoided cost on the sum of several factors. These factors represent an estimate of the costs that the utility company avoids by not having to generate the amount of power provided by the wind system or by not having to maintain the capacity for the power supplied by the system. The estimate includes the avoided fuel cost, operating and maintenance costs, fuel inventory carrying costs, the capacity factor, and adjustments for forced outages and peak loads. The capacity factor may be eliminated if the utility already has excess power-generating capacity, or if the P.U.C. rules that the wind system does not represent any firm savings in generating capacity to the utility because of the variable nature of the wind. In any event, if the utility uses a very expensive fuel, such as oil, or if the utility is short on power-generating capacity, the avoided cost may well be greater than the regular retail rates charged to utility customers.

Billing

Once the avoided-cost rate structure has been determined, there are two possible methods for metering your use and production of electricity to establish your monthly bill. In *simultaneous purchase and sale billing,* also called *differential billing,* the customer sells all of the power produced by the wind system to the utility at the avoided cost rate and buys back all of the power used on the homeowner's premises at the regular retail rate. This involves installing a kilowatt-hour meter to register the output of the wind system. A second, one-way or ratcheted meter is installed to register all power used from the utility (discussed later in this chapter). Thus, at the end of the month, your account is credited at the avoided-cost rate for all of the energy produced by the wind system and debited at the retail rate for all of the energy that you used. For example, if you use 900 kwh. per month and the retail rate is $.06 per kwh., the retail charge would be $54. If your wind system produces 300 kwh. per month and the avoided cost rate is $.07 per kwh., then your account would be credited $21 ($.07 × 300). Your resulting monthly bill (neglecting metering charges) would then be $54 – $21, or $33. In general, simultaneous purchase and sale billing is to your advantage if you live in an area where the avoided cost rate is greater than the retail rate, and significant amounts of electricity are fed back to the utility. If only small amounts of electricity are involved, the monthly charge for the extra meter will offset the value of the electricity sent back.

In *parallel operation billing,* also called *net energy billing,* you are reimbursed for only the amount of electricity that you send back to the utility, and you pay only for the amount of electricity that you use from the utility. This involves connecting two ratcheted meters between the utility transformer and the house's service entrance panel (see Illus. 3-1), one to register power use from the utility company and the other to register the amount of electricity fed back to the utility. Using the same 900 kwh. per month example as before, assume that 50 kwh. of the 300 produced by the wind system are fed back to the utility. The one-way meter will register 50 kwh. and you'll be credited for $3.50 at the avoided cost. The other meter will register the 650 kwh. used from the utility company (the other 250 of the total 900 consumed from the wind system) for a debit of $39. Your net bill is then $35.50 ($39.00 less $3.50). Utilities such as Southern California Edison (S.C.E.) have instituted a simultaneous purchase billing policy throughout their area for a three-year trial period.

Either of these billing plans require two meters and, therefore, involves an extra monthly charge by the utility company for reading the extra meter and calculating the bill. In most cases with a home-size wind system, the value of the wind-generated electricity will be largely offset or completely eliminated by this meter-reading charge. Thus, if the avoided cost exceeds the retail cost of your utility power, it may be advantageous to attempt to negotiate for the installation of a single-meter system. In a single-meter system the meter runs backward when the wind generator sends power back to the utility, and forward when the utility company power is being used. In effect, you will be reimbursed at the retail rate for backfed electricity instead of at the avoided cost rate, but the elimination of the extra meter charge makes it worthwhile for both you and the utility company.

In cases where the avoided cost is less than the retail cost, and the wind system is backfeeding only small amounts of power, it may be best to negotiate the use of one ratcheted meter that registers only the power you use from the utility without recording any power sent back to the utility. You lose reimbursement for any power sent back to the utility but also avoid any extra meter charge, and your bill is still reduced by the amount of energy you use from the wind system. Actually, for most home-size wind systems the question of the rate paid for backfed power should not be the major concern. Although it may be satisfying to sell power back to the local utility, the amount of energy sold to the utility will not be significant for most home-size wind systems. If the wind system is sized to your needs, only about 10 to 20 percent of the wind system's electricity should be fed back to the utility. The main advantage of the utility-connected system is that it eliminates the need for battery storage and thus reduces costs.

In my own case, my local utility company (which was not subject to FERC regulations) gave me the option of selling electricity from the wind system at the imbedded cost rate of $.025 per kwh. and paying for a ratcheted meter, or to receive nothing for the backfed power and eliminate the extra meter charge. The charge for the extra meter was $2.50 per month so that the wind system would have to send 140 kwh. per month back to the utility just to cover the meter charge.

As more wind systems are installed, their effects on the utility grid and their contribution to energy savings for both the utility and the individual will be assessed. Studies are being conducted now by the U.S. Department of Energy (DOE) as well as by the Electric Power Research Institute (E.P.R.I.) of Palo Alto, California, which conducts research for private utilities in the United States.

Issues such as zoning, building codes, and utility rates are real factors that must be considered in installing a wind system, but once you've become familiar with these factors in your area and have checked with the appropriate agencies, you can move on to considering the wind equipment itself.

Chapter 4

In the Air: Generators, Rotors, and Towers

Once you know your site has enough wind for a system, and you've sorted out the legal and financial requirements, you've reached the point at which it's wise to learn some specifics about the actual equipment available for wind systems. While it is not necessary to be familiar with all of the engineering specifications of wind system components, knowing some of the concepts behind their design and use is necessary before you can choose the best system for your needs.

As mentioned before, the fundamental components of most systems are a rotor and generator mounted on a tower, with batteries, inverters and a control panel operating on the ground. We'll begin with the equipment in the air, starting first with generators.

Generators

Electrical generators used in wind power systems are not substantially different from any other type of generator. In fact, many manufacturers of wind systems use "off-the-shelf" generators and

The Nature of Electricity

Electricity from a battery is called *direct* or *direct current* (d.c.) because it flows in one direction, just as water flows in a pipe. On the other hand, electricity supplied by the utility company is called *alternating* or *alternating current* (a.c.) because it flows first in one direction and then in the other. It takes 1/60 of a second to complete each alternating cycle, so we say that the electricity has a *frequency* of 60 cycles per second (or 60 H., in which 1 cycle per second = 1 Hertz = 1 H.). If you were to connect a televisionlike device known as an oscilloscope to an a.c. outlet in your home, the picture would show a pulsating wave called a *sine wave.*

Just as water flowing in a pipe has a pressure, the "pressure" in an electrical circuit is called *voltage* and is measured in *volts* (v.). Similarly, just as the flow of water in a pipe is measured in gallons per minute, the flow of electricity in a wire is called *current* and is measured in *amperes,* usually shortened to *amp.* The amount of power in an electrical circuit is measured in *watts* (w.) and is found by multiplying the current by the voltage in the circuit (w. = v. × amp.).

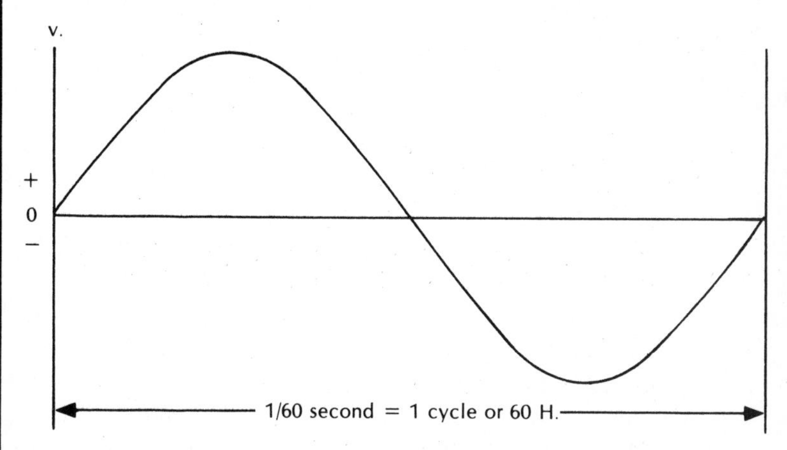

Illus. 4-1 The Wave Form of One Cycle of A.C. Electricity—The frequency of a.c. electricity is 60 cycles per second, which means that it takes 1/60 of a second to complete a full pulsation of power.

make minor modifications to them for use with their product.

All electrical generators work on the same basic principle. If a wire is moved through the magnetic field around a magnet (or if a magnet is moved past the wire), a small electrical current is "induced" or produced in the wire. This sounds like magic, and in a way it is, since the wire does not come in physical contact with the magnet and there is no force that we can see.

If you consider motors, generators, and electricity to be mysterious subjects and wish to better understand them, try building some of the toy motors often described in school science manuals, science kits, or Scout manuals. I built some of these motors with my seven-year-old son using some copper wire, a few nails, and the small magnets used to hold pot holders onto refrigerators and ranges. They worked very well, and demonstrated the basics of electrical generators quite well.

Types of Generators

There are two basic types of generators that can be used with a wind system: a d.c. generator or an a.c. generator. As the names imply, a d.c. generator puts out direct current electricity and an a.c. generator puts out alternating current electricity. The reason for using one or the other is largely a matter of the manufacturer's judgment of the expense and maintainability of a particular generator.

D.C. Generators

The most basic d.c. generator consists of a coil or loop of wire that rotates between the north and south poles of a magnet. As the coil of wire rotates, the magnets, called *field poles,* induce an electric current in the coil of wire. Copper bars at the ends of the coil of wire make contact with carbon brushes which connect the electricity to a storage battery (see Illus. 4-2).

The current in the coil of wire is strongest when the coil is perpendicular, or nearest, to the field poles and weakest when the coil is parallel to the field poles at the greatest distance from them. As shown in Illustration 4-2, when the coil is perpendicular to the field poles at position A, an electrical current flows from commutator segment 1 to segment 2, and thus from carbon brush 2 to brush 1. When the coil has rotated 90 degrees and is parallel to the field poles at position B, the coil is said to be at the neutral plane and no electrical current flows. When the coil has rotated another 90 degrees to position C, current again flows in the same direction, from brush 2 to brush 1. Since the current always flows through the brushes in the same direction, the generator is called a direct current generator. The direction of this current flow is determined by the magnetic polarity of the field poles.

In a practical d.c. generator there are many coils of wire rotating rather than just one as described above. These coils of wire are wound around a metal armature. Increasing the number of turns of wire in each coil raises the voltage that the generator can develop, and increasing the size of the wire in each coil adds to the amount of current that the generator can generate (see Illus. 4-3).

More power is induced in the armature coils as they move faster through the magnetic field produced by the field poles. The voltage and current from a d.c. generator are both directly propor-

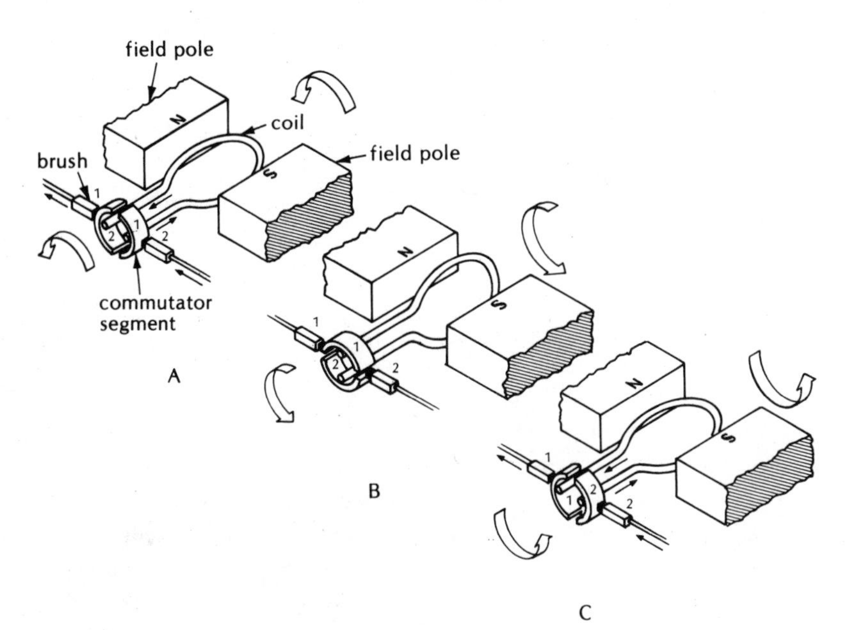

Illus. 4-2 A Simple D.C. Generator—In position A, the current flows at its strongest from the copper commutator segments 1 to 2, and, thus, from carbon brush 2 to 1. At position B, the coil is at the neutral plane, and no current flows. In position C, the current flows again, in the same direction as A.

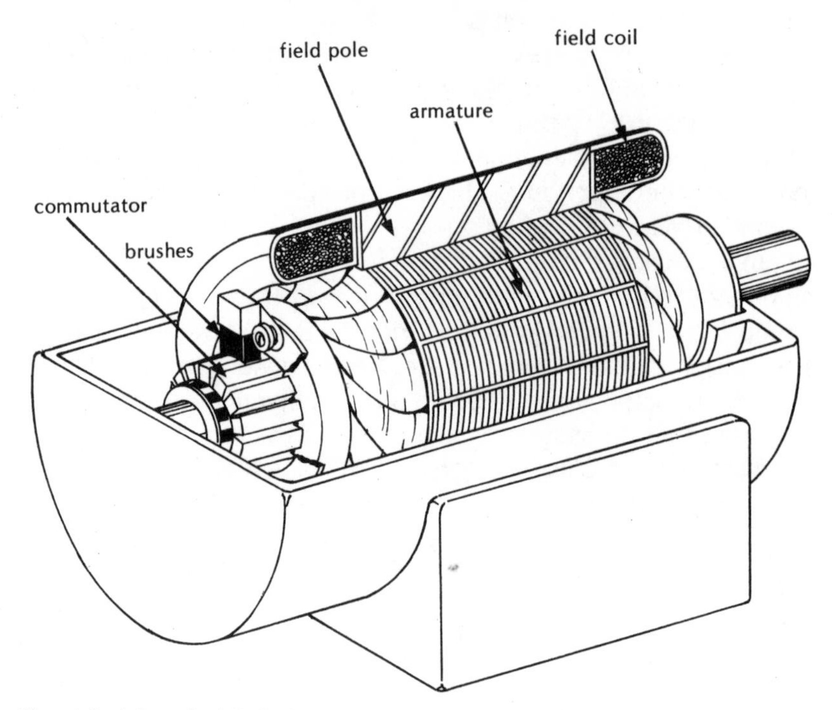

Illus. 4-3 A Practical D.C. Generator—Many coils wound around a metal armature of wire rotate to produce power. When the armature rotates faster, the amount of available power increases. The size of the wire and its number of turns around the armature are also directly proportional to voltage and current.

tional to the speed of rotation of the armature, and, thus, the amount of electrical power (voltage × current) available from the generator increases as the armature is rotated faster. The faster the armature turns, the more power it can generate, until a point is reached where the windings start getting hot from carrying too much current. At this point, the generator has reached what is called its *rated power,* and increasing its speed beyond this will cause it to burn out its windings.

The copper bars or segments attached to the ends of each coil of wire are arranged in a circle and separated from each other with mica insulation to form a commutator. In a small generator, such as the Winco (formerly Wincharger) generator, the commutator may be about 2 inches in diameter and have about 20 segments or bars. In a larger generator, such as the R.E.A. Jacobs type, the commutator can be up to 10 inches in diameter and have almost 200 commutator segments. I have constructed a commutator for my own Jacobs generator, and it involved arranging 188 wedge-shaped bars that were separated from each other by pieces of flat mica into a circle 10 inches in diameter. Constructing a commutator of this size requires considerable labor and is one reason why most manufacturers use a.c. generators.

A d.c. generator has 2, 4, 6 or even more field poles, each consisting of a coil of wire, called a field winding, that is wound around an iron core to form an electromagnet. In this way the electricity generated by the armature's movement through a magnetic field is also directed into the coils around the iron cores, thereby creating the magnetic field itself. In some units, the field poles are made from permanent magnets. Permanent magnet (p.m.) generators are easier to manufacture since no coil of wire has to be wound around a form. A disadvantage is that the magnetic pull of the p.m. fields must be overcome before the armature can begin rotating in low wind speeds. With a wound field, the strength of the magnetic field builds up gradually and can be controlled by varying the current flowing through it.

Favorable features of d.c. generators are that they are rugged, efficient, and can be made self-exciting, which means that the current for the field poles is supplied from their residual magnetism rather than from an external source.

Another useful property of d.c. generators is that they can act as d.c. motors as well. This means that a wind-powered d.c. generator can easily be "checked out" to see if it is running properly by connecting the storage battery to the generator while the wind is not blowing, thereby running a current through it. This is called "motoring" the generator.

On the other hand, d.c. generators cost more than other types because they use more copper and are more difficult to manufacture. The carbon brushes in a d.c. generator need periodic replacement, although they are not as much of a problem as they are sometimes made out to be. For instance, the brushes on the pre-R.E.A. Jacobs Wind Electric units were made large for long wear. They were 1½ inches wide and ¼ inch thick and lasted for at least five years.

A.C. Generators

The most basic a.c. generator is very similar to the d.c. generator just described. The difference is that, instead of having a commutator segment at the end of each coil of wire, an a.c. generator operates by using copper rings called *slip rings.* Each carbon brush makes continuous contact with the same respective slip ring.

As with the d.c. generator, the current is always flowing in the same direction in the coil when the coil is perpendicular to the field pole. This means that, as shown in Illustration 4-4, the current first flows from slip ring 2 to 1 (and thus from brush 2 to 1) and then flows from slip ring 1 to 2 (and thus from brush 1 to 2). The result is that the current changes direction with each "cycle" or, in this example, 180 degree rotation of the coil. Hence the term "alternating" current.

In a practical a.c. generator, the armature is made stationary and the field poles are rotated instead. With the stationary armature, called a *stator,* the connection to the outside electrical circuit can be direct. The connection to the rotating fields, or *rotor,* is made

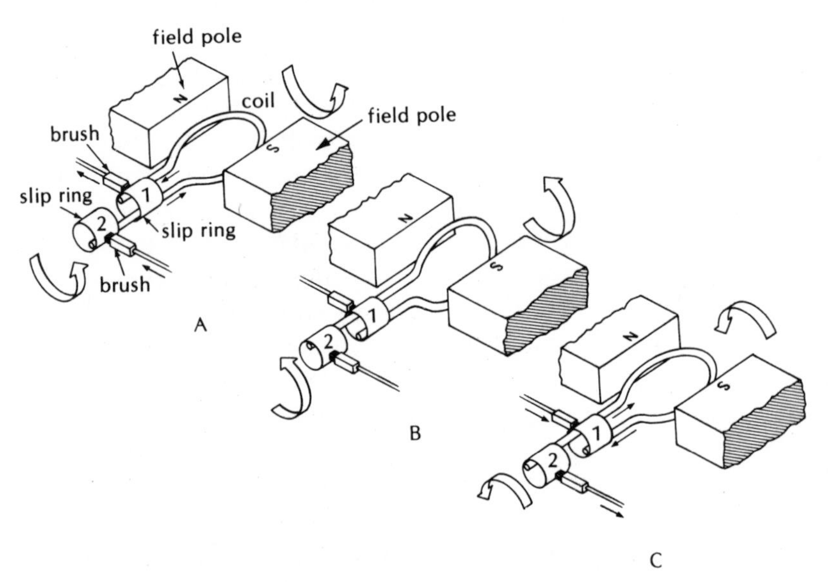

Illus. 4-4 An A.C. Generator—Current flows from slip ring 2 to 1 at position A, stops at position B, and then flows from slip ring 1 to 2 at position C, which completes one cycle.

through relatively small brushes and slip rings since the field current is much smaller than the stator current.

The a.c. generator still uses d.c. current to energize its field windings. For this reason, an a.c. generator is not inherently self-exciting since the a.c. current from the stator cannot be supplied to the rotor, which needs d.c. current. In some a.c. generators, that direct current is supplied from a small d.c. generator mounted on the same shaft with the a.c. generator. In an alternator, such as an automobile alternator, the a.c. current from the stator is converted to d.c. via electronic diodes. Diodes are simply electrical valves that allow current to flow in one direction only and thus convert a.c. to d.c. The d.c. provided by the diodes goes to the rotor so that the alternator can be self-exciting.

Those a.c. generators without brushes or slip rings to supply current to the rotor are called *brushless.* In these generators an exciter winding induces an electric current in the rotor without any physical connection between the two. Since the exciter uses diodes to provide d.c. electricity to the rotor, carbon brushes and slip rings, in effect, are replaced with diodes.

Other a.c. generators use permanent magnets for the field poles just like some d.c. generators. Permanent magnet a.c. generators eliminate any problems associated with carbon brush wear or the possibility of diode failure. However, as with permanent magnet d.c. generators, the field current does not vary with the wind speed, and the pull of the magnets on the generator at low wind speeds must be overcome to start the system up. Both of these factors can be dealt with by designing the wind rotor and the charging control circuit to be compatible with the permanent magnet generator.

Although a generator may employ carbon brushes, electronic diodes, or permanent magnets, proper design is important to insure low maintenance for all of these models. With the proper design of the carbon-brush holding device and the correct sizing of the brushes, brush replacement should be limited to a routine of only once every several years.

With the proper choice of diodes and with good lightning protection, diode replacement should be rare. In one instance, an early model wind system needed frequent diode replacement because of inadequate lightning protection. Since the diodes were located in inaccessible positions inside the generator housing, repairs were difficult. To deal with the problem, one owner I talked to mounted all of the diodes on a panel at the rear of the generator for easier maintenance. Reports of diode malfunctions are becoming an exception, however, as designers understand the potential problems associated with wind-system generators.

As with d.c. generators, the voltage and current produced by a.c. generators increase in proportion to the speed of the generator rotor. In addition, the *frequency* of the a.c. increases with the speed of the generator rotor. The frequency, as a term of measurement, is unique to a.c., because the flow of electricity is intermittent, unlike the constant flow of d.c. Many a.c. generators are designed to produce a voltage with a frequency of 60 cycles per second (60 H.) at a speed of either 1,800 or 3,600 revolutions per minute (rpm). When an a.c. generator is used in a wind power system, the speed at which the shaft of the generator turns varies with the speed at which the wind turns the wind system rotor, and thus the frequency of the voltage produced varies. This makes the power coming from the generator incompatible with regular 60 H. utility power. To make the output of the generator compatible with utility power, a device called a synchronous inverter must be used (see Ch. 5). For battery storage systems that must use d.c. to operate, the a.c. is converted to d.c. by the use of diodes, and the frequency of the power from the a.c. generator is not a concern.

Many a.c. generators are designed with three separate electrical windings called *phases.* The phases are spaced 120 degrees apart so that each phase develops its maximum current in equally spaced intervals. Whereas the single phase generator described above produces one pulse per $1/120$ second interval, the three phase generator produces three pulses of current in the same time interval. The current from each phase is converted to d.c. with diodes and then combined together. This combination produces a closer approximation to d.c. than does an a.c. generator with just one phase.

Automobile Alternators

Automobile alternators have been the subject of much misunderstanding concerning the difference between d.c. and a.c. generators. When the automobile industry converted from d.c. generators to alternators for charging batteries, the commonly understood reason was that alternators were more efficient because they could produce more current at low engine speeds, a desirable quality in modern stop-and-go traffic. The alternator is an a.c. generator that uses electronic diodes to change from a.c. to d.c. While automobile alternators do produce more current at low engine speeds, it is not because they are more efficient than the old d.c. automobile generators, but because they are designed to turn at speeds 2 to 3 times faster than automobile generators. Automobile generators that produce d.c. cannot be designed easily to rotate as fast as an alternator's 6,000 to 10,000 rpm because the windings would fly out of the armature at such speeds. Because of their high speed of rotation, then, automobile alternators produce more current than d.c. generators even when the engine is idling.

Actually, off-the-shelf automobile alternators are ill-suited for use as wind generators, as many experimenters have discovered. They are inefficient and must be geared up considerably (since the wind rotor turns relatively slowly) to rotate at a high enough speed to generate a usable amount of electricity. Also, either a wind-actuated switch or an rpm sensor must be used to connect the battery current to the rotor when wind speeds are high enough for the alternator to start producing power. This is necessary because usually car alternators are not self-exciting.

Induction Generators

A third type of generator that can be used in wind systems is the a.c. *induction generator.* The induction generator has a stator and a rotor just like a standard a.c. generator. However, there is no physical connection between the two and, thus, no need for slip rings, brushes, or even diodes. This is because the rotor's magnetic field is "induced" through an electromagnetic interaction with the stator.

The electrical windings of the stator are arranged so that when a 60 H. current is applied to it, the stator produces a magnetic field. This magnetic field rotates around the stator in synchronization with the 60 H. electricity. In turn, this rotating magnetic field induces a similar magnetic field in the rotor. The rotor "locks in" to this rotating magnetic field and turns at a very constant speed (see Illus. 4-5). Thus, although the electrical current increases as the rotor turns faster, the voltage and frequency of the induction generator change very little.

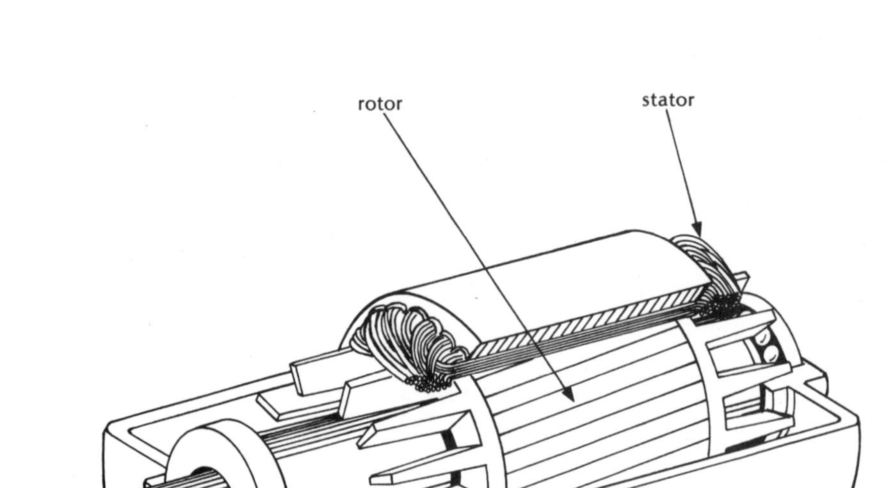

Illus. 4-5 A Practical A.C. Induction Generator—The magnetic field of the rotor is induced by electromagnetic interaction with the stator. No slip rings or brushes are needed.

The motors found on furnace blowers, water pumps, washing machines, and similar appliances are usually induction types. Induction motors are essentially the same as induction generators. When used as a generator, the only necessary difference is that the rotor must be turned about 5 percent faster than the speed the rotor turns when used as a motor—usually 1,800 rpm.

When employed in a wind power system, the induction generator rotates very close to one speed even though the wind speed changes. The Enertech model "1800" is an example of a wind system that uses an induction generator. As soon as wind speeds are averaging 10 mph, the Enertech hooks in to utility power and, for about one second, the generator acts as a motor and actually spins the system rotor until it is up to the speed it should be. Then the generator begins producing its own power. Because of the nearly constant speed of the induction generator, the wind system rotor speed changes from 160 rpm to a maximum of only 170 rpm.

Induction generators are well suited for utility-connected wind systems. As a matter of fact, they cannot generate power if utility power is not connected to them, a safety factor that utility companies appreciate. So induction generators cannot be used in stand-alone wind power systems. The main reason for using induction generators is that they can generate a.c. electricity without the use of additional electronic circuits to make the generated power compatible with utility power.

Gearing

The power output of a generator is approximately proportional to its speed. Most generators are designed to operate at steady speeds of either 1,800 or 3,600 rpm. These are the standard speeds of the industry, and if such generators are used in wind power systems, a problem arises. Most wind rotors turn at speeds of only 200 to 400 rpm. This is not fast enough to turn most generators at the speed for which they were designed, so a gearbox must be used to speed up the rotation. For example, if a generator has a power rating of 1,000 w. at a speed of 1,800 rpm and is used with a rotor that has a top speed of 360 rpm, a gearbox with a 5 to 1 (5:1) step-up ratio must be used. The gearbox has two gears inside it, one gear being 5 times the diameter of the other. The rotor drives the larger gear which in turn drives the smaller gear connected to the generator shaft. In the given example then, if a gearbox were not installed, the top speed of the generator would be only 360 rpm and its power output would be only ⅕ of 1000 w. or 200 w. The gearbox ratio on gear drive wind systems varies from 2 to 1 up to 100 to 1.

Another way to match the speed of the rotor to that of the generator is to use a generator designed to operate at lower speeds and eliminate the need for a gearbox. This is called a *direct drive* system. A direct drive generator that gives its maximum power at a very low speed, say 180 rpm, is much larger and heavier than an equivalent gear drive unit. The advantage of using a direct drive generator is that the lower speeds mean longer life and less wear on the generator bearings. Also, there are no gears to wear out or

Photo. 4-1 A Gearbox for Wind Generators—The large gear shown is 5 times the diameter of the smaller gear. When the gearbox is placed upon the wind system, it speeds up the rotation of the generator to the proper number of revolutions to produce the system's rated power.

gearbox oil to change periodically. On the other hand, a direct drive generator would be very large and expensive for systems rated above about 5 kw.

The most famous examples of direct drive systems are the Jacobs Wind Electric generators made from the 1930's to the 1950's. These units, rated at 3,000 w. of power when turning at about 225 rpm, have often been put back into service by wind power buffs, with a minimum of repairs to the generator. This is not to say that a gear drive unit cannot last long, however. With regular maintenance and good design, any wind system should last at least 15 to 20 years.

Rotors

The rotor is the business end of the wind power generator. Without a properly designed rotor (or turbine or propeller, as it is also called) the generator would develop ineffective amounts of power. The rotor must turn the generator at a speed fast enough for the maximum amount of power but not so fast that it presents any danger to the system itself from overspeeding.

A rotor is like an airplane wing traveling through the air, except that it goes in a circle instead of in a straight line. (Vertical axis rotors, discussed later in this chapter, have different aerodynamic shapes, however.) In fact, the aerodynamic shape (or airfoil) of the rotor is very similar to that of an airplane wing. The rotor of a wind power system should not be confused with the propeller of an airplane, however. The airplane propeller is made to pull the plane through the air, whereas the wind rotor is made to be moved by the wind.

Like an airplane wing, the leading edge of the rotor blade is thick and rounded, but it tapers down to a very thin trailing edge. The side facing the wind is flat or slightly curved and the side away from the wind is curved. Each blade is set at a slight angle in relation to the wind, called the *angle of attack.* The smaller the angle of attack, the faster the rotor will tend to turn. Increasing the angle of attack slows the rotor down, just as changing the position of the sail on a sailboat changes its speed (see Illus. 4-6).

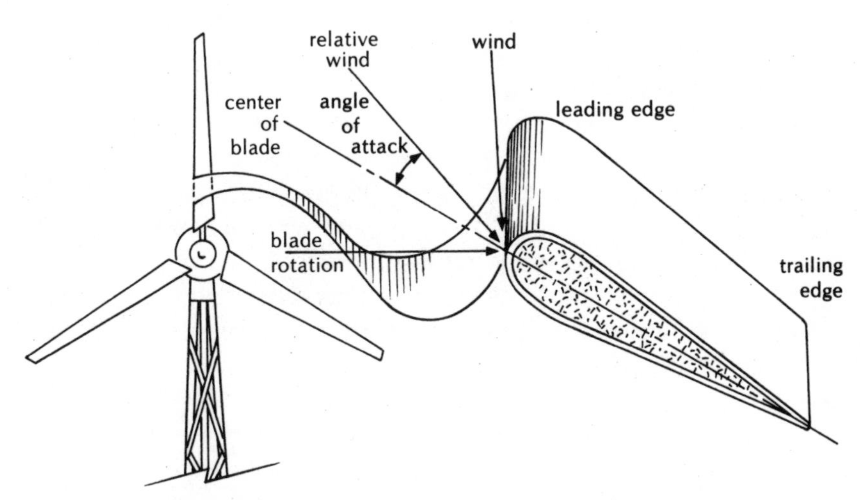

Illus. 4-6 Cross Section of a Wind System Rotor Blade—Most rotor blades have a thick leading edge tapering to a thin trailing one. The wind strikes the flat side of the blade which is set at a slight angle called the angle of attack.

Tip Speed

Because of its aerodynamic shape, the outer tip of the rotor or turbine actually turns faster than the speed of the wind—5 to 8 times faster. The ratio of the blade tip speed to that of the wind is called the *tip speed ratio.* A tip speed ratio of 5, for example, simply means that the tip of the blade moves 5 times faster than the wind speed. Thus, when the wind is blowing at 20 mph, the tip of the blade would travel in a circle at 100 mph.

The maximum rpm at which a rotor is designed to turn depends in part upon the length of the blades. Very short blades, such as the 6-foot rotor on the Winco, will turn at speeds of up to 900 rpm at wind speeds of 23 mph. At the other extreme, Energy Development's model "440," with a 40-foot diameter rotor, turns at a maximum of 60 rpm in a 40-mph wind.

A tip speed ratio between 5 and 8 is desirable for wind power generators since the rotor must turn the generator at a high enough speed to generate a useful amount of power. In a well-designed system, when the rotor turns at its maximum speed, the generator produces its maximum power.

Solidity

The tip speed ratio of a rotor is determined in part by the solidity of the rotor. The solidity is the amount of rotor surface-area facing the wind in proportion to the total area swept by the rotor. The higher the solidity of a rotor, the lower is its tip speed ratio. Thus, a rotor with a large number of wide blades will have a lower tip speed ratio than a rotor with only a few narrow blades.

For better starting (higher torque) in low wind speeds, a rotor with high solidity is used. Water-pumping windmills usually have several blades with tip speed ratios of less than 3. They are made to start up in low winds, since they don't need the high rotor speeds necessary for turning a generator. But for wind power generator use, rotors generally have from two to six blades, with two, three, or four being the most common.

On small wind power generators, two blades are often used because they are easier to make and the cost is lower. The one disadvantage of the two-blade rotor is that its performance is "choppy." As the wind changes direction, the tail vane on these

Illus. 4-7 Rotors with Different Numbers of Blades—The number of blades on a wind system rotor range from 1 to 48, though two-, three-, and four-blade rotors are most common for electrical generators.

systems moves the rotor so that it faces the wind head on. When the two-blade rotor is straight up and down, it offers little resistance to this orienting motion. However, when the rotor is parallel to the horizon, the blades act as a gyroscope and the wind system resists the force to turn into the wind. Thus, the wind system tends to move rather unevenly when it's orienting itself into the wind.

A smoother action is provided by using three blades. This is similar to adding more pistons to an engine. Most medium-size wind systems use a three-blade rotor even though adding the third blade increases their complexity and cost.

An exception to the two-or three-blade design of most wind power systems is the four-blade designs with 40-foot diameter blades built by innovator Terry Mehrkam of Hamburg, Pennsylvania. Mehrkam likens his designs to the four-blade designs built by the Wincharger Company in the 1930's. Mehrkam feels that the reason for using four and even six blades on the larger-size wind systems is that the wind rotor will turn more slowly with better balance, thus keeping stresses down to a minimum and lengthening the life of the machine. This is the same design approach applied to the old Dutch windmills, which had four straight blades with 60- and 100-foot diameters.

Photo. 4-2 Four-Blade Rotors Designed by Terry Mehrkam—Mehrkam models his rotors after the 1930's Wincharger systems. He also builds six-blade rotors.

In yet another design approach, both the Altos and the American Wind Turbine units use multi-blade "bicycle wheel" turbines. The Altos design uses 24 aluminum blades and the American Wind Turbines use up to 48 blades each. These rotors are designed to be low cost, but they do not provide as much power at high wind speeds as rotors with fewer blades.

Materials

Rotor blades for wind systems are made from wood, aluminum, steel, and plastic. Wood is the most tried and proven material for blades, particularly for small- to moderate-size rotors. The ideal blade material is both light and strong at the same time, and certain woods serve quite well. The blades should be light because the heavier they are, the more centrifugal force tends to throw the blades outward as they spin. A typical wooden blade on a 12-to 15-foot diameter rotor will weigh from 5 to 8 pounds. At the same time, wood is strong and can take the slight flexing that it must withstand in the wind.

The kind of wood considered ideal for wind rotor use is Sitka spruce. It is very light, strong, and does not warp due to moisture, a very important property. Unfortunately, Sitka spruce, which is also used in musical instruments and airplane propellers, is fast becoming scarce and cannot be used on any large scale. Douglas fir is very close to Sitka spruce in desirable properties, though it weighs about 1½ times as much.

Aluminum is also used extensively for rotor blades. It too is light and strong, although it is more difficult to form into the complicated shapes of an aerodynamically-designed blade. Aluminum blades either can be formed from sheets of aircraft-grade aluminum over struts, like an airplane wing, or they can be formed from extruded aluminum. The latter is done by forcing hot aluminum through a die which forms the aluminum into the shape of the blades as it comes out.

Galvanized or stainless steel is also used for some blades, such as those of the Dunlite rotors. This is a good choice for an ocean environment where salt can attack wood or corrode aluminum blades.

Blades of composite materials, including fiberglass coatings over plastic foam, are also used and are potentially low-cost replacements for wooden rotors. Composite blades are used on Astral/Wilcon's model "AWIO-B" and Wind Power Systems "Storm-/Master 10" model.

Types of Wind Turbines

The type of wind system with a propellerlike rotor is more properly called a *horizontal-axis* system because the generator shaft—the axis of rotation—is parallel to the horizon. In recent years, interest has grown in *vertical-axis* wind systems, in which the generator shaft or axis of rotation of the rotor is perpendicular to the ground. Vertical-axis machines are of interest because, potentially, they can be constructed for lower cost. Vertical-axis turbines also do not need to be oriented into the wind as does a horizontal-axis turbine. How much of an advantage this feature is in practice has yet to be determined.

Horizontal-Axis Rotors

Horizontal-axis wind systems can be either upwind or downwind. In an upwind design, the wind first passes over the rotor and then passes by the generator behind the rotor. A tail vane acts like the rudder on a sailboat and keeps the rotor facing the wind. The tail vane may also be used to shut the plant off by folding at right angles to the generator. Most small- and moderate-size wind systems are of the upwind design.

In a down-wind design, the wind first passes over the generator and then the rotor. No tail vane is used, so a mechanical or electrical brake is necessary to shut off the plant. The tower can block some of the wind before it gets to the blades in downwind designs, causing vibration problems. This is called *tower shadow.* The relationship between the tower design and the wind plant is so important to operating downwind designs that Enertech Corporation, for example, guarantees its wind systems only with towers recommended for their downwind machines.

On larger downwind machines, such as the Mehrkam Energy Development Company model "440," a motorized mechanism called a yaw control is used to swing or "yaw" the turbine into the wind. When wind speeds are high enough to generate electricity, such a large turbine would not swing into the wind by itself, which is why a yaw control is necessary.

Speed control

All rotors must have some type of speed control system to prevent them from overspeeding in high winds. Most wind systems are designed to operate in winds of about 8 to 40 mph. The wind speed at which the rotor is turning fast enough for the generator to start putting out useful amounts of electricity is called the *cut-in speed.* The speed at which the rotor is stopped during high winds is called the *cut-out speed,* usually at around 40 mph. There are many different methods used to limit the speed of the rotor, including a variable-axis control, coning, spoiler flaps, blade-pitch control, and brakes.

Variable-axis control relies upon wind pressure on the rotor to turn the rotor sideways or upwards, away from the wind. One possible problem with this mode of speed control is that it depends upon the pressure on the rotor while the electrical generator is connected to a load. If the electrical wires are broken at any point, the generator is free to spin without any magnetic force to restrain the speed of the rotor or armature. In such an occurrence, the pressure to turn the wind rotor sideways or upwards does not build up, but the rotor simply speeds up faster. Wind pioneer Marcellus Jacobs reports that pre-R.E.A. units built by the Paris-Dunn Company, which used a tilt-up design, sometimes damaged their blades due to overspeeding this way.

With side-tilt control, the generator is set slightly off-center on its mounting and the tail vane is hinged at the point where it connects to the generator. The rotor and generator start turning away from the wind while the tail vane stays in the direction of the wind at high speeds. A spring is used to restrain the tail vane until the wind pressure on the rotor starts to overcome it at speeds of 25 mph or so. This type of speed control is used effectively on the Sencenbaugh Wind Electric units.

The generator mounting with the tilt-up design can be hinged so that the rotor and generator tip upwards instead of sideways. This design was used on the Paris-Dunn wind generators made during the 1930's. The North Wind Power Company model "HR2" uses

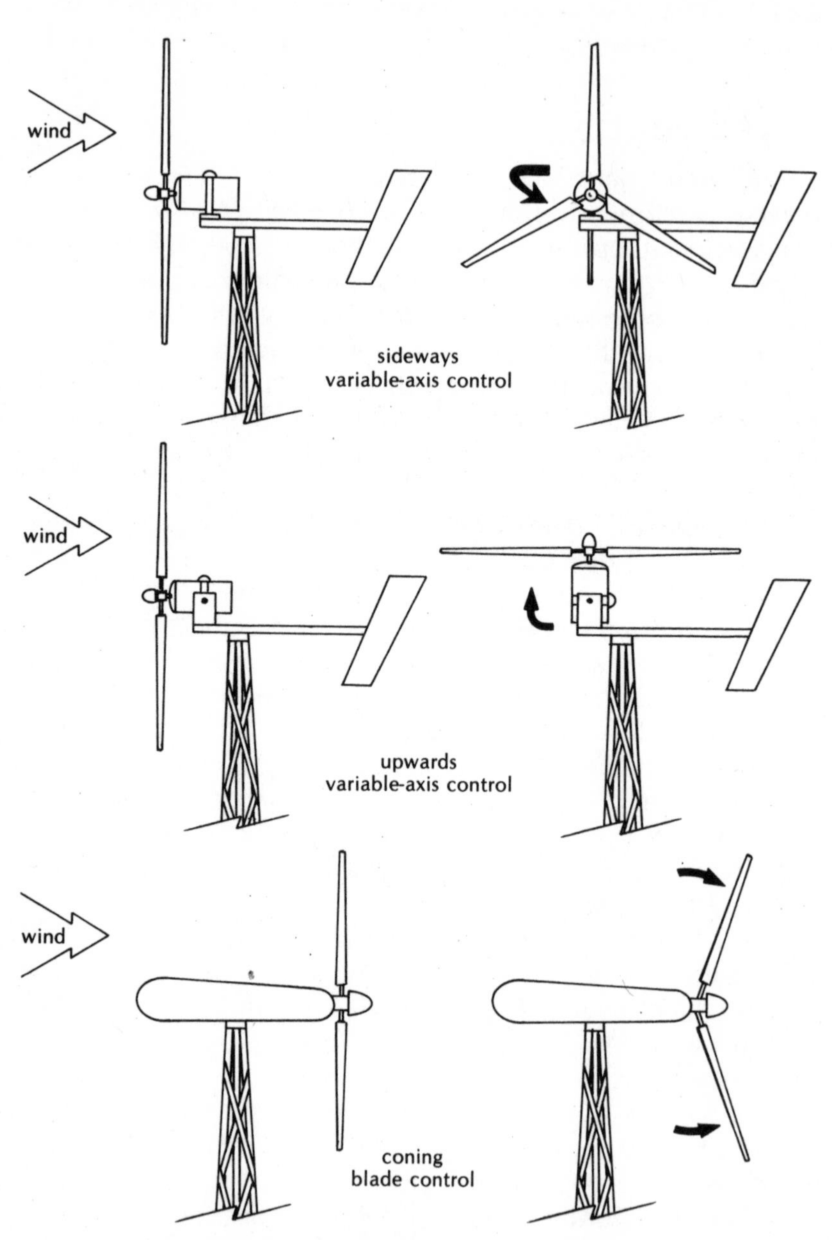

Illus. 4-8 Variable-Axis Control and Coning—In a variable axis system for limiting the speed of the rotor, gyroscopic controls tilt the rotor either up or to the side, away from the wind. In coning, the blades themselves bend away from the wind.

a tilt-up design with a restraining spring to limit the tilting action until the pressure on the rotor overcomes the force of the spring.

In coning speed control, the blades actually bend away from the wind as the wind speed increases. This design is feasible only on downwind machines since the blades would hit the tower in an upwind design.

Spoiler flaps can also slow rotor blades by disrupting the smooth airflow over them. The small Wincharger design has used this method for many years. As the wind speed picks up and approaches the cut-out speed, a spring restrained air-brake, or flap, begins to extend. This causes turbulence on the leading edge of the blades and slows them down.

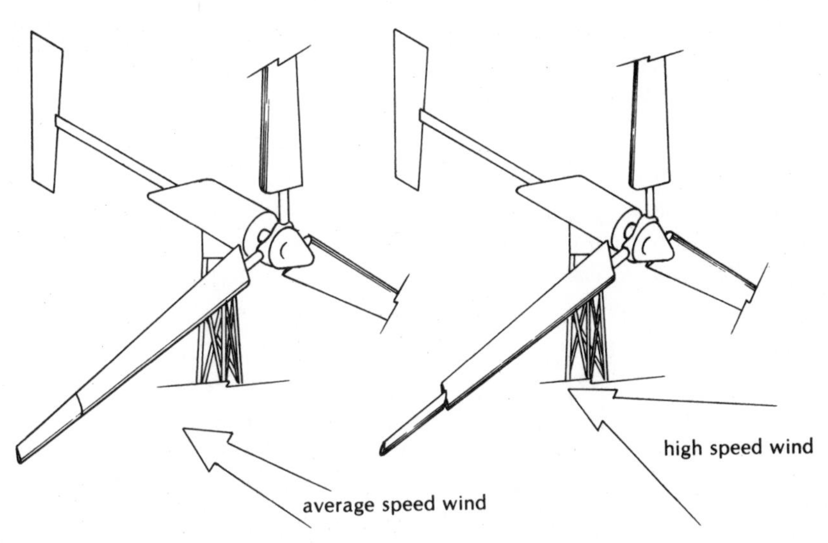

Illus. 4-9 Tip Flaps—An air-brake or tip flap extends from the blade tips to create drag that slows down the rotor.

Tip flaps, movable flaps at the tips of the blades, work well on very large machines. Most of the power occurs at the outer 20 to 30 percent of the blade. If the outer end of the blade is made movable (which increases the angle of attack), the blade will slow down without having to change its position.

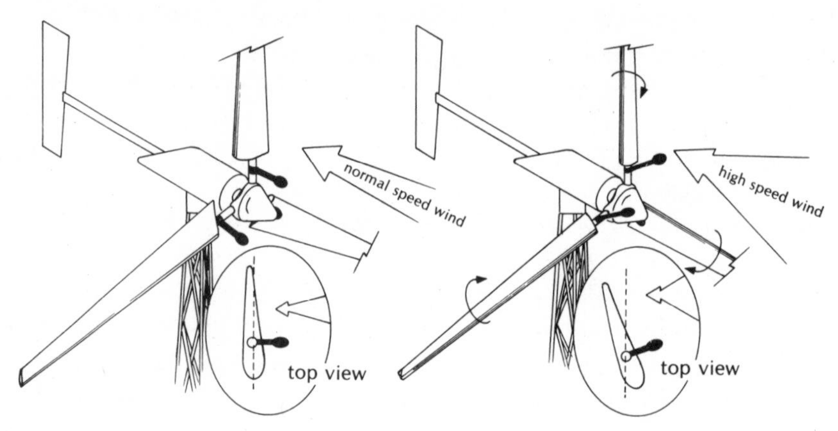

Illus. 4-10 Blade-Pitch Control—The rotor speed is controlled by the action of weights that change the pitch of the rotor during high-speed winds.

Blade-pitch control is one of the most commonly used methods of speed control on moderate-size rotors, 10 to 15 feet in diameter. As the wind speed starts exceeding about 22 to 25 mph, the pitch, or angle of attack, is increased. By changing the blade pitch just a few degrees, the rotor can be kept at a safe speed even in very high winds.

The hub to which the blades are attached has a governing device that changes the blade pitch as the speed increases. The most common type is a flyball, a weight usually restrained by a spring that moves out as speed increases. A lever arm attached to the weight rotates the blade slightly, thus slowing it down. The speed at which the weight starts moving or "feathering" the blades is determined by the tension of the restraining spring. Variations on this design are used in such wind systems as the Dunlite, Elektro, pre-R.E.A. Jacobs and also Kedco.

In the blade-controlled version, developed by the Jacobs Wind Electric Company, the blade itself is the weight. As the wind speed picks up, the blades are allowed to move up and out in a helical motion to change the pitch. This motion is controlled by special ball joints and a synchronizing mechanism that make the blades move in unison and control the helical motion. The weight of the blade itself moves the blade out as speed picks up and a spring restrains each blade at high speeds. Of all the pitch control designs, the blade-controlled design is one of the simplest and has the least number of moving parts. Variations of this design are used on the Aeropower, Independent Energy, Dakota Wind, and Jacobs wind systems.

In all governing mechanisms for rotors, it is important to keep wear to a minimum. Fail-safe shutdown features are important also, so that if a part of the system does fail, the blades should turn out of the wind to their maximum-angle, full-feather position, thus preventing overspeeding. This is not the case with all designs.

Is it possible for a rotor blade to be thrown off during a storm? Nearly all wind systems are designed with speed-limiting and shutdown features that protect the rotor from high speeds. Winds of 60 to 100 mph are certainly probable during the lifetime of any wind system, but most wind systems begin governing speed at wind speeds of 22 to 30 mph.

In addition, many systems have mechanical devices preset to turn the rotor completely away from the wind if the speeds are averaging between 40 to 60 mph. Down-wind systems such as those manufactured by the Enertech, Mehrkam Energy Development Company, and Whirlwind Power Company use a mechanical or electrical brake to stop the rotor from turning at a certain wind speed. Dunlite governor mechanisms use a magnetic latch that locks the blades into a fully feathered position at wind speeds of 32 mph and then releases them when speeds have dropped back to 20 mph or under.

The pre-R.E.A. Jacobs wind systems have been known to survive very high winds over the years. Wind pioneer Martin Jopp of Princeton, Minnesota reported that two tornadoes have passed through his area during the 25 years he had used his wind system. No damage to the wind system occurred in either instance, even though during the second tornado a tree was thrown into Jopp's water-pumping windmill, toppling the tower.

Changing the mechanical settings on a governing device is definitely not recommended. Philip Moore of the Upland Hills Ecological Awareness Center, a school and community center in Oxford, Michigan, related how he tightened the spring tension bolts on the school's flyball governor. He intended to increase the generator's output during high wind speeds. The result was that all three blades were thrown from the governor during a violent storm. One blade flew almost 300 feet! Fortunately, no other harm was done and the blades themselves were not significantly damaged.

In another case, Douglas Brooks, also of central Michigan, tightened the tension bolts on his flyball governor to their maximum setting in order to increase the generator's output. The output did increase, but during a storm with 80 mph winds the overstressed governor flew apart. Two blades landed at the foot of the tower and one landed 25 feet away. Again, there was no other damage. Needless to say, Brooks now cautions wind power users not to push the

Photo. 4-3 and Photo. 4-4 Vertical-Axis Rotors—The Savonius or "S" rotor and the Darrieus or "Eggbeater" rotor are two types of vertical-axis rotors. The "S" rotor is traditionally used for mechanical applications, and the Darrieus is used for both mechanical applications and electrical generation. (Photographs by Joe Carter.)

wind equipment beyond the limits for which it was originally designed. When a wind system is operated within the limits set by the manufacturer, it's just as reliable as any other machine.

Vertical-Axis Rotors

The vertical-axis rotor, in the form of various paddle vane mills, has been used for centuries. Its recent application to electricity-producing wind systems is due to advances in aerodynamics and the development of materials available for the turbines. Vertical-axis rotors are designed as either *drag* types or *lift* types. The drag types, also called panemones, rely on the difference in wind pressure on the paddle vane to cause rotation.

The Savonius rotor is a type of rotor which is interesting due to its extreme simplicity. Also called the "S" rotor because of its shape, many experimenters have built such rotors from 55-gallon oil drums cut in half and welded together into an "S" configuration. While such designs are useful for water-pumping applications, the tip speed of the "S" rotor is too low for electrical generation. Also, the efficiency of the rotor is only about 60 percent of that of a horizontal-axis machine. But for many, the Savonius rotor has been a valuable teaching device.

Typical is the experience of Bob Bartlett, who lives in northwestern Pennsylvania. Bartlett, who has extensive mechanical experience, built a three-tiered Savonius rotor in his backyard and connected it with a belt-drive to a car alternator. Bartlett says that it seldom produced enough electricity to be useful and then only at very high wind speeds. The rotor still sits in Bartlett's yard, but is now surrounded by two working wind systems that power his home.

Lift-type vertical-axis machines use aerodynamically shaped blades similar to those on a horizontal-axis system to create an aerodynamic "lift," just like that of an airplane. Lift-design vertical-axis machines are about as efficient as horizontal-axis rotors. One version, called either the Darrieus rotor after its inventor or the "Eggbeater" because of its shape, shows great promise. A commercially available version of the Darrieus, built by the Dynergy Corporation, has a 15-foot diameter rotor made of extruded aluminum. For speed control, the Darrieus rotors rely upon blade stall, along with an electric brake to shut them down completely. In blade stall, a mechanism slows the rotor at a designated high wind speed so that the rpm of the blades do not increase as fast as in a normal range of wind speeds.

Another lift-design vertical-axis turbine, called the H-rotor, cycloturbine, or gyromill, uses straight blades rather than curved ones. Unlike the Darrieus rotor, the H-rotor can be made to be self-starting. The straight blades of the H-rotor enable the pitch to be varied for self-starting, and for speed control as well. Both the H-rotor and the Darrieus rotor can be used with a variety of electrical generators or for other applications, such as pumping water or compressing air. (For more information, see Ch. 10.)

Towers

As discussed in Chapter 1, the main function of the tower is to get the wind turbine up high enough so that it is free from turbulent winds near obstructions such as trees and buildings. Other considerations in choosing a tower are its strength, cost, and aesthetics.

The tower must be strong enough to withstand the force of the wind against it. Also, the force of the wind on the rotor (the lateral thrust) acts in such a way as to push the tower over. Proper anchoring of the tower is important to withstand these forces, and the greater the size of the turbine, the greater the lateral thrust. Of course, the tower must be strong enough to support the weight of the wind system.

Towers are subject to vibrations due to interactions between the wind, the motion of the turbine, and the electrical "hum" sent up by the generator. For this reason, it is not recommended that towers be installed on the roofs of buildings. The vibrations could be detrimental to the building, the tower itself, and to your nerves. Towers have a natural frequency or point at which they vibrate most, just as a violin string resonates at a definite pitch. If the manufacturer of a wind system recommends a particular type of tower to use with his equipment, seriously consider it. Some manufacturers have found that their systems work best with certain towers and that other types tend to vibrate more. Experienced tower manufacturers or suppliers can also recommend tower specifications given the size of the wind turbine, the weight of the system, the maximum wind speeds to be encountered, and whether the rotor is an upwind, downwind, or a vertical-axis model.

The two basic types of towers are the self-supporting tower and the guyed tower. Guyed towers for wind systems are similar to those used for antennas in that guy wires hold the towers up. The guy wires extend out almost as far from the base of the tower as the tower is high, about 80 percent of the height. Guyed towers have the advantage of being lower in cost than self-supporting towers, but they take up more ground area since the guy wires must be anchored. Also, the tension on the guy wires must be properly maintained. If one wire becomes loose, or is cut, the tower will be severely weakened and can come down. Potential users of guyed towers should consider whether they can properly maintain guy wires over a period of years and if nearby trees or an unaccounted-for vehicle could cause an accident.

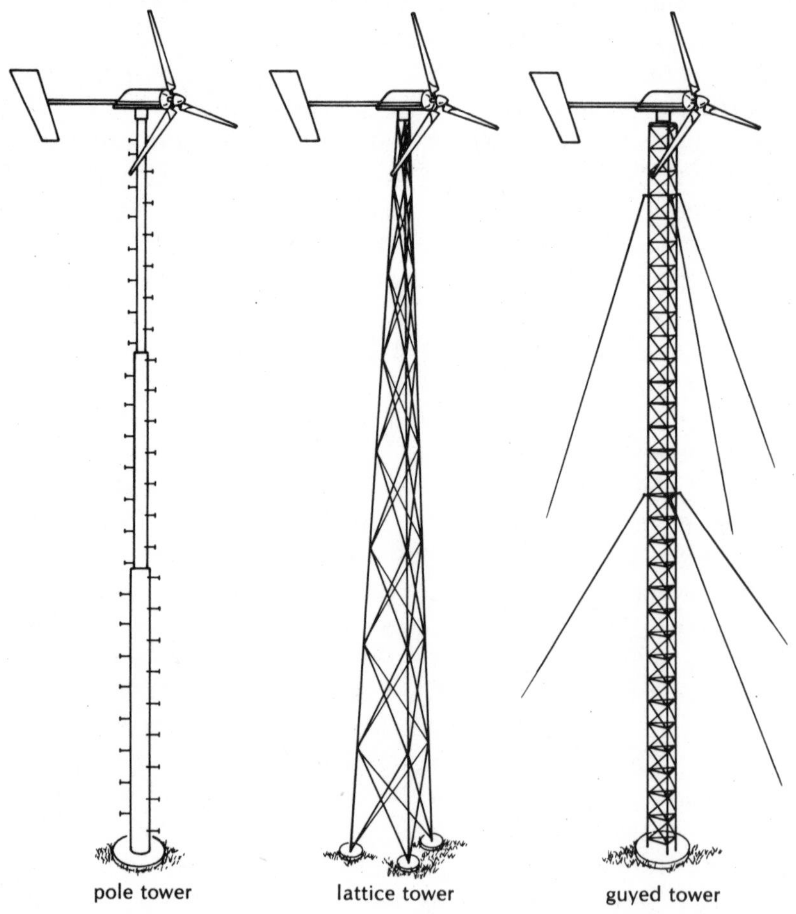

Illus. 4-11 Wind Power System Towers—Whether they are self-supporting or braced by guy wires, wind system towers perform the same function; they position the generator as far from obstacles and ground turbulence as possible.

Aluminum towers designed for TV antennas should *not* be used for wind systems because they are not strong enough to support generators.

Self-supporting towers of the type used for water-pumping windmills are suitable for use with some of the smaller wind systems. The generator is mounted on an adapter at the top of the tower. Most water-pumping windmill towers have an angle of about 76° between one leg and the ground. This means that the base of the tower is about 1/5 of the tower height in width. Thus, a 60-foot tower is about 12 feet wide at the base. The wide base of this kind of tower provides a good safety factor.

Rotor diameters of the water-pumping tower are limited to about 15 feet since the blades would hit the tower if they were longer. Narrow-based latticed towers need considerably more concrete volume to anchor the base than the wide-based towers. Self-supporting towers come in three or four leg models.

A relatively new type of self-supporting tower, called the *octahedron* tower, employs the principles of triangulation used in geodesic domes for strength. Octahedrons cost more to manufacture but have a greater aesthetic appeal to many people.

Treated utility poles serve well as towers for some of the smaller wind units. Poles are probably the lowest cost of the available towers but they are definitely limited in lifetime, since they eventually rot in the ground. Pole towers also require guy wires for additional support.

Tubular steel towers tend to be more expensive than other types but they conceivably are the most aesthetically pleasing in appearance. One design sold by Solargy Corporation consists of interlocking sections that are easy to raise into place.

Chapter 5

On the Ground: Batteries, Inverters, and Controls

The wind does not always blow when you are using electricity, and so, in stand-alone wind systems, batteries are extremely useful to store electricity until it is needed. An analogy is that if the electrical current flowing through wire is considered to be similar to water in a pipe, then a battery is similar to a water tank.

The batteries in a wind system also serve as a voltage regulator. As the wind turns the generator at different speeds (from second to second), the electrical voltage varies, as does the current. Electrical voltage is like water pressure in a pipe, and, since the voltage varies widely with changes in wind speeds, this greatly varying electrical "pressure" could damage lights, motors, and other appliances connected directly to the wind generator. Although the voltage of a battery changes somewhat as it is charged and discharged, it does maintain its voltage within a certain tolerable range for those electrical appliances powered by the batteries.

A number of storage batteries connected together to provide the desired voltage and amount of storage capacity constitutes a battery bank. Each battery consists of one or more cells. In the type of batteries generally used in automobiles, for standby power, and with wind systems, the voltage of each cell is 2 volts (v.) A 12-volt car battery, then, is made up of six 2-volt cells, while industrial batteries for standby power often consist of only one 2-volt cell per battery.

Each cell in a battery is made up of a number of positive and negative plates that are kept apart from each other by separators. The plates are made of lead compounds and the separators are made of porous, nonelectrical materials such as rubber, wood, or plastic. The plates are submerged in an electrolyte, usually dilute sulphuric acid, and water familiarly known as battery acid. The separators prevent the plates from touching each other but allow the electrolyte to seep through it. At present, this lead-acid battery is the most common and generally the lowest cost battery for automotive and industrial-storage use.

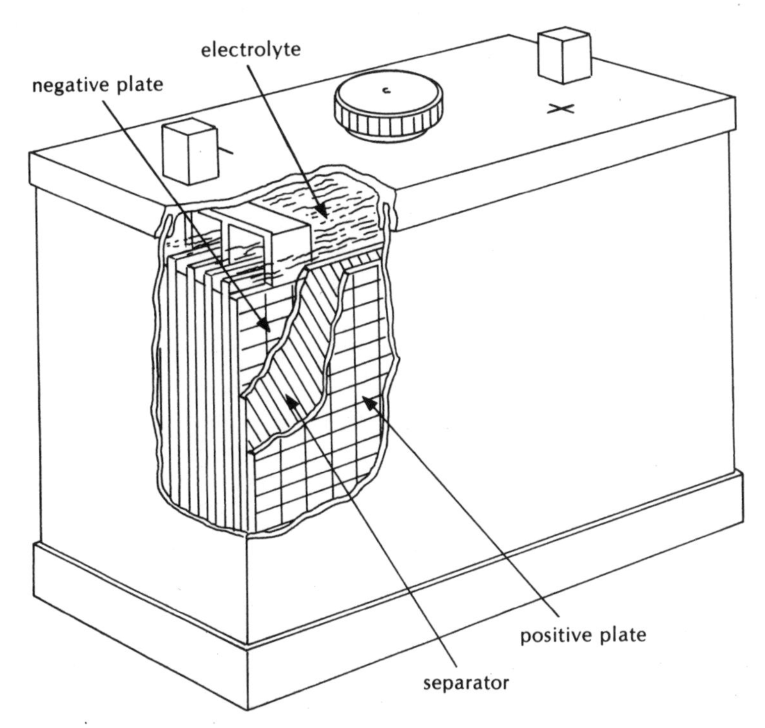

Illus. 5-1 The Parts of a Typical Battery—Plates made of lead compounds interact with the electrolyte to store electricity. Separators keep the plates from touching while still allowing free passage by the electrolyte.

Battery Chemistry and Types

When electrical pressure or voltage is applied to the plates of a battery cell, a chemical reaction takes place. This reaction is reversed as current is drawn from the cell, such as when a light bulb is connected to the battery. In a fully charged cell, the negative plates are made of sponge, or soft lead which is gray in appearance. The positive plates are made of lead peroxide which is dark brown in appearance. When electrical current is drawn from the battery to discharge the battery cell, the sulphuric acid breaks down into hydrogen ions and sulphate ions. At the negative plates, the sulphate ions combine with the sponge lead to form lead sulphate which is whitish gray in appearance. At the positive plates, the lead peroxide loses its oxygen ions to the hydrogen ions that came from the sulphuric acid. These combine to form water. In turn, the lead ions in the positive plate combine with sulphate ions to form lead sulphate. Thus, as discharging takes place, both the negative and positive plates change from sponge lead and lead peroxide, respectively, to an increasing percentage of lead sulphate. Also, the electrolyte changes from a high concentration of sulphuric acid to a lower concentration, with more water being formed.

The electrical pressure or voltage of the cell changes as it is discharged, too. A fully charged cell indicates about 2.2 v. when a voltmeter is connected to it. When the cell is considered to be completely discharged, the voltmeter reads 1.75 v.

As the battery is recharged, the chemical action is reversed. During charging, water in the electrolyte breaks down—is electrolyzed—into hydrogen and oxygen ions. The hydrogen ions attract sulphate ions from the lead sulphate that has formed on both the positive and negative plates during discharging. The hydrogen and sulphate ions combine to form sulphuric acid. As this occurs, the negative plate changes back to sponge lead and, at the positive plate, lead ions combine with oxygen ions from the electrolyzed water to form lead peroxide once again.

During charging, small amounts of hydrogen and oxygen are given off as a result of the chemical action. These gasses escape through the vent plugs in the top of the battery. Since hydrogen and oxygen can be explosive, the batteries should be installed in a ventilated room and kept away from sparks (see Ch. 9). Explosion-proof vent plugs that trap these gases so they cannot ignite are available from battery suppliers at about $5 per cell.

The hydrogen and oxygen that escape during charging come from the electrolyzed water. So, as they escape, the amount of water in the electrolyte decreases. Periodically (every several months or so), distilled water must be added to the cells to replace the water lost through gassing.

Lead-Acid Plates and Grids

Lead-acid batteries are classified by the types of plates within them. Most common batteries have Faure-type plates in which the sponge lead or lead peroxide is pasted onto a lead grid. For heavy-duty, long-life batteries, Plante model plates are used, which are basically heavy sheets of lead. One variation, the Manchester grid, is a plate made with spiral buttons of lead forced into a lead grid. This exposes a large area of lead to the electrolyte. If kept close to a full charge, with a minimum loss of water, lifetimes of up to 30 years have been reported for this type of cell. Batteries with Plante cells cost up to 50 percent more than those with Faure plates, however.

Battery grids in lead-acid batteries are made of either lead-antimony or lead-calcium alloys. The antimony or calcium is present in small amounts in order to make the grid stronger than if it were made of pure lead. The grid holds the sponge lead or lead peroxide in place and must resist damage due to physical shock or temperature change. In addition, the alloy used has an effect on the lifetime and maintenance requirements of the battery.

As a battery with lead-antimony grids gets older, the antimony tends to migrate out of the grid and onto the negative plate material. This alters the chemistry of the battery and causes the battery to partially discharge itself as well as to increase the amount of gassing while the battery is being charged. Thus, as the battery ages, it requires increasing amounts of charging current to maintain a steady charge and it also requires more frequent additions of water to the electrolyte.

On the other hand, lead-calcium grid batteries require less current to maintain a steady charge, which is important for floating service use, discussed later in this chapter. They also require less frequent additions of water to the cells. Automotive batteries classified as "maintenance-free" use lead-calcium grids, for instance.

Alkaline Batteries

Another type of battery is the alkaline battery, invented by Thomas Edison. The oldest type of alkaline battery is the Edison, or nickel-iron cell. In an Edison cell, the positive plate is made of nickel dioxide and the negative plate is made of iron. The electrolyte is a solution of potassium hydroxide and water. The voltage of an Edison cell is only 1.2 v. compared to the 2.2 v. of a lead-acid cell. This means that more cells are needed per battery bank. Also, it is difficult to tell the state of charge of an Edison cell until it is completely discharged. Their main drawback, though, is that they are much more expensive to manufacture than lead-acid batteries.

Edison cells have some very interesting characteristics, however, which make them suitable for either electric vehicle or wind energy use. They have extremely long lifetimes—4 times or more than that of lead-acid batteries. They are also extremely rugged, being relatively unaffected by vibration or cold weather. Heavy charging or discharging does not damage Edison cells either, a characteristic particularly valuable in wind energy use.

Ampere-Hour Ratings

Batteries are rated according to their storage capacity in units called ampere-hours (a-h.) A battery rated at "200 a-h. at the 8-hour rate" can be discharged at the rate of 25 amp. for 8 hours. In theory, the battery can also be discharged at a rate of 200 amps. for 1 hour; 50 amp. for 4 hours; or 12.5 amp. for 16 hours. Actually, experience has shown that at the heavier discharge rates, the battery loses its charge proportionately faster than at the slower rate. For example, the "200 a-h. at the 8-hour rate" battery may actually be capable of discharge at 200 amp. for only ¾ hour instead of the anticipated 1 hour.

Usually, ampere-hour ratings are given as just so many a-h. A battery is rated as a 100 a-h. for example, with the hour rate not given except in the manufacturer's literature. Basically, ampere-hour ratings are useful for making comparisons between different batteries of the same general style. They are also useful in "sizing" a system—choosing the particular model number of battery offered by manufacturers that will match the storage requirements most closely.

The weight of a battery also plays an important role, because the amount of lead in a battery is an indication of its capacity and expected lifetime. A 50-pound electric vehicle battery rated at 225 a-h. will not last as many years as a 150-pound industrial truck battery with a 225 ampere-hour rating.

The total amount of kwh. of energy that can be stored in a battery bank can be found by multiplying the ampere-hour rating of the batteries by the total voltage of the battery bank and dividing by 1,000. For example, a 120-volt battery bank with batteries rated at 180 a-h. each can store 21.6 kwh. (120 v. × 180 a-h. ÷ 1,000). The number of days of storage capacity for the battery bank can be found by dividing the kilowatt-hour capacity by the average daily use of electricity in kwh. If the average daily use is 6 kwh. (180 kwh. per month) in the example just given, then the storage capacity would be about 3½ days (21.6 kwh. ÷ 6 kwh. per day).

In practice, it is not necessary to provide for more than three to five days of storage capacity. If the area has sufficient wind, the few times a year when the wind is inadequate for over a five-day period can be taken care of by an auxiliary power source. Typically, an ampere-hour rating of 150 to 350 is used with 120-volt wind systems. For 12-volt or 32-volt wind systems, a battery bank of 350- to 1,000-ampere-hour ratings is common.

Lifetime

Whenever a battery goes from full charge to full discharge, one cycle of its lifetime is considered to have passed. After many cycles, the battery starts losing its ability to turn the lead sulphate on its plates back into sponge lead or lead peroxide. A portion of the lead sulphate becomes hard and brittle, causing a loss in the ampere-hour capacity of the battery. This process, called *sulphation,* is irreversible and eventually causes the battery to become useless. Manufacturers test their various batteries for the number of full charge-to-discharge cycles that the batteries can take. Unfortunately, the cycle lifetime of batteries is not usually included in the sales literature, but you might be able to obtain the figure from the manufacturer.

The state of charge in a battery bank used in a wind system will fluctuate from day to day according to the amount of wind available and how much energy is being drawn from the battery bank. A wind system battery, however, is usually only partially charged or discharged each day. In contrast, a battery in a warehouse forklift truck goes through a complete cycle each day, being almost completely discharged every day and then recharged every night.

The wind system owner is at an advantage in this, because if a battery is never discharged below 50 percent of its ampere-hour capacity, it will last considerably longer than one that is completely discharged and recharged each time. For example, if the manufacturer estimates the lifetime of the battery at 2,000 cycles, the battery may actually last for 5,000 partial cycles. In other words, it takes about 2½ to 3 partial cycles (the battery is not discharged below 50 percent of capacity) to age the battery as much as one complete charge-discharge cycle. The battery rated at 2,000 cycles would last 5½ years (2,000 ÷ 365 days) if cycled every day. If the battery averages one partial cycle per day, however, it would last 14 years (5,000 ÷ 365).

Undercharging and overcharging also have an effect on battery lifetime. If a battery is allowed to sit for long periods at a low charge, say 80 percent discharged, the plates begin to sulphate and become more difficult to recharge. A completely drained battery that sits for only a month will begin to deteriorate. After six months, the battery would sulphate completely and thus be ruined. Undercharging can also occur if the battery is never charged to its full capacity. If a battery's charge is never over 80 percent, for example, it will eventually lose about 20 percent of its capacity. It is best to have the batteries in a wind system brought to a full charge once, or preferably twice a month, and they should be discharged below 50 percent of capacity as infrequently as possible.

If the batteries are consistently undercharged, either the wind system has too small a rating for the size of the battery bank, or the amount of energy used from the batteries is too much in excess of the energy provided by the wind. In the latter case, either a larger wind system is necessary or an auxiliary power source must be used to maintain the proper battery charge.

Whereas undercharging of a battery leads to sulphation, overcharging physically damages the plates. Once the battery is fully charged, any further charging drives off water from the electrolyte in the form of hydrogen and oxygen. If heavy overcharging occurs, the rapidly escaping gas bubbles can loosen plate material, mostly on the positive plate, and this material settles as a brownish red (lead peroxide) sediment at the bottom of the battery. Also, as the electrolyte loses water, the concentration of sulphuric acid increases. A heavy concentration of acid beyond the normal range attacks and corrodes the plates. As overcharging occurs, the battery heats up and, if the battery gets hot enough, the plates can actually buckle.

Overcharging is not commonly a concern, however, with a properly designed wind system. As we will discuss later in this chapter, a regulator in the wind system control panel slows or "tapers" the rate of charging current to a "trickle" charge as the batteries near full capacity. Actually, a slight overcharge about once a month is good for the batteries. It has the effect of ridding the plates of any slight sulphation and also of bringing the sulphuric acid concentration up to the same level in each cell. Such a maintenance overcharge is called an *equalizing charge.*

Connecting Batteries

Batteries must be connected together to obtain the required voltages or ampere-hour capacities for different applications. If a 120-volt battery bank is required, for example, and 6-volt batteries are being used, then twenty 6-volt batteries must be connected in *series* to obtain the required voltage. Connecting in series means to connect the positive terminal of one battery to the negative of the next one, and so on. When connected in series, the voltage of the battery bank is the sum of the individual voltages of each battery or cell. The ampere-hour rating of the battery bank remains the same as that of the individual batteries, however. For example, if sixteen 2-volt cells with an ampere-hour rating of 100 a-h. each are connected in series, the voltage of the battery bank would be 16 $\times$ 2, or 32 v., and the capacity would be 100 a-h. The kilowatt-hour storage capacity would be 32 $\times$ 100 $\div$ 1,000, or 3.2 kwh.

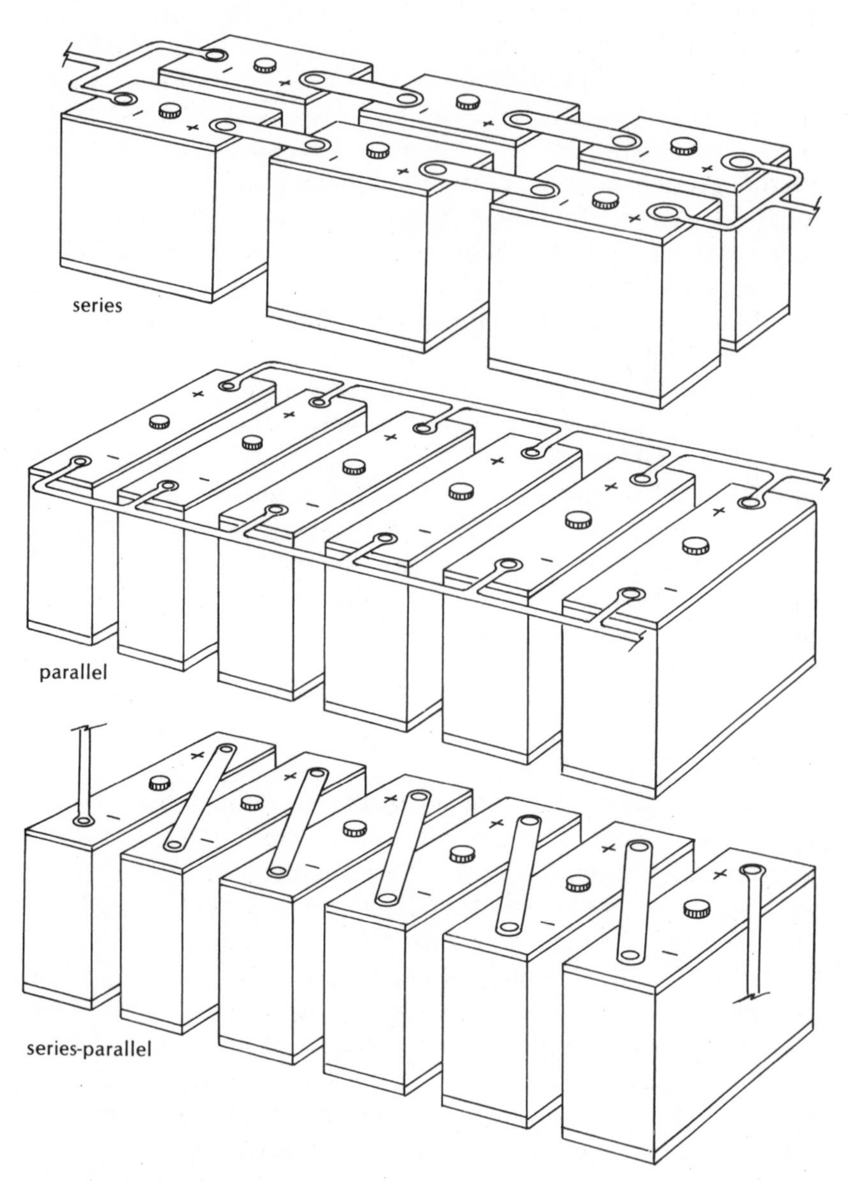

Illus. 5-2 Connecting Batteries—Batteries can be connected in series, to increase voltage; in parallel, to increase the number of ampere-hours in a battery bank; or in a combination of each called series-parallel.

To increase the storage or ampere-hour capacity of a battery bank while maintaining the same electrical "pressure" of voltage, the batteries are connected in *parallel.* That is, the positive terminal of one battery is connected to the positive terminal of the other battery and the negative terminal of the one battery is connected to the negative terminal of the other. This is the same connection used to connect cables from one car battery to another when "jump-starting" a car. If two 12-volt batteries rated at 180 a-h. each are connected in parallel, the resulting battery bank would have a voltage of 12 v. and a capacity of 2 × 180, or 360 a-h.

Batteries connected in series should always be of the same ampere-hour capacity. When a battery bank of mixed capacities connected in series is charged, the smallest capacity battery will be overcharged and the larger battery will probably not be charged enough.

If one battery in a series of batteries is connected in the wrong direction, or "polarity," that battery will soon become discharged. In fact, if it is left in that condition long enough, the battery will even start recharging in the opposite direction. The negative plates begin turning to brownish lead peroxide and the positive plates begin turning to sponge lead! Although such a reverse charge certainly does not extend the lifetime of a battery, it can be brought back by discharging the battery and then recharging it slowly in the correct polarity.

Batteries connected in parallel must be of the same electrical voltage. If you connect a 6-volt battery in parallel with a 12-volt battery, the voltage between the two batteries will quickly equalize; that is, the 12-volt battery will discharge itself by heating up the cable connecting the two batteries until its voltage declines to 6 v. also.

If a series battery bank is connected in parallel with another series battery bank of the same voltage, the arrangement is called a *series-parallel* connection. For example, if a series battery bank of 32 v. consisting of sixteen 2-volt, 200 ampere-hour cells did not have enough capacity, the capacity could be doubled by connecting another series battery bank of sixteen 2-volt, 200 ampere-hour cells in parallel with the first. One possible drawback with such an arrangement is that any weak cells in one side of the battery bank will tend to draw down the voltage of the other battery bank. It is best to select batteries with the capacity you desire so that series-parallel connections aren't necessary. If you must connect your batteries in a series-parallel arrangement, isolating diodes in series with each bank will prevent either bank from discharging the other while still allowing the flow of electricity to its use in your home.

Measuring the State of Charge

Since it's impossible to look at a battery and know if it's full, as you can with a bank of water, instruments that indirectly monitor the battery's state of charge must be used. These include the hydrometer, the voltmeter, and the electronic battery indicator gauge.

Hydrometers

A hydrometer measures the *specific gravity* of the electrolyte. The specific gravity is the weight of the electrolyte compared to the weight of the same volume of water. The specific gravity of water is 1.000 and that of pure sulphuric acid is 1.840, meaning that the acid is 1.840 times heavier than water. Since the electrolyte in a battery is a mixture of water and sulphuric acid, the specific gravity of the electrolyte will always be greater than 1.000 and less than 1.840. In actuality, the specific gravity of the electrolyte is 1.110 when the battery is completely discharged and about 1.300 when completely charged. The concentration of sulphuric acid and thus, the specific gravity, increases as the battery is charged.

Hydrometers are inexpensive instruments and can be obtained at any auto supply store. To measure the specific gravity of a battery cell, remove the battery cap from the cell and take a sample of the electrolyte by gently squeezing and then releasing the rubber bulb on the hydrometer while holding the tip of the hydrometer tube in the electrolyte. The float inside the hydrometer tube will rise to a given level and the number on the float even with the liquid's level in the tube will indicate the specific gravity of the cell. For accuracy, the hydrometer should be held straight up and down while being read.

The rule for knowing the state of charge of the battery based upon a hydrometer reading follows: each change of .030 (30 points) from the maximum specific gravity of the cell indicates a change of 25 percent in the state of charge of the battery. For example, if the hydrometer reads 1.300 when the battery is fully charged, it will read 1.270 (1.300 minus .030) when the battery has lost 25 percent of its charge. In other words, it is at 75 percent of full capacity. Similarly, a reading of 1.240 would indicate 50 percent and a reading of 1.210 would indicate that the battery is at 25 percent of full charge. A reading of 1.110 is considered to be a 100-percent discharge condition for any battery.

Actually, the maximum hydrometer reading for a fully charged battery varies with the type, design, and manufacturer of the battery. It is good to know this maximum number when first choosing and using a set of batteries. Typically, the maximum specific gravity for deep-cycle batteries is 1.275 to 1.300; for automobile batteries, it is about 1.260; and for stationary or floating-type batteries, it is around 1.210.

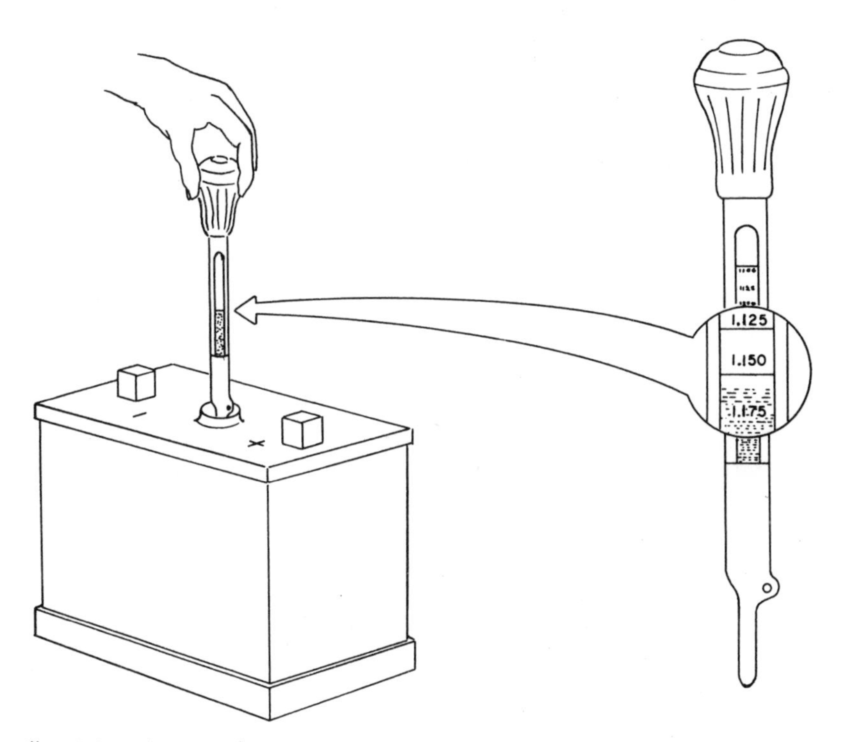

Illus. 5-3 Using a Hydrometer to Measure Battery Charge—Hydrometers gauge the specific gravity of the battery's electrolyte. If mostly water is in the electrolyte, the battery's charge is low.

To use a hydrometer accurately, the reading must be corrected according to the temperature. Most hydrometers include a thermometer to read the temperature and a scale giving the number to add or subtract from the hydrometer reading for the given temperature. The ideal temperature for a battery is 77°F. The capacity of the battery is reduced somewhat at any temperature above or below 77°F. The correction for the hydrometer reading amounts to .001 (1 point) for each 3-Fahrenheit-degree change in the temperature. For example, in Illustration 5-3 the hydrometer reads 1.160, but the temperature is 42°F, so that a correction of .012 ($[77-42] \div 3 \times .001$) is subtracted from the reading. The corrected hydrometer reading is thus 1.148 (1.160 – .012). To make things simpler, a new type of hydrometer is now available with readings that are automatically corrected for temperature. They have an easy-to-read dial instead of a float and are available from many auto supply stores.

Temperature not only affects the hydrometer reading, it also significantly affects battery capacity. Battery capacity is considered to be 100 percent at 77°F. In other words, if a battery is rated at 100 a-h., that rating is true at 77°F. At freezing, 32°F, the capacity of a battery is only about ¾ of its maximum. Temperatures higher than 77°F actually increase the capacity of a battery, but this is not desirable since it also shortens the battery's life. Ideally a battery should be kept between 60° to 85°F for longest life and greatest capacity.

Fortunately, the electrolyte in a battery acts as an antifreeze, which is why a car battery does not freeze unless it's completely discharged. If the hydrometer reading is 1.280, the electrolyte will not freeze until it reaches −90°F, which is rather unlikely. At a hydrometer reading of 1.230, the solution will not freeze until temperatures of −40°F are reached, but at a reading of 1.110, or complete discharge, the battery will freeze at 16°F.

Hydrometer readings give the "true" state of charge of a battery when corrected for the temperature, but it should be pointed out that readings are accurate only when the battery is not being charged and has not been for several hours. This is because there is a lag time between the electrolyte reading and the state of the battery cells due to stratification of the electrolyte during charging.

Reading a hydrometer is relatively simple, but it is an inconvenient way of making a daily check on the state of the battery charge.

TABLE 5-1
Freezing Point of Battery Acid

Hydrometer Reading	Freezing Point, °F
1.300	−95
1.280	−90
1.270	−82
1.260	−74
1.250	−63
1.240	−50
1.230	−40
1.220	−32
1.210	−23
1.110	+16

As discussed in Chapter 10, the best use of the hydrometer is in making periodic checks of each cell in the battery bank. If the hydrometer reading of a cell varies from the other cells by more than .050 (50 points) several hours after the battery bank has been given an equalizing charge, then the cell is probably bad.

Voltmeters

A voltmeter that is permanently connected to the battery bank is the simplest and most convenient way of monitoring the state of the batteries. As the battery is charged, the voltage of each cell increases just as the specific gravity does. A voltmeter connected to a completely discharged battery cell will read 1.75 v. (as long as the cell is out of use, and no appliance is drawing any remaining current). The voltage of a fully charged cell depends upon the specific gravity of the electrolyte, but it generally falls between 2.1 and 2.2 v. When a battery is being charged, the voltage of each cell will rise as high as 2.5 v. This is temporary, and after charging stops for a few hours, the voltage settles back to the 2.1 and 2.2 volt range. Between complete discharge and full charge, the voltage of each cell can be found by adding .84 to the specific gravity of each cell. For example, if the hydrometer reading is 1.200, the voltage of the cell would be 1.200 plus .84, or 2.040 v. This applies only when the battery is not being charged at the time the hydrometer reading is taken.

The voltage of the complete battery bank is just the number of cells multiplied by the voltage per cell. Consider a battery bank consisting of 6 cells, for example. Nominally, this is a 12-volt battery bank. When the battery is completely discharged, a voltmeter connected across its terminals will read 10.5 v. (6 × 1.75 v. per cell). When completely charged, the voltmeter will indicate 13.2 v. (6 × 2.2 v. per cell). Similarly, a 16-cell battery bank that is nominally a 32-volt battery will indicate from 28 to 35.2 v. on the meter when going from empty to full charge. A 60-cell battery bank, which is nominally a 120-volt battery, will range from 105 to 132 v.

In practice, a voltmeter is useful in indicating whether a battery is fully charged or low in charge. An accurate reading of the state

TABLE 5-2
Battery Charge Measured by Hydrometer and Voltage

	State of Charge	Hydrometer Reading	Volts per cell
Battery A	100%	1.260	2.10
	75	1.230	2.07
	50	1.200	2.04
	25	1.170	2.01
	Discharge	1.110	1.75
Battery B	100%	1.280	2.12
	75	1.250	2.09
	50	1.220	2.06
	25	1.190	2.03
	Discharge	1.110	1.75
Battery C	100%	1.300	2.14
	75	1.270	2.11
	50	1.240	2.08
	25	1.210	2.05
	Discharge	1.110	1.75

of charge is difficult, however (and generally not necessary), since for the sake of accuracy the reading must be adjusted for temperature changes. Also, if the battery has been charged within the last few hours, the reading will be higher than normal.

A good rule of thumb is never allow the battery bank voltage reading to go below its nominal value of 2.0 v. per cell. For instance, say a battery has a specific gravity of 1.260 and a voltage per cell of 2.10 when full. At ¾ of full capacity, the hydrometer reads 1.230 and the voltmeter reads 2.07 v. per cell. Likewise, at ½ capacity the readings are 1.200 and 2.04 respectively, and at ¼ of capacity the readings are 1.170 and 2.01 respectively (see Table 5-2). If the battery bank consists of 60 cells, a voltmeter connected across its terminals will indicate 126, 124.2, 122.4, and 120.6 v. respectively, in going from full, to ¾ to ½ to ¼ capacity. These are only small changes in the meter scale. However, once the meter drops below 120 v. it is safe to assume that the battery is nearly empty; if it is not at least partially charged within a day or so, its voltage will quickly drop to its low value of 105 v.

Battery Indicator Gauges

For automatic indication of the state of the battery charge, an electronic battery indicator gauge can be used. These gauges are used on forklift trucks and other electric vehicles and have a meter dial that indicates empty to full just like an automobile gas gauge. The indicator meter is operated by an electronic circuit that monitors the state of the battery. Some activate a light to indicate low charge and others operate an electrical relay or switch that can sound an alarm, or start an auxiliary power source. Such gauges are convenient but they cost over $100 each and are not essential.

Battery Design for Different Service

Batteries are generally divided into two types, commercial (automotive) and industrial. Automotive batteries are the most common type of lead-acid battery in use today. They are designed to provide large amounts of electrical current in a very short time, as much as 400 amp. just to start a car's engine. To give the battery this high charge and discharge capability, the cells are made with relatively thin and closely spaced plates. This leaves a large amount of plate area exposed to the electrolyte and allows for rapid discharging to take place when the engine is being started. Automotive batteries cannot take the continuous charging and discharging that a battery must undergo when used in a wind system, however. If used in a wind system, they would only last a year or two at most.

Industrial batteries are made with fewer but thicker plates than automotive batteries so that they can provide lesser amounts of current, but over longer periods of time. Industrial batteries are designed for either floating service or deep cycles.

Floating Batteries

Floating batteries are kept fully charged, or "floating," with an electronic battery charger until they are needed for some emergency use. Also called stationary batteries, they may be designed for communications (as in telephone offices), for emergency lighting and engine starting (to provide emergency power in buildings), for uninterruptible power service (U.P.S., such as those used to operate computers during power outages), or for switchgear and control in power stations.

Floating batteries are designed to require a minimum amount of current to maintain the battery at full charge, as well as to lose a minimum amount of water from the electrolyte while being charged. Typically these batteries are guaranteed for 20 years and may last for 25. This guarantee assumes that the batteries are constantly maintained at full charge except for relatively brief discharge periods.

In floating service use, a lead-calcium battery will last 20 to 25 years compared to 12 to 14 years for an equivalent lead-antimony battery. Because of their lower maintenance requirements and longer lifetime, lead-calcium batteries are increasingly being used for floating service even though they cost 20 to 30 percent more than lead-antimony batteries. Batteries designed for floating service, however, are not made to be cycled repeatedly. Consequently, they are *not* recommended for use in wind power systems.

Deep-Cycle Batteries

Deep-cycle batteries are designed to withstand a large number of charge and discharge cycles and for this reason are best suited for use with wind systems. Types of deep-cycle batteries include industrial truck batteries, such as those used in electric forklifts, and electric-vehicle batteries, such as those used in golf carts.

Industrial truck batteries are designed for a longer cycle lifetime than electric-vehicle batteries, but cost 2 to 4 times more. A typical industrial truck battery may be designed for a lifetime of 1,800 cycles. If such a battery is never discharged below 50 percent of capacity, it may last for 5,000 partial cycles. Assuming one partial cycle per day, the batteries would last about 14 years (5,000 partial cycles ÷ 365 days per year). At one partial cycle per 1½ days, the batteries would last 20 years.

When comparing batteries then, assess not only the cost per a-h. of capacity but also the cost per cycle of lifetime. For example, consider two models of 6-volt, deep-cycle batteries. One costs $180.00, is rated at 350 a-h., and has a lifetime of 500 cycles. Another battery costs $350.00, is also rated at 350 a-h., and has a lifetime of 1,500 cycles. The first battery has a cost of $.51 per a-h. ($180.00 ÷ 350 a-h.) and a cost of $.36 per cycle ($180.00 ÷ 500 cycles). Similarly, the second battery costs $1.00 per a-h. and $.23 per cycle. Even though the second battery costs more for its rated capacity, $1.00 as opposed to $.51, it costs less per cycle, $.23 compared to $.36 per cycle, and so is the better buy if you can afford the extra initial cost.

Electric vehicle batteries are physically smaller than industrial truck batteries. They are also designed for a lower number of cycles. Because they are used in golf carts and similar vehicles, weight is a primary consideration. The ampere-hour capacity-to-weight ratio is more of a concern than cycle lifetime with such batteries.

An electric-vehicle battery is designed for a lifetime of about 500 cycles. A 500-cycle, electric-vehicle battery may last about 1,300 partial cycles. Thus, at one partial cycle per day, the battery would last 3½ years. At one partial cycle every 1½ days it would last 5½ years, and it would last 10 years at one partial cycle every three days.

There are other types of batteries being developed for electric vehicle use that possibly have merit for wind energy use, but they are only in the research stage and may well prove to be too expensive because of their reliance on costly metals (see Ch. 11).

Among the types of grids found in deep-cycle batteries, lead-calcium grids are also available for deep-cycle use. However, they will not last any longer than deep-cycle, lead-antimony batteries. The fact that they require less frequent replacement of water is a possible convenience in industrial forklift use but is no real advantage when used in a wind system, given the extra cost. In short, batteries with lead-calcium grids are not recommended for wind system use because their extra cost is not justified by the application. Thus, the most suitable and cost-effective type of battery for use with wind systems is the industrial, deep-cycle, lead-acid battery with lead-antimony grids. Both industrial truck and electric-vehicle, deep-cycle batteries of this type will do well in wind systems.

Used Batteries

Buying used batteries can be one way of cutting the initial cost of a wind system. Of course, it is risky to buy used batteries since their condition and remaining lifetime is in question. However, if you obtain them for their price as scrap, or slightly above that value, then your risk is minimized. Batteries that prove to be worn out can always be sold again as scrap. If the batteries are uncharged when purchased, you will have no way of knowing how long they sat that way, and whether or not the plates have sulphated. It is best to purchase used batteries only if they are fully charged when you get them.

If the battery has a transparent case, check for whitish areas on the plates. If these areas are large, it could mean that the plates may be irreversibly sulphated. If so, the battery cannot hold a charge, and is useless. Also, check for a large amount of sediment in the bottom of the case. At the end of the battery's lifetime the sediment, which comes from the plates, builds up high enough so that it often "shorts out" the positive to the negative plates.

Batteries cannot be rebuilt in the sense of revitalizing the plates. Compounds that are sold to revitalize or restore batteries should definitely be avoided. Some may temporarily perk up the battery but they considerably shorten the cell's lifetime. A battery can be rebuilt by replacing a defective cell with a new one if the battery case has a removable top. This is only worthwhile if the battery is new.

Keep in mind also that the most commonly available batteries from telephone companies and power stations are most likely designed for floating service and not for deep-cycle use. If they are in reasonable shape they will still last a few years but it is not advisable to pay much more for them than their scrap price.

Inverters

You may wonder how a.c. appliances in a stand-alone type wind system can be operated when the electricity from the batteries is d.c. In fact, many appliances will operate on d.c. electricity, which we will discuss in Chapter 7. However, there are some that always require a.c., and some wind power users do not want to bother with a dual a.c./d.c. electrical system in their home. In these instances, inverters are necessary to convert the d.c. electricity from the batteries to a.c.

Inverters come in all sizes, from small ones—units big enough to power a television or stereo—to home-size units that can run all the appliances in the house. They have many different features that make them suitable for specific applications but not for others. Furthermore, inverters are expensive, costing around $1 per w. of rated power, so it is imperative to choose the size of unit that is geared to the appliances you'll be using.

Efficiency

An inverter uses power to convert d.c. to a.c. and the efficiency of the inverter, or the percentage of power that is converted, increases with the amount of power used. For example, an inverter rated at 200 w. may be 89 percent efficient when full power is being used. In other words, 225 w. are taken from the battery bank to deliver 200 w. (.89 × 225 w.) to the appliances in use. However, at 100 w. the inverter may be only 83 percent efficient, meaning that 120 w. are taken from the battery bank to supply the 100 w. to the appliance load.

Inverters also use a small amount of power, called the *idle current,* even when no appliance is connected to it. If the inverter is left on for considerable periods of time, this power drain on the storage batteries can be significant. In a low-wind area the idle power drain of the inverter could equal 10 to 30 percent of the wind system's output. The idle current is not always stated in the manufacturer's literature, so it should be determined before a purchase is made. The idle power drain can be as low as 20 w. and as high as several hundred watts depending upon the size and design of the inverter. Some inverters are designed to avoid this problem by automatically switching to a "standby" mode when not in use, or when only small appliances are in service.

Rotary and Vibrator Inverters

Rotary inverters are low-cost, rugged inverters. Each consists of a d.c. motor that turns an a.c. generator. They are only 60 to 70 percent efficient but cost half as much as other types of inverters. They're available in many different models for different power ranges and for different d.c. input voltages.

Another low-cost inverter is the vibrator type. These inverters use an electrical "vibrator" similar to a doorbell buzzer to convert the d.c. electricity to a pulsating a.c., or "square wave." The square-wave voltage is sufficient to operate certain small appliances but may cause electrical "hum" in televisions and radios. Also, vibrator mechanisms need to be replaced every few years.

Electronic Inverters

Electronic, or "solid state," inverters use electronic circuits containing transistors or silicon-controlled rectifiers (S.C.R.'s) to convert the d.c. to a.c. as well as transformers to increase the a.c. voltage to 120 v. These are generally the most efficient type of inverters and they come in a wide variety of power ranges and input voltages. Most electronic inverters in the low-cost range are of the square-wave type, and electrical hum, or an audible hum due to vibrators in the transformer, may be a problem with this type of inverter, too. The solution is to use an electronic *sine-wave* inverter that produces a voltage the same as or very close to utility company voltage. Tape recorders, for example, operate best on sine-wave inverters.

There is no hard and fast rule for knowing which type of inverter will cause undesirable electrical "hum" on home entertainment appliances. For example, I occasionally use a Wilmore 300-watt inverter that is designed for equipment requiring sine-wave voltage. However, when I tried it with our radio and stereo, it caused a considerable hum. On the other hand, Jim Sencenbaugh, wind power user and distributor of wind products, reports that his square-wave, Topaz inverter creates very little hum or radio frequency interference (RFI) on his Sony color TV and MacIntosh FM receiver. Sencenbaugh points out that in some cases a poorly filtered power supply on the radio or television can be the source of the hum or noise rather than the inverter. So, whenever possible, see a particular model of inverter in operation before purchasing one.

In many cases the hum or RFI problem can be reduced or eliminated by putting an electrical device called a *line filter* in the line cord of the affected appliance. You can obtain these from electronics shops, and an electronics repairman may be helpful in selecting the proper filter. Other features that inverters of good quality have are:

- *Short circuit protection,* which protects the inverter from harm if an electrical "short" occurs in the wiring or appliance being used.
- *Frequency regulation,* which keeps the frequency close to the 60 H necessary for operating such appliances as phonographs, electric clocks, and microwave ovens.
- *Reverse polarity protection,* which protects the inverter from harm if it is connected to the batteries with the wrong polarity.

Most manufacturers of inverters have small ones in the 100- to 500-watt power range that are suitable for small appliances. When used to power refrigerators, compressors, and other appliances with motors, however, the inverter must be capable of handling "inductive loads," that is, electric motors. Whenever a motor is started, it requires a very large current, called "surge," to start it. Only inverters that are advertised as able to handle motors should be used to start them. Otherwise the surge current, which may be 5 to 10 times larger than the normal current put out by the inverter, will damage the inverter.

Before purchasing an inverter for starting motors, ask the manufacturer to supply information on the largest-size motor that the inverter can start without damaging itself. Typically, the largest motor in horsepower (h.p.) that can be started by an inverter is about one-fifth the rating of the inverter in kw. This is because of the large surge of current required to start the motor. Best Energy Systems for Tomorrow inverters have one of the better ratios of h.p. to kw. For example, the 1-kilowatt model can start a ¼-horsepower motor and the 5-kilowatt model can start a 2-horsepower motor. These inverters also automatically adjust the "power factor," the relationship between the voltage and the current, while the motor is running, thus causing it to run more efficiently.

Synchronous Inverters

An inverter that not only converts the d.c. from the wind system to a.c. electricity but also works in conjunction with utility power is called a synchronous inverter. A synchronous inverter can eliminate the need for battery storage in a wind system. When the wind system is generating electricity, the synchronous inverter converts it to a.c. for use by any household appliances that are operating at the time. Any power not supplied from the wind system is supplied by utility power. If the wind system generates more power than that used at the time, the excess power is "backfed" to the utility. As discussed in Chapter 3, in order to get paid for the energy returned to the utility, the system must be metered; several different metering arrangements are possible.

S.C.R.'s inside the synchronous inverter are turned on and off like switches by an electronic control circuit. If the d.c. voltage from the wind system is greater than the a.c. line voltage during the first half of the a.c. voltage sine wave, this control circuit functions (see Illus 4–1). During such an occurence, the control circuit turns on the S.C.R.'s, which temporarily connect the d.c. generator to the a.c. line and a pulse of electric current flows on the a.c. line. The process

Photo. 5-1 A Synchronous Inverter—A synchronous inverter converts varying d.c. electricity to a.c. electricity that is compatible with utility power.

is repeated during the first half of each sine wave of the a.c. voltage.

The synchronous inverter is designed so that when utility power is unavailable, such as during a storm, the inverter ceases to function. This is a safety feature to protect any lineman working on a downed line.

Another concern of the utility company is the effects the wind system might have on its waveform. It's possible that the synchronous inverter could add "harmonics," or distort the sine-wave shape of the electrical current from the utility lines. This should be a minor concern, however, because the synchronous inverters are designed to distort the waveform by less than 2 percent. However, sensitive electronics, such as a home computer system, may require electronic filtering to prevent errors due to any small distortions caused by a synchronous inverter. But, in many instances, this filtering is required whether powered by a wind system or not.

Wind systems with power ratings on the same order of magnitude as the power rating of the line transformer, such as a 15-kilowatt synchronous inverter or induction generator that sends power through a 15-kilowatt line transformer, can also cause distortion. Consult a power company engineer on any such possibilities when installing a 15-kilowatt or larger wind system.

At this time only a few manufacturers offer synchronous inverters to wind power users. The Gemini Synchronous Inverter, sold by Windworks, Incorporated, is used with many types of wind systems as well as in hydroelectric and photovoltaic systems. It is available in 4-, 8-, and 20-kilowatt sizes to match the inverter size with the power rating of the wind system. Proper interfacing between the wind system, the Gemini, and the utility power is important, and the recommendations of the manufacturer should be followed. Generally, the interfacing consists of electrical elements called inductive reactors and electrolytic capacitors. These elements are of the right specifications to match the electrical characteristics of the wind system to those of the Gemini. This interfacing not only increases the efficiency of the system, but can prevent damage to the wind system should a failure of a component occur in the Gemini.

Wind generator developer Terry Mehrkam installed a 45-kilowatt wind system with a Gemini inverter, but without the reactor and electrolytic capacitors. The failure of an electronic component, an S.C.R., in the Gemini caused the generator to "shock load" when it generated power into a short circuit. This caused the turbine to slow down almost instantly, and the top of the tower swayed in an almost 2-foot arc. The only damage was a broken shear bolt on the turbine. Although such a dramatic occurrence probably will not happen with most smaller, home-size systems, if you're thinking of using a synchronous inverter, check first with the manufacturer before installing a unit.

Another synchronous inverter, the Sine-Sync inverter, manufactured by the Real Gas and Electric Company, is more versatile than the Gemini inverter but is also more expensive. This unit can act not only as a synchronous inverter while connected to utility power, it can also act as a regular inverter when utility power is not available. If utility power fails, the unit will automatically switch from one mode to the other. Other companies, such as Power Group International, have developed their own synchronous inverters that are part of their wind systems.

Control Panels, Auxiliary Power, and Standby Plants

Nearly all wind systems include electronic panels designed to control the rate of charge of the generator, among other functions. Besides meters to indicate the charging rate of the generator and the voltage level of the batteries, a control panel generally includes a cut-out circuit and regulator. The cut-out, which may be as simple as an electronic diode, prevents electricity from flowing from the batteries to the generator when the wind is not blowing. The regulator decreases the electrical current supplied to the field or rotor of the wind power generator as the battery voltage rises. This has the effect of cutting down the generator's rate of charge as the battery bank nears full charge.

Another type of regulator, a load control switch, controls the amount of power used rather than limiting the power generated by the wind system. If the battery voltage gets too high due to overcharging by the wind system, the load control switch can connect power-using appliances, such as an electric space or hot-water heater. Thus, if the batteries are full, the excess wind energy can be put to use rather than simply limiting the output of the generator.

Auxiliary Power

When the batteries are low in power and there is no wind, some type of auxiliary power must be used to charge the batteries. How often an auxiliary power source is needed can vary from a few hours per year to several hours per week depending upon the size of the system, the power used, and the amount of wind energy available. For example, during the period of September, 1977 through June, 1978 my own standby generator was used for a total of 14½ hours of operation. Of this, 13½ hours occurred in February and one hour in March.

Standby Plants

Engine-driven generators, or standby plants, are the most common source of auxiliary power. For home and farm use, standby generators come in sizes of from 1,000 to 30,000. For use with a wind system, the standby generator needs to be of a size to charge the batteries at a reasonable rate, but does not have to be large enough to supply all of the electrical needs of the home at once. For most wind systems, this means that the generator should be in the 2,000- to 6,000-watt size. Units for home emergencies often have 3,600-rpm engines instead of the slower-speed, longer-life, 1,800-rpm engines. The emergency units are not meant for extensive use, as are the slower-speed units.

A good rule of thumb is to use an auxiliary generator that has about the same power rating as the wind powered generator. For example, a standby plant charging at a rate of 20 amps of current would run for ten hours to completely fill a 200-ampere-hour battery bank. (Actually, the plant would run longer because of losses while charging, and because the battery voltage rises as it is filled, thus cutting down on the charging rate.)

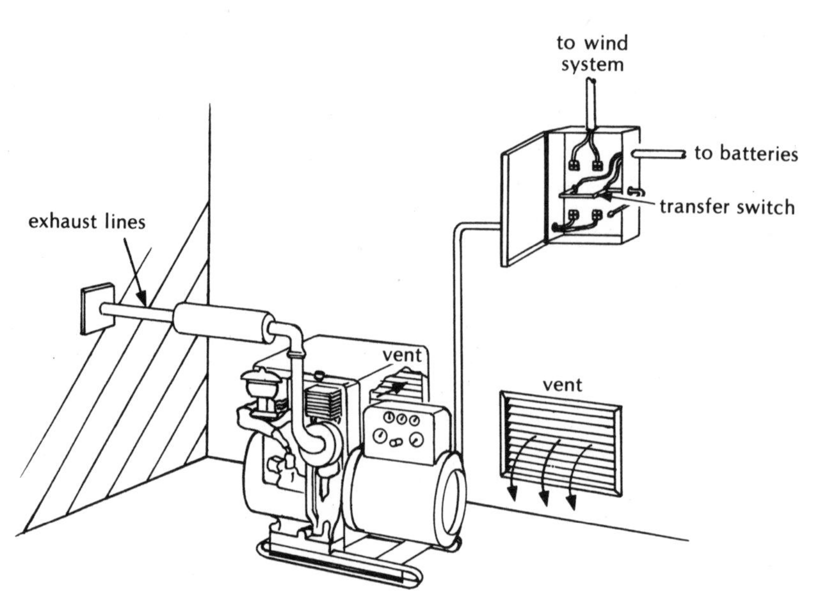

Illus. 5-4 Standby Engines—When the wind doesn't blow, an auxiliary engine will maintain the battery charge. Ventilation and good exhaust-valving are important, as well as a transfer switch to change from charging by the wind system to charging by the auxiliary engine.

Since most modern standby plants have a.c. generators on them, the a.c. must be changed to d.c. with a rectifier or battery-charging unit. A full-wave bridge rectifier, available at any electronics shop, is sufficient to change the a.c. to d.c. electricity. Choose a bridge rectifier with a current rating sufficient to handle the largest charging current it will have to withstand, and a voltage rating of at least twice the a.c. voltage rating of the generator. A battery charger circuit gives greater control of the charging current than a bridge rectifier, but it is much more expensive, too.

For safety reasons, the exhaust of the engine should always be vented outdoors. Also, a transfer switch should be used to prevent the standby generator's electricity from being sent back to the wind power generator. A transfer switch is a double-pole, double-throw switch mounted inside an electrical box (see Illus. 5–4).

Where utility power is available, a battery charger similar to types used for car batteries can be used to charge the battery bank instead of an engine-driven generator. A battery charger with electronic regulation can charge the batteries with a heavy charge and then reduce to a "trickle" charge as the batteries become full.

Finally, auxiliary generators can be powered from gasoline, diesel, or liquified petroleum gas engines. They can be started by hand with a pull rope, with an electric starter, or with an automatic or remote starter. Electric or remote starters are preferable.

Chapter 6

Choosing a Wind System

Deciding which wind system to install involves not only picking the right size system for your needs, it also involves considering its use. Your wind system can be used with storage batteries, connected to utility power, or a combination of both, with several variations of each. Each possibility has its own advantages that may suit your own situation best.

A whole range of manufactured wind systems exist today, with several dozen different models from which to choose. Typical costs for these systems are in the range of $1.50 to $3.00 per w. of rated power exclusive of a tower, storage batteries, and other equipment. Larger systems usually cost less per w. than smaller ones. Many of the currently available wind systems and the characteristics of their rotors and generators (discussed in Ch. 4) are listed in Appendix E. Although these models are subject to design changes by their manufacturers, the appendix gives an idea of the range of features available. Each model in Appendix E is listed according to the size of the wind turbine. The units range in size from the 20-inch-diameter Sencenbaugh model "24–14," which is big enough to power a small TV or radio, to the 40-foot-diameter "445" made by the Mehrkam Energy Development Company, which can power several homes at once.

The rated power and rated wind speed for each model is also

listed in Appendix E. Note that the rated power of each model does not necessarily increase along with increases in rotor size as you might expect. The reason for this is that manufacturers specify the wind system's rated power at different speeds. Power ratings of the listed models are made at wind speeds from 17 to 33 mph. When comparing two models with the same size rotor, the one rated at a higher wind speed will most likely have a higher rated power, since the power in the wind is proportional to the cube of the wind speed. Also, the efficiency of wind systems varies to some extent, so that a model with a larger rotor can have less power output than one with a smaller rotor if the larger one is less efficient.

Range of Size

For purposes of comparison, the different models of wind systems can be grouped into three size ranges: "cabin-size," "home-size," and "all-electric-size" systems. These size classifications are arbitrary, of course, but are useful in considering the approximate size of wind systems used for different applications. Keep in mind that the appropriate size of the wind system for your site also depends upon your average wind speed.

Cabin-Size Systems

Wind systems with rotors from 6 to 12 feet in diameter can be classified as "cabin-size," in that they are suitable for minimal energy-use situations. They are typically used in a small home or cabin that is remote from utility power and where the objective is to power a few lights, a stereo or television, and a few other low-power appliances. They usually have low voltage (12 to 24 v.) generators and produce in the range of 25 to 350 kwh. of electricity per month (depending upon the size of the rotor at an average wind speed of 12 mph.). Their advantages are that they are rugged, low-cost, easy to install by one or two people, and easy to repair. The combination of features that best allows for a simple, rugged design in a cabin-size system is, in my opinion, a direct-drive d.c. generator or three-phase alternator, and a horizontal-axis, upwind, three-blade rotor with a simple rotor speed control, such as a side-

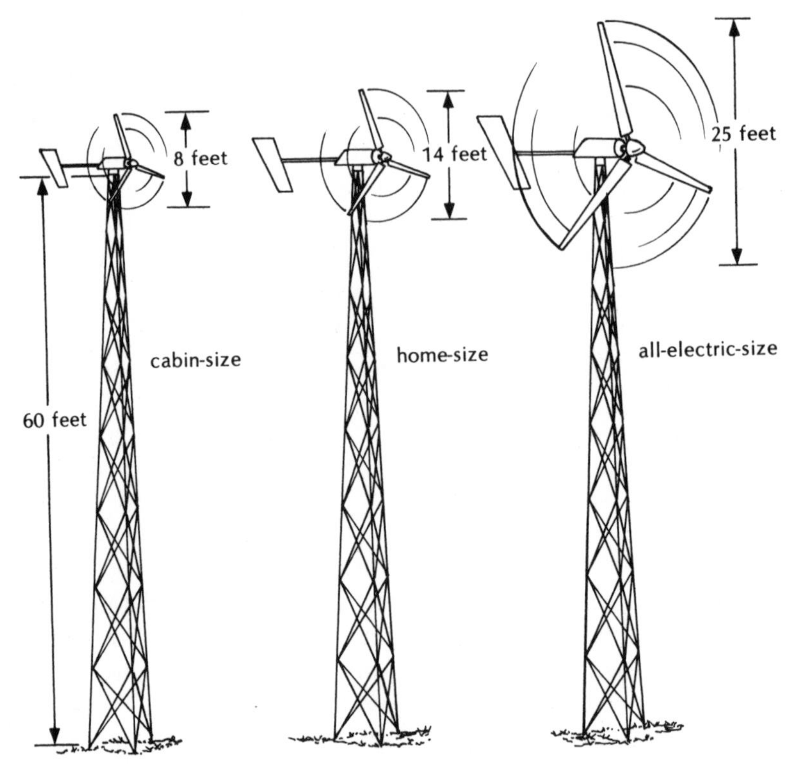

Illus. 6-1 Ranges of Wind Power System Sizes—Cabin-size systems have 6- to 12-foot rotors, and are big enough to power a few lights and appliances in a remote area. Home-size systems utilize 12- to 16-foot rotors to power most or all of the electrical needs of a small home. All-electric-size wind systems depend upon 16- to 40-foot rotors to power an all-electric home or a small business.

tilt control design. Some typical systems in the cabin-size category are the Winco model "1222H," the Sencenbaugh model "500," and the Whirlwind "Model A." These and other systems are described in Appendix E.

Home-Size Systems

Wind systems with 12- to 16-foot rotors can be called "home-size" in that they are capable of powering all or a good percentage of the electrical needs of a home. They can produce in the range of 300 to 500 kwh. per month at an average wind speed of 12 mph. This is the size range in which the greatest number of models are available.

Models with 12- to 13-foot rotors provide overlap between the "cabin-size" and "home-size" range. Most of them have low voltage (12 to 24 v.) generators for use with battery storage, and an optional inverter for converting the d.c. to 120 v. a.c. The Aero Power model "SL 1500," the Sencenbaugh model "1000," and the Dunlite model "2000" are examples of wind systems that fit into this category.

Several of the larger home-size units with 13- to 16-foot-diameter rotors use d.c. generators or alternators that are designed either for charging a 120-volt d.c. battery bank or for use in conjunction with a synchronous inverter. The Dakota Wind and Sun model "BC4" and the Power Group International "Hummingbird" model are examples of systems of this type.

As with cabin-size systems, a home-size unit should be as simple as possible. Three-blade rotors with blade-pitch control, a side-tilt, or a tilt-up design have proven to be reliable over the years. Systems with direct-drive generators are simpler and start easier at low wind speeds, but they do tend to cost more.

Still other home-size units, such as the Enertech model "1800," are designed around the use of an induction generator intended for utility interconnection. With this type of system, the generator itself limits rotor speed. However, at high wind speeds or if utility power is disconnected, an electric brake is commonly used to stop the rotor.

Several home-size units are designed around either a vertical-axis rotor or a "bicycle-wheel" rotor connected to a belt-driven generator. This arrangement allows for the use of different types of generators with the rotor. These systems are also easily adaptable to nonelectrical use, such as running a compressor or a water pump. In fact, they are better suited for nonelectrical applications than horizontal-axis units, since a shaft can be extended easily from the rotor to a mechanical device on the ground. Vertical-axis systems in the home-size category include the Pinson Energy Corporation "Cycloturbine C2E3" model and the Dynergy "5 Meter" model. Systems with bicycle-wheel rotors include the Altos and the American Wind Turbine models.

Photo. 6-1 Bill Gallea's Home-Size Wind System—Gallea's Hummingbird generates enough energy to run most of his Minnesota household's appliances.

All-Electric-Size Systems

Systems with rotors over 16 feet and up to 40 feet in diameter are large enough to power an all-electric home, a small business, or even a farm. This size of wind system can produce 500 to 5,000 kwh. of electricity per month at an average wind speed of 12 mph. Besides the obvious benefits of producing more energy than smaller systems, wind systems of this size cost only about ⅓ to ½ as much per w. of rated power as smaller home-size systems do, although the total initial cost is more, of course. These large systems are more complex than home-size units, however, and installation and maintenance is more involved, too.

Most systems of this size are designed for utility connection through a synchronous inverter or an induction generator. Rotor speeds are slower for the larger systems, typically from a maximum

Photo. 6-2 William Sheperd's All-Electric Wind System—At 25-mph average wind speeds, this Carter "Mod 25" system generates 25 kw., enough energy for most of Sheperd's needs in his home in Henrietta, Texas. In this view the system may appear to be blocked by the trees, but in fact the tower clears the surrounding obstacles by the correct margin.

of 250 rpm for a 16-foot rotor to a maximum of 60 rpm for a 40-foot rotor. Because of their large size, centrifugal forces on the blades are greater if the rotor overspeeds, and most large systems use more than one speed control method, including blade-pitch control, tip flaps, the side-tilt design, or electric brakes. A three-blade, downwind rotor is most commonly utilized in these systems, though notable exceptions are the four-blade Coulson Wind Electric system and the six-blade Mehrkam systems. The use of more than three blades results in slower and, thus, more stable speeds in higher winds (although three-blade rotors can also be designed to "stall" or slow down at any desired speed by the proper choice of airfoil shape). The Dakota Wind "10 KW" model, with a 28-foot-diameter rotor, is the largest unit to use wooden blades. Most other systems use aluminum blades, although composite (fiberglass plus urethane foam) blades are increasingly being tried, with varying success.

Warranties, Service, and Reliability

By carefully comparing the features of each model of wind system that relies upon a well-engineered rotor and generator package, you'll find that few differences in power and efficiency exist between most models of similar size. The real differences in wind systems come in the quality of the materials, workmanship, and service. A wind system must work in all kinds of weather, including wind speeds of up to 70 mph in most locations, and up to 100 mph or more in others. Your wind generator may make more revolutions in one year than a car alternator makes during the total lifetime of the car! On top of that, a quality wind system must be designed for a 15- to 20-year lifetime to make it a worthwhile investment.

Most wind system manufacturers are relatively new, so there is no consumer's "frequency of repair" record available yet. Ongoing testing of many models under extreme wind conditions is being done at Rockwell International Corporation's Rocky Flats test site in Golden, Colorado, under contract to the U.S. Department of Energy. But these tests will have to continue over a period of several years in order to show meaningful differences in designs. At this time, then, the best insurance in choosing a reliable wind system is to consider thoroughly the quality of construction, the effectiveness of the speed-limiting mechanisms, and the quality of service.

Good construction is most important in the rotor and speed control mechanisms. A particularly important feature in any wind system is the shutdown device that stops the rotor in very high winds, between 40 and 60 mph. At most sites, these winds occur so infrequently that the total amount of energy generated at these speeds is not worth the bother. But even the best speed control mechanism can overspeed from a sudden gust of 60 to 90 mph in a single second or less during a violent storm. This overspeeding causes wear on the rotor, gears, and bearings, which also must be of good quality and of more than adequate size to assure a wind system's long life.

Warranties on most wind systems extend for one year, including parts and workmanship. Some warranties are contingent upon having the system installed or inspected by a qualified dealer. The availability of parts and the ability of the manufacturer or installer to give service or advice is perhaps just as important as any warranty. If the manufacturer has a dealer network, the dealer should be capable of complete service and installation. Some companies provide service contracts that, at a minimum, offer a yearly inspection of the wind system. Also, a dealer will sometimes lend or rent out installation and siting equipment.

Before purchasing a wind system, ask how you can obtain repair parts and how quickly. Find out the charge for a service call. Although most companies are concerned about giving good service, I have talked to several wind power users who were frustrated by the lack of parts or help from the company that sold them the unit. Usually, the problems are relatively minor, but having to wait a month or more to get a small part keeps the system from working just the same.

Richard Davis of Virginia experienced a typical situation. Lightning caused a component failure in Davis's control panel, and he waited for over three weeks for the part when simple instructions from the manufacturer could have shown him how to keep the system going in the meantime despite the malfunction.

Another sign of a service-conscious manufacturer is the usefulness of the owner's manual. If the manual contains step-by-step, troubleshooting, and maintenance procedures, this is a good indication that the manufacturer is prepared to give good service.

Estimating Wind Energy

Having reviewed the features of the different wind systems available in the appropriate size range for your needs, the next step is to obtain an estimate of how much energy in kwh. per month or year that the different wind systems you're considering will generate at your location. With this estimate, you can make a judgment of what percentage of the electricity you use (see Ch. 2), or will need, can be supplied by a particular wind system.

After doing a site analysis and measuring your average wind speed, obtaining the estimate is a simple matter. You can estimate the amount of energy in various ways, including referring to the manufacturer's estimate, using a wind monitor, using the power curve method, or using the standard distribution method. Keep in mind that each of these methods is based upon assumptions that simplify reality; once the system is installed, the actual results for a particular month may vary by 50 percent above or below the estimate.

Manufacturer's Estimate and Wind Monitors

As discussed in Chapter 1, many manufacturers publicize their own energy output estimates for their wind systems. Since you have already determined the average wind speed at your site, you simply have to refer to the manufacturer's estimate of the energy that the wind system generates at that average wind speed. The manufacturer's energy output estimates for many wind systems are listed in Appendix F. These estimates are helpful for considering a particular wind system, but it is difficult to compare different wind systems since the manufacturers use different methods to arrive at their estimates. Some use more optimistic data than others to arrive at their system's energy output.

Using a wind monitor to estimate the energy available at a site is also a simple matter. As discussed in Chapter 1, the monitor indicates the number of kwh. that a particular wind system would have generated during a specific time period while operating at your site. If you are interested in comparing several systems, however, a wind monitor is of limited usefulness because it is programmed to simulate only one wind system at a time and does not record the average wind speed.

The Power Curve Method

The energy available at a site can also be estimated by what I call the power curve method. Using the power curve method, you multiply the number of hours that each wind speed occurs by the power output of the generator at that wind speed. The resulting kwh. are added up to give a total estimate. This is done by using the power curve for a particular wind system in conjunction with the wind frequency chart for your location, derived from a Percentage Frequency or Frequency of Occurrence chart (see Ch. 1).

Manufacturers make available the power curve of their wind systems at different speeds in their sales literature. In Table 6–1 you'll find the Dakota Wind model "BC4" power curve as an example of a manufacturer's power curve. The graph indicates that this generator starts charging at 8 mph—the cut-in speed of the rotor. The power output increases with the wind speed until winds reach 25 mph. At speeds above 25 mph, the speed control mechanism, a blade-actuated governor in this case, limits the power output to a constant level. Some wind systems are stopped completely at a shutdown speed of around 40 mph, and no power is generated beyond that speed.

The power curves for many of the commercially available wind systems are summarized in tabular form in Appendix G in the order of decreasing rotor size. Although the power curve for a particular model may change as a manufacturer makes design changes, this appendix is useful for making comparisons between systems of a similar size.

The results of using this method are only as accurate as the manufacturer's estimate of his system's output at various speeds. A manufacturer typically uses a moving test bed to determine the power curve for his wind system. The wind system is mounted on a short tower attached to the bed of a truck. To simulate a steady wind, the truck is driven at a constant speed on a windless day. The generator's power output at that speed is then noted.

A more sophisticated but expensive method to establish a

TABLE 6-1 Power Curve for a Dakota Wind Model "BC4"

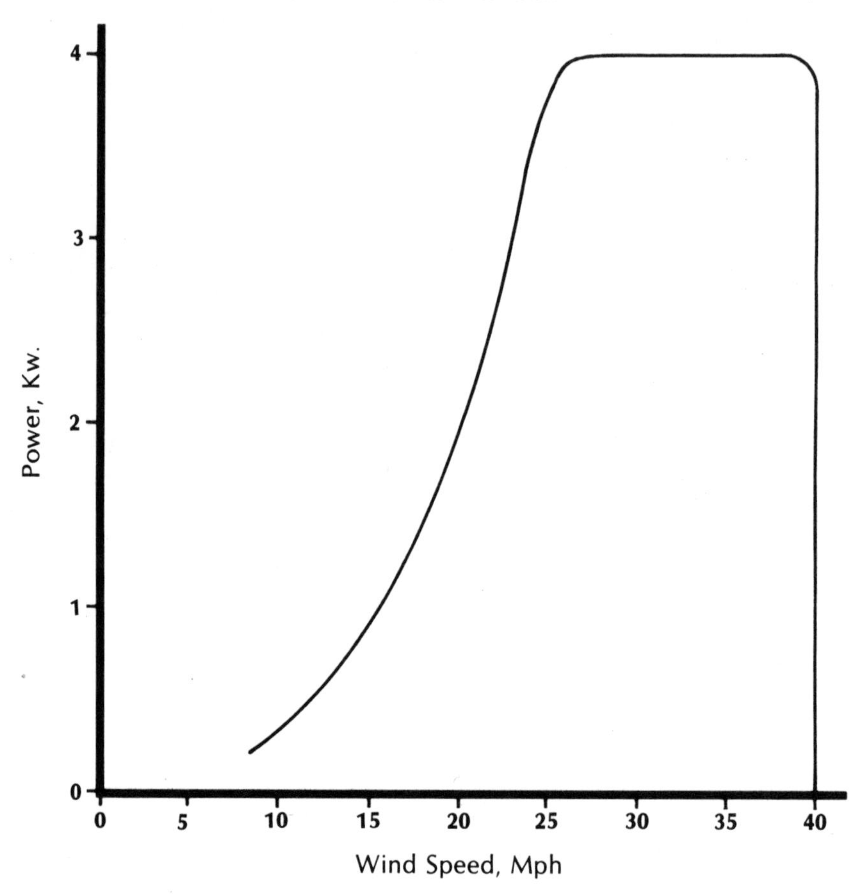

power curve is to use a computer to simultaneously monitor the power output of the wind system as well as the wind speed during actual operation. Such testing is conducted at the U.S. Government's Rocky Flats test site. This testing is called the "method of bins," since each time that a particular wind speed is recorded, the generator's power output is also noted, or figuratively put into a "bin." Later on, all of the power readings from each "bin" are averaged to give the typical power output of the wind system at each wind speed.

Some manufacturers test their wind systems inside a wind tunnel. A fan causes steady air movement past the rotor. Because actual winds are more complex than wind tunnel simulations, these tests tend toward optimistic estimates of the generator's output.

A Sample Power Curve Calculation

To understand how the power curve method works, let's consider the following example based upon my own wind system, a restored 2-kilowatt pre-R.E.A. Jacobs wind system. The Frequency of Occurrence Chart for the month of March at my nearest airport, St. Cloud, Minnesota, is shown in Table 6–2. The bottom line of the chart shows the frequency of observation, with the calculated percentages added for each speed class of 0 to 3 mph, 4 to 7 mph, and so on.

Photo. 6-3 Moving Test Bed—Manufacturers often determine a wind system generator's power curve by mounting the generator on a truck or a railroad flatcar, then running it at a constant speed on a windless day.

TABLE 6-2
Frequency of Occurrence Chart Directions by Speed Groups

U.S. Dept. of Commerce

St. Cloud, Minnesota #14926 (Station Name) — Month of March, 1953–1957 (Period)

Surface Winds
Job No. 7019

Speed / Direction	0–3 Mph	4–7 Mph	8–12 Mph	13–18 Mph	19–24 Mph	25–31 Mph	32–38 Mph	39–46 Mph	Total No. of Observations	
									Percent	Observations
N	27	44	53	45	9	. . .	2	1	4.9	181
NNE	20	29	27	19	. . .	. . .	. . .	. . .	2.5	95
NE	22	67	50	28	1	. . .	. . .	. . .	4.5	168
ENE	13	46	38	19	3	. . .	. . .	. . .	3.2	119
E	50	83	56	55	5	. . .	. . .	. . .	6.7	249
ESE	43	76	67	50	8	. . .	. . .	. . .	6.5	244
SE	45	110	111	39	6	. . .	. . .	. . .	8.4	311
SSE	11	47	61	30	3	. . .	. . .	. . .	4.1	152
S	21	38	51	37	2	. . .	. . .	. . .	4.0	149
SSW	17	38	31	21	. . .	. . .	. . .	. . .	2.9	107
SW	22	39	24	4	. . .	. . .	. . .	. . .	2.4	89
WSW	15	26	12	8	. . .	. . .	. . .	. . .	1.6	61
W	34	63	54	43	21	5	2	. . .	6.0	222
WNW	60	115	147	161	67	13	8	3	15.4	574
NW	74	117	170	135	27	4	. . .	. . .	14.2	527
NNW	25	48	71	66	4	1	1	. . .	5.8	216
CALM	256	. . .	. . .	. . .	. . .	. . .	. . .	. . .	6.9	256
Totals	755	986	1023	760	156	23	13	4	100	3720
Percent (Author's addition)		20.3	26.5	27.5	20.5	4.2	.6	.3	.1	

In Table 6–3, the percentage figures from the Frequency Chart have been converted to numbers of hours per month, based on 744 total hours per 31-day month. The power output of my wind system at the midpoint of each speed class is also listed in Table 6–3. For example, the midpoint of the 8- to 12-mph speed class is 10 mph and the power output at that speed is 0.2 kw. The amount of energy produced at each speed class is then calculated by multiplying the number of hours of occurrence times the power output. For the 8- to 12-mph speed class, the energy produced is 204.6 hours × 0.2 kw., or 40.9 kwh. For the example given in Table 6–3, the total for all the speed classes gives an estimated energy of 195.1 kwh.

Site Correction

In this example, no adjustment was made for differences between the wind site and airport speeds or for differences in the height of the towers. To correct for differences between the site and the nearest airport, you must compare the average wind speed at the site with the concurrent average recorded at the airport, preferably using a three month or more average. If the average measured at the site for February, March, and April is 9.2 mph, for instance,

TABLE 6-3
Power Curve and Estimated Energy Output for Don Marier's Pre-R.E.A. Jacobs Wind Power System

Wind Speed, Mph (Speed Class Midpoints*)	Hours per Month	1.8-Kw. Pre-R.E.A. Jacobs Power Output, Kw.	Estimated Energy, Kwh.
1.5	151.0	0	0
5.5	197.2	0	0
10.0	204.6	0.2	40.9
14.5	152.0	0.6	83.6
21.5	31.2	1.8	54.6
28.0	4.6	2.0	9.2
35.0	2.6	2.0	5.2
42.5	0.8	2.0	1.6

Total: 195.1 Kwh.

*See Table 6-2

and the average measured at the local airport during the same time period is 8.6 mph, then the site correction factor would be 9.2 mph ÷ 8.6 mph, or 1.07.

Similarly, if the airport measurements were made at a typical height of 20 feet above the ground, and the proposed wind system is to be installed on a 60-foot tower, the height correction factor would be 1.17 (see App. C). The combined correction factors would then be 1.07 × 1.17, or 1.25. The midpoint of each speed class in Table 6–3 is then multiplied by the correction factor of 1.25 and the corresponding output of the generator (from Table 6–3) at the wind speed is noted. The height- and site-corrected data for the example given in Table 6–3 are shown in Table 6–4. The corrections lead to a new estimate of 302 kwh. instead of 195.1 kwh. produced at the site.

The actual energy produced at the site will be less than these estimated amounts because of losses in the system, such as through the use of storage batteries or inverters. To be realistic, assume that from 50 to 70 percent of the estimated energy will be obtained when the wind system is actually installed and operating.

TABLE 6-4
Height and Site-Corrected Estimated Energy Output for Don Marier's Pre-R.E.A. Jacobs Wind Power System

Wind Speed, Mph	Hours per Month	1.6-Kw. Pre-R.E.A. Jacobs Power Output, Kw.	Estimated Energy, Kwh.
2.0	151.0	0	0
7.0	197.2	0	0
12.5	204.6	0.4	71.6
18.0	152.0	1.0	152.0
27.0	31.2	2.0	62.4
35.0	4.6	2.0	9.2
44.0	2.6	2.0	5.2
53.0	0.8	2.0	1.6

Total: 302 Kwh.

Standard Distribution Method

The standard distribution method for estimating energy output is more simple to use than the power curve method. In standard distributions, the power curve and the wind frequency chart are reduced to mathematical formulas in order to allow for the comparison of several different wind systems on a similar basis. Using the standard distribution method, the power curve of many of the commercially available wind systems are summarized for average wind speeds of 8 to 31 mph in Appendix G. Once you have measured the average wind speed at your site and know how much energy will be required, refer to Appendix H to indicate which size and models of wind systems will produce enough energy for your needs.

It is worthwhile to compare the manufacturers' estimates shown in Appendix F with the standard distribution estimates shown in Appendix H. If the manufacturer's estimate is much higher, it may indicate that the manufacturer used some more optimistic method of estimating. If the estimates are similar, it is probably because the manufacturer used the standard distribution method to arrive at the estimate printed in his sales literature. If the manufacturer's estimate is much lower, it may indicate a poor match between the rotor and generator design so that maximum efficiency is not obtained, or it may simply indicate that the manufacturer relied upon rather conservative data in arriving at the published estimate.

Rayleigh Distributions

To make the estimates shown in Appendix H, a standard wind frequency distribution called the Rayleigh distribution was used in conjunction with the manufacturer's estimated power curve (App. G) to find an estimated monthly energy output in the same manner as used in the power curve method (see Table 6–3). The Rayleigh distribution is simply a mathematical description of what a wind frequency chart would look like for an average site with a given average speed. A mathematical formula called the Rayleigh formula is commonly used for this purpose. The standard distributions derived from the Rayleigh formula are shown in Table 6–5 for average wind speeds of 8, 10, 12, and 14 mph.

Maximum Energy Available

The maximum amount of energy that can be produced by any wind system of a particular rotor size can also be determined by using the standard Rayleigh distribution method. This is done by assuming that the wind system is 100 percent efficient and, thus, that its power curve is the same as the theoretically available power (see Betz's Law, App. A) for that size of rotor. Table 6–6 shows the estimated maximum energy output for wind systems with rotor diameters of 2 through 40 feet at average wind speeds of 8 through 14 mph. Thus, if only the rotor size of a wind system is known, its energy output can be estimated by using the maximum energy figure from Table 6–6 for the appropriate rotor size and average wind speed, and then reducing that figure by 40 to 60 percent to correspond with the efficiency range of most real wind systems.

To make the estimates shown in Table 6–6, the power output of the wind system is approximated by taking the theoretically available power as determined by the Betz rules (see App. A) and multiplying that power by the power coefficient of the wind system under consideration.

TABLE 6-5
Rayleigh Wind Distributions for Average Wind Speeds of 8, 10, 12, and 14 Mph

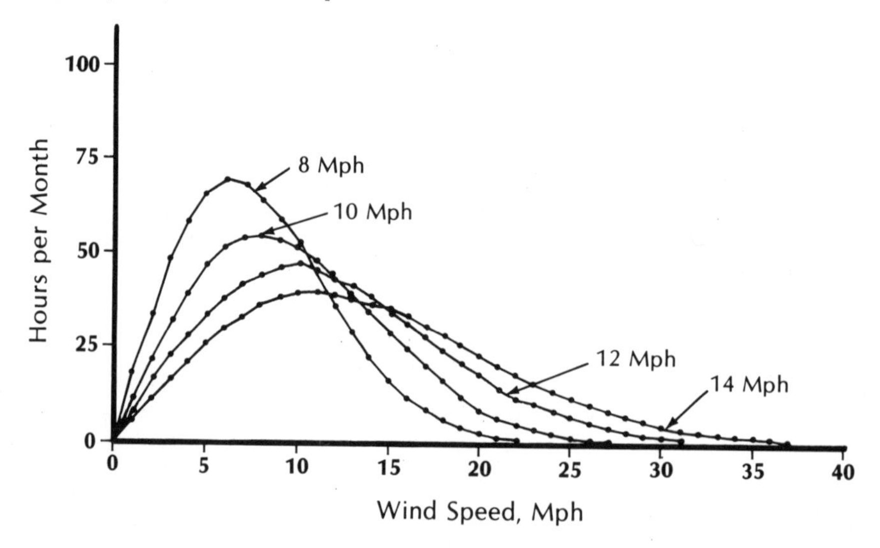

TABLE 6-5—Continued
Data Points for Rayleigh Wind Distributions for Average Wind Speeds of 8, 10, 12, and 14 Mph
Number of Hours per Month at a Site with an Average Wind Speed of:

Wind Speed, Mph	8 Mph	10 Mph	12 Mph	14 Mph
1	17.5	11.2	7.8	5.8
2	33.7	21.9	15.4	11.4
3	47.5	31.6	22.4	16.7
4	58.1	39.9	28.8	21.7
5	65.0	46.5	34.3	26.1
6	68.2	51.2	38.7	30.0
7	67.8	53.9	42.1	33.2
8	64.5	54.7	44.3	35.7
9	58.9	53.9	45.4	37.5
10	51.8	51.6	45.5	38.7
11	44.0	48.1	44.7	39.1
12	36.2	43.8	43.0	38.9
13	28.9	39.0	40.6	38.1
14	22.3	34.0	37.8	36.8
15	16.8	29.0	34.5	35.1
16	12.2	24.2	31.1	33.1
17	8.7	20.0	27.6	30.8
18	6.0	16.0	24.2	28.4
19	4.0	12.6	20.8	25.8
20	2.6	9.8	17.7	23.2
21	1.7	7.4	14.9	20.7
22	1.0	5.6	12.3	18.3
23	0.6	4.1	10.1	15.9
24	0.4	2.9	8.2	13.8
25	0.2	2.1	6.5	11.8
26	0.1	1.5	5.1	10.0
27	. . .	1.0	4.0	8.4

TABLE 6-5—Continued

Wind Speed, Mph	8 Mph	10 Mph	12 Mph	14 Mph
28	. . .	0.7	3.1	7.0
29	. . .	0.4	2.3	5.8
30	. . .	0.3	1.7	4.7
31	. . .	0.2	1.3	3.8
32	. . .	. . .	. . .	3.1
33	. . .	. . .	. . .	2.4
34	. . .	. . .	. . .	1.9
35	. . .	. . .	. . .	1.5
36	. . .	. . .	. . .	1.2
37	. . .	. . .	. . .	0.9

Power Coefficients

If the rated power of the wind system under consideration is known, its efficiency or power coefficient can be calculated. The power coefficient is found by dividing the rated power (see App. E) by the theoretically available power at that wind speed (see App. A). An example is the Kedco model "1620," which has a 16-foot-diameter rotor and is rated at 3.0 kw. of power in a 25-mph wind (see App. E). From Appendix A the maximum amount of power that could be generated by a system with this size rotor at 25 mph is 9.87 kw. The power coefficient is then 3.0 ÷ 9.87, or .304, indicating that the system extracts 30.4 percent of the available energy from the wind. The power coefficient for each model of wind system in Appendix G is listed along with its power curve. These power coefficients vary from .185 to .680, indicating the percentage (20 to 60 percent) of the theoretically available power that the system puts out at its rated wind speed.

The power coefficient used with Betz's Law gives an estimated power curve that assumes that the wind system has the same efficiency at all wind speeds. While this is not the case, it is a mathematically easy way to define the characteristics of the wind system

TABLE 6-6
Theoretical Maximum Estimated Energy in Kilowatt Hours per Month

Rotor Diameter, Feet	Wind Speed, Mph 8	10	12	14	16
2	6.7	13.0	22.5	35.8	53.4
4	26.7	52.1	90.1	143.1	213.5
6	60.1	117.3	202.7	321.9	480.3
8	106.8	208.6	360.4	572.3	853.9
10	166.8	325.9	563.1	894.2	1,334.2
12	240.3	469.3	810.9	1,287.6	1,921.2
14	327.0	638.7	1,103.7	1,752.6	2,615.0
16	427.1	834.2	1,441.5	2,289.1	3,415.5
18	540.6	1,055.8	1,824.4	2,897.1	4,322.8
20	667.4	1,303.5	2,252.4	3,576.6	5,336.7
22	807.5	1,577.2	2,725.4	4,327.8	6,457.5
24	961.0	1,877.0	3,243.5	5,150.4	7,684.9
26	1,127.9	2,202.9	3,806.6	6,044.5	9,019.1
28	1,308.1	2,554.8	4,414.7	7,010.2	10,460.0
30	1,501.6	2,932.8	5,067.9	8,047.5	12,007.7
32	1,708.5	3,336.9	5,766.1	9,156.2	13,662.0
34	1,928.7	3,767.0	6,509.4	10,336.5	15,423.2
36	2,162.3	4,223.2	7,297.8	11,588.4	17,291.0
38	2,409.2	4,705.5	8,131.2	12,911.7	19,265.6
40	2,669.5	5,213.9	9,009.6	14,306.6	21,346.9

for purposes of comparison.

The power coefficient by itself is not a complete indicator of the effectiveness of a wind system's design. Some utility-connected wind systems have rotors designed to turn at constant speed to allow for better control of the system. These designs show a low power coefficient at their rated wind speed, but this is offset by their more simple overall design.

Watt-Hour Meters

After the wind system is installed and operating, it may be helpful to measure the actual amount of energy that the wind system produces. For wind systems that generate a.c. power, you can use the same type of meter that the utility company uses to measure your electrical consumption. These are actually kilowatt-hour meters. For wind systems that generate d.c. for use with storage batteries, a d.c.-type of kilowatt-hour meter must be used.

Ampere-hour meters such as those used on electric vehicles and fork lifts can also be used with d.c. systems. They are lower in cost but less accurate. To convert the ampere-hour reading to watt-hours, it is necessary to multiply the reading by the voltage of the batteries used in the system.

I have used kilowatt-hour meters to measure both the energy generated by my own wind system and the energy used from it after being stored in batteries. Table 6–7 shows the actual measured monthly energy output of my wind power generator, "energy in,"

TABLE 6-7
Energy Generated vs. Energy Used for Don Marier's Pre-R.E.A. Jacobs Wind Power System

Date	"Energy In," Kwh.*	"Energy Out," Kwh.†	Estimated Energy, Kwh.‡
Nov. 1977	98.1	77.0	164.7
Dec.	79.5	70.3	114.7
Jan. 1978	74.8	57.1	138.8
Feb.	40.6	57.9	147.8
Mar.	44.0	50.0	184.0
Apr.	78.8	70.8	247.3
May	51.5	58.6	190.6
June	61.9	63.4	141.6

*Output of the wind power generator
†Energy used from the storage batteries
‡Based on wind frequency data for St. Cloud, Minnesota airport and for the power output curve for a 1.8-kilowatt pre-R.E.A. Jacobs wind system.

and the actual measured monthly energy used from the storage batteries, "energy out," for the period November 1977 through June 1978.

Battery Banks Compared to Utility Power

What is the best way to use wind power—with storage batteries, connected to utility power, or for electric heat? As the question implies, the best way depends upon your application and the situation. As mentioned before, the two basic choices are between a utility-connected system or a stand-alone system, in which the wind power is stored in batteries or is used for electric heat. But these are not "either-or" choices, because there are ways to combine these options into hybrid systems to suit different circumstances.

Battery Storage

It is entirely possible to power a home completely from storage batteries with d.c. electricity. If your goal is to become energy self-sufficient, this is the type of system to use. You should be aware that a stand-alone system has definite limitations, however. The capacity of both the batteries and the wind system itself determine how much energy can be used. For instance, the battery bank necessary for an all-electric home that requires 1,000 or more kwh. per month to function would be quite expensive because of the large amount of storage such a house would demand.

A battery-storage system also involves more attention by the user than a utility-connected system. Although it is relatively easy to learn how to watch the state of charge in the batteries to see how "full" they are, the effort does have to be made. Wind systems with battery storage cannot be treated such as "set it and forget it" appliances.

In choosing a battery storage system, you should decide whether appliances will operate directly from the battery bank or from an electronic inverter. As you'll see in Chapter 7, there are many appliances that operate on d.c. electricity, but many others rely solely upon a.c. If you are interested in limiting your energy use and the number of appliances that you have, such considerations should not be bothersome. On the other hand, using an electronic inverter allows you to operate just about any available a.c. appliance. Be sure to check on the idle power drain of the inverter you choose since power lost in this way may be a significant percentage of the power output of a small wind system (see Ch. 5). If the idle power drain of an inverter is 50 w., for example, then the inverter would require 36 kwh. of electricity per month to operate 24 hours per day. If the wind system's output is only 200 kwh., then the inverter would be using 18 percent of the system's output.

Remote Areas

In a remote area where utility power is not available or is too expensive to bring in, a wind system with battery storage is usually cheaper than a diesel-driven generator system, and involves much less upkeep. Many who live in such an area first choose a small "cabin-size" wind system that charges a 12- or 24-volt battery bank. The amount of electricity that these systems provide is small compared to the requirements of the "average" home. However, for someone going from no electricity to just enough to power a water pump, a few lights, and a home entertainment appliance, these small wind systems just fit the bill.

Elizabeth and Stephen Willey use two Winco units to supply the electrical needs of their energy-efficient Idaho home. They do not live in a windy area, and periods without wind last up to two weeks, so they average only 5 to 10 kwh. per month from the system. This is 1/100th of the amount of electricity most people use! Yet, with this system the Willey's have sufficient power to operate lights, a TV and stereo, a ham radio and CB, an intercom to a neighbor who lives 3/4 mile away, wood stove fans, shop tools, and several other small 12-volt appliances. Needless to say, when lights are not needed, they are not left on by the Willey family.

Generally, battery-storage wind systems are not large enough to provide sufficient power for heating and cooking with electricity. The Willeys use a wood stove for heating and for providing hot water as well as for cooking. A propane tank provides back-up water heating and cooking fuel. This arrangement is typical for people with wind power systems in remote areas. I have met very few wind power users with battery-storage systems who do not also use wood for fuel.

Chris Chomiak lives at the top of a 6,000-foot ridge in Montana, and his only source of power is a Sencenbaugh model "500" and a 350-watt wind system that he built himself. Although they are limited in output, Chomiak feels that these systems allow the user to become involved with their workings. He says, "I have a good time with my wind system. I built my own blades, electronic cut-in regulators, and so on."

Sites near a lake, on an island, or near an ocean are often very favorable for the use of wind power because of the increase in wind speeds over a body of water (see Ch. 1). Irv Benson, who lives on a lake island in northern Minnesota, switched from a gas-driven generator and propane lighting to a Jacobs 32-volt, 2,500-watt wind system. Before the switch, the fuel for the generator had to be hauled seven miles by boat.

Benson notes that the wind blows at least four to six hours per day across the island, with the longest windless period being three to four days in October. The Bensons use the wind-generated electricity for lights, a 12-volt TV, and a Citizen's Band radio system. Benson uses the CB to keep in touch with home while walking his trap-lines. This means that the base set is often on for up to 20 hours per day.

Henry Wood had absolutely no room for an ordinary wind system on his island in Narragansett Bay near Rhode Island—the house on the island is as big as the island itself! After consulting with an engineer, Wood shored up the house's stone fireplace chimney with reinforcing bars and filled the chimney with concrete to support the wind system. The 32-volt, rebuilt pre-R.E.A. Jacobs wind system runs unattended most of the time. Wood notes that the Edison batteries that he uses fit well with his application because of the Edison cells' ability to take considerable overcharging.

Nonremote Sites

For sites where utility power is available, it is less expensive to use utility-connected systems rather than a stand-alone battery-storage wind system. Utility power is generally lower in cost than the electricity from a battery-storage wind system, which costs on

Photo. 6-4 Henry Wood's Wind Power System—Having no extra room at all on his small island off of Rhode Island, Wood shored up his cabin's chimney and mounted a wind system on top of it. (Photograph by Henry A. Wood.)

the order of $.10 to $.20 per kwh. However, there are other reasons for installing a battery-storage system in these areas. For some people, disconnecting from the utility company and using a pollution-free power source is a chance to make a philosophical statement concerning the effects of unlimited use of fossil fuel energy on our environment. Another factor is the uncertainty of the future—who can guarantee the reasonable pricing and availability of fossil fuel energy? Electricity rates already exceed $.10 per kwh. in several parts of the United States. With brownouts and blackouts an increasingly probable part of the future, a battery-storage wind system may well prove to be a very good investment.

Photovoltaics

Living in a remote area of Idaho where over 400 houses are not serviced by utility power, the Willey family helped organize a group of interested neighbors to buy a number of wind systems at one time—a consumers' wind cooperative! The Willeys also plan to do the same group-buying to obtain photovoltaic cells.

Photovoltaic cells, which convert solar energy into electricity just as the wind system converts wind into electricity, provide a good complement to a small wind system. A photovoltaic panel provides most of its power during the sunny, and usually less windy, summer months, while the wind system provides the bulk of its power during the rest of the year.

Photovoltaic panels are rated in peak watts, which is the power they can generate in bright sunshine. For example, a common-size unit will generate 1.2 amp. of current at 14 v. (to charge a 12-volt battery) in direct sunlight. The cost per peak watt is about $20 compared to $3 per watt of rated power for a small wind system.

Joel Davidson used a 30-watt photovoltaic panel for several months before installing his wind system. Having lived for two years without electricity on their Arkansas homestead, Davidson and his wife, Sherri, point out that the 30 w. of power "changed our lives" by providing power for lights, a water pump, and a TV.

Another wind and solar power user, Caleb Scott, uses three 24-watt photovoltaic panels to complement his Sencenbaugh model "1000" wind system. Both systems charge a bank of three 221-ampere-hour, 12-volt batteries. Scott says that "the two make a good system, as our slow wind times are mid- and late summer when there's lots of sun." Scott notes that in three years of use, two of the solar panels had to be repaired by the manufacturer. Leads connecting individual photovoltaic cells broke when the panels expanded in the hot sun.

Some battery manufacturers offer batteries that are designed specifically for photovoltaic systems. These should not be used if you plan a combined wind and photovoltaic system, however, because they are designed as floating batteries. Again, it's best to use only a deep-cycle battery bank with wind systems (see Ch. 5).

(continued on page 158)

Photo. 6-5, Photo. 6-6, and Photo. 6-7 Caleb Scott's Cabin-Size Wind/Solar Power System—Scott, who lives in Connecticut, uses three 24-watt photovoltaic panels to fill in when his wind system isn't active.

Electric Heat

Using wind energy for electric heat is an attractive possibility since it eliminates the use of batteries. This is not always a good use of wind energy, however. To heat a house electrically takes from 1,000 to 2,000 kwh. of electricity per month, while an electric hot water heater takes from 300 to 600 kwh. per month to operate. Thus, to heat a house with wind energy necessitates a wind system with a 20- to 40-foot rotor, and to provide hot water would take all of the output of a 13- to 16-foot rotor system. Generally, heat can be provided at lower cost by other sources such as oil, gas, wood, or solar heating. In fact, any form of electric heat is very expensive.

In cases where excess wind energy would otherwise be wasted, however, converting this excess to heat makes use of a normally lost resource. Furthermore, an advantage of converting all of the wind energy to heat is that the need to use storage batteries is virtually eliminated.

How much heat can the wind provide? Consider a 40-gallon water tank that is heated by wind power. At 8.4 lbs. per gal. of water, the tank contains 336 lbs. of water. If a wind system that generates 6 kwh. per day (180 kwh. per month) is connected to the tank's heating element, it would generate 20,484 British Thermal Units (BTU's) of heat (6 kwh. × 3,414 BTU's per kwh. = 20,484). Since one BTU of heat raises each lb. of water one degree F, the water temperature will rise by 61° F (20,484 BTU's ÷ 336 lbs., or 61° F). This is assuming that no water is taken out of the tank and that no heat is lost through the tank's walls. This example shows that, on a windy day, a home-size wind system can supply a significant increment of a household's hot water needs.

Vernon Orf of western Wisconsin found that using the wind energy for heat was a good way to start out. Orf first installed two rebuilt pre-R.E.A. Jacobs systems, but he didn't have any batteries. Orf connected the wind systems to a resistance coil heater and circulated the heat with the regular furnace fan. This system provided enough electricity to heat the Orf home on windy days. Later, he added yet a third wind system and a set of used telephone batteries. He now uses any excess wind energy to heat hot water in an auxiliary hot water tank. According to Orf, on windy days this tank provides all of the hot water he needs.

Besides using wind energy for direct heating, another possibility is to incorporate a wind system along with a solar hot water system. The water would be heated by either system, thus assuring a more constant supply of energy than with solar heat alone, since the winds tend to be greatest on cold, cloudy days (see Ch. 10).

Utility-Connected Systems

In areas where utility power is available, a utility-connected wind system can provide significant savings over a battery-storage system. Utility-connected systems also provide the greatest amount of convenience compared to battery-storage systems. On the other hand, they offer little incentive to save energy since more power can always be drawn from the utility grid as it's desired.

A synchronous inverter costs the same, or less, than a set of deep-cycle storage batteries. Thus, initial costs of either type of system can be about the same, depending upon the size of the battery bank. However, a wind system using batteries plus a large inverter to run all of the household appliances costs significantly more than a synchronous inverter system. The synchronous inverter costs only about a quarter as much as a regular electronic inverter and is not limited to a certain surge current for starting motors. An induction generator system is even lower in initial cost since the conversion of the wind power from d.c. to a.c. is done by the generator itself and no additional electronics are needed.

From the standpoint of efficiency, utility-connected systems can be more efficient than battery-storage systems, but not in all cases. A d.c. battery wind system supplies 75 percent of the electricity from the generator to the appliances being used. If an inverter is added, only 55 to 60 percent of the electricity is made available to the appliances. A synchronous inverter is about 95 percent efficient when supplying its maximum rated power. At lower power levels, this efficiency drops off to about 50 percent.

Craig Toepfer, who repairs and installs wind systems, used a Gemini synchronous inverter with a pre-R.E.A. Jacobs wind system near Ann Arbor, Michigan for four years. At that location, the power output averaged about 70 kwh., with a low of 40 kwh. and a high of 230 kwh. during the four-year period. Toepfer estimates that the inverter used between 10 and 20 kwh. per month in powering its continuously operating sensing circuit.

In another test, Toepfer installed a 1,800-watt Jacobs wind system with a set of new Exide "Ironclad" batteries in the same area as the system using the synchronous inverter. Although precise

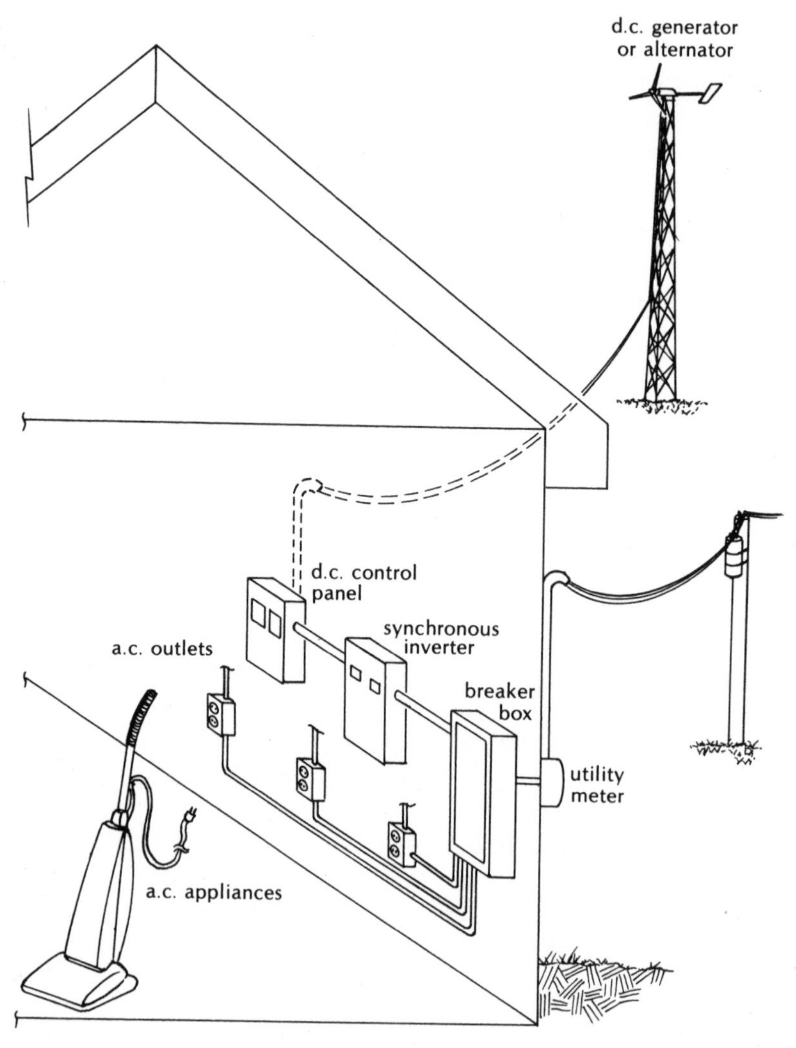

Illus. 6-2 A Utility-Connected Wind System—Electricity from the d.c. generator or alternator is converted to a.c. by a synchronous inverter. Local utility power supplies the house's electricity when the wind doesn't blow.

calculations are difficult to make, Toepfer noted that in periods of low wind speeds, the battery wind system put out twice as much power as the inverter system. In high winds, the battery system generated one-half again as much as the inverter system. In this case, at least, the battery system generated more power than the inverter system. In a high wind speed area, though, the results may not be as dramatic since the synchronous inverter would be operating at its peak efficiency most of the time.

One factor that must be taken into account when using utility-connected wind systems is the amount of power that is backfed to the utility company. When the electrical use in the house is less than the output of the wind system, the excess is fed back to the utility company. The amount of power that is sent back to the utility by the wind system varies with the use patterns in each home. If, for example, 15 percent of the power at a particular installation is backfed to the utility company, the overall efficiency of the system would be about the same as a d.c. battery system (95 percent — the 15 percent sent to the utility company), but still better than the battery-plus-inverter system.

The utility-connected systems rate high, then, in compatability with existing a.c. power in efficiency and in low cost compared to battery systems. They do have the disadvantage of the wind system becoming inoperable whenever utility power lines are down, but this does not happen often. Also, it should be kept in mind that the utility-connected wind system is most economical when it provides only a percentage of the household power needs, up to 50 or 60 percent. If the system is sized to provide all or most of the power for the house, then there will be a number of periods when the system backfeeds power to the utility. This represents wasted capacity for the user unless the utility company offers favorable rates for buying back the energy. These rates should equal or exceed the retail of the utility's energy (see Ch. 3).

Although synchronous inverters are designed to convert any variable d.c. electricity into 60-hertz a.c. power, low voltage generators (such as 32-volt) are not recommended. Bruce Hilde of Moorehead, Minnesota connected a Gemini synchronous inverter to a restored 32-volt pre-R.E.A. Jacobs generator. He obtained only about 100 kwh. per month from the wind system, even though the estimated yield had been 300 kwh. Because of the design of its electronic circuits, the unit does not operate efficiently with wind system generators rated below 120 v. Hilde's solution was to install a transformer between the synchronous inverter and utility power to change the voltage connected to the inverter to 32 v. Since the inverter functions by comparing the voltage level of the a.c. power

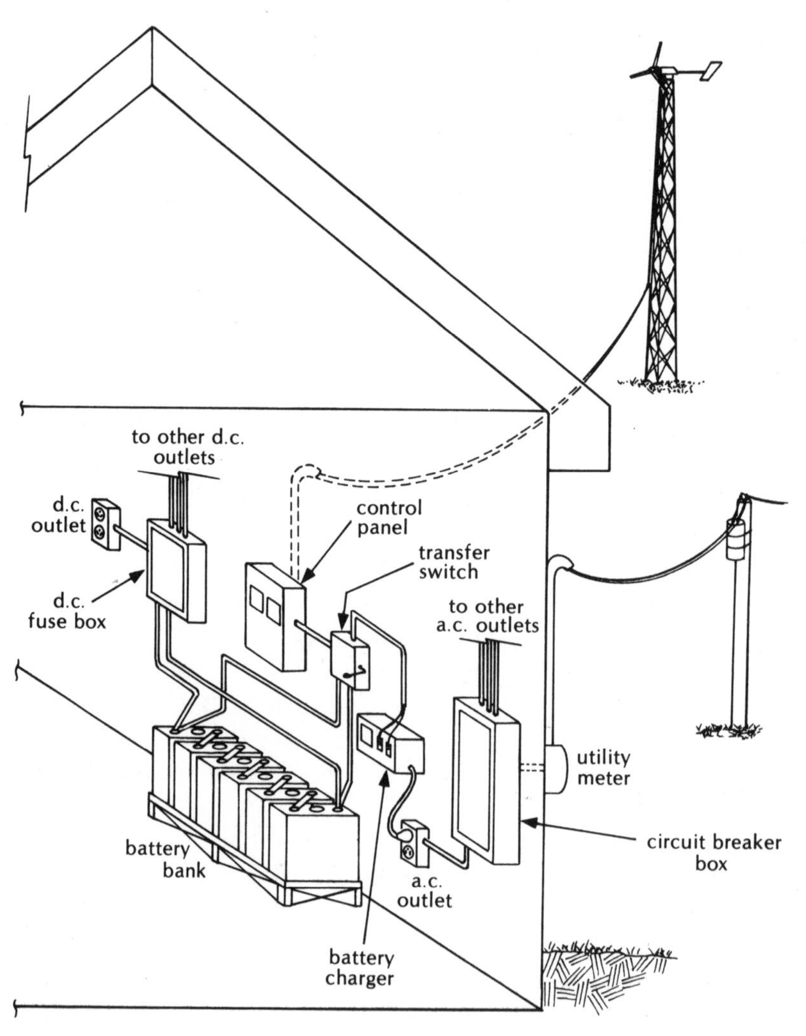

Illus. 6-3 Hybrid Set-Up Using A.C. to Charge Batteries—This arrangement takes the place of a standby generator.

from the utility company to the d.c. power of the wind system, this increased the efficiency of the operation.

Hybrid Systems

A hybrid wind system using battery storage in conjunction with utility power allows you to utilize the best features of both. The possibilities include using utility power to charge a battery bank or using a battery bank in conjunction with a utility-connected wind system to directly power your appliances.

Instead of using a gas or diesel-driven generator to charge the battery bank during periods of low wind, you can use utility power. A battery charger with standard-charge and trickle-charge features can keep the batteries from going below the 50- to 60-percent charge level. This arrangement not only extends the life of the batteries, it also reduces the size (and thus expense) of the battery bank needed. Instead of allowing for 4 to 7 days storage, you only need allow for 2 to 3 days storage if you charge your batteries by utility power during periods of low wind. Although a good battery charger costs almost as much as an engine-driven generator, it costs less to operate, since the cost of electricity from the engine-driven generator is at least 4 times that of utility power, even at the minimum rate charge.

You can also use a split circuit arrangement in which some of the household circuits are connected to the d.c. battery bank and others to the a.c. utility power. Lighting and appliances that work on d.c. electricity can be connected to the battery bank circuits, and other appliances that would normally require an inverter can be connected to the a.c. circuits (see Ch. 8 for details). Obviously this arrangement is suitable only for a system that is intended to provide just a part of the household electrical needs.

Ken Preston of central Michigan installed his wind system with a d.c. fuse box next to the a.c. circuit breaker panel. He began by changing one or two of the lighting circuits from a.c. to d.c., simply by moving the branch circuit from the a.c. box to the d.c. box. As he became aware of how much of his house's electricity could be supplied by the wind system, Preston added more circuits, leaving only the kitchen appliances that normally require a.c. on the a.c. circuits.

In another situation, Bob Bartlett of western Pennsylvania powers his all-electric home with two pre-R.E.A. Jacobs wind systems using electronic inverters for five days per week. He then throws a transfer switch to operate the house circuits on utility power for the other two days.

Using battery storage in conjunction with a synchronous in-

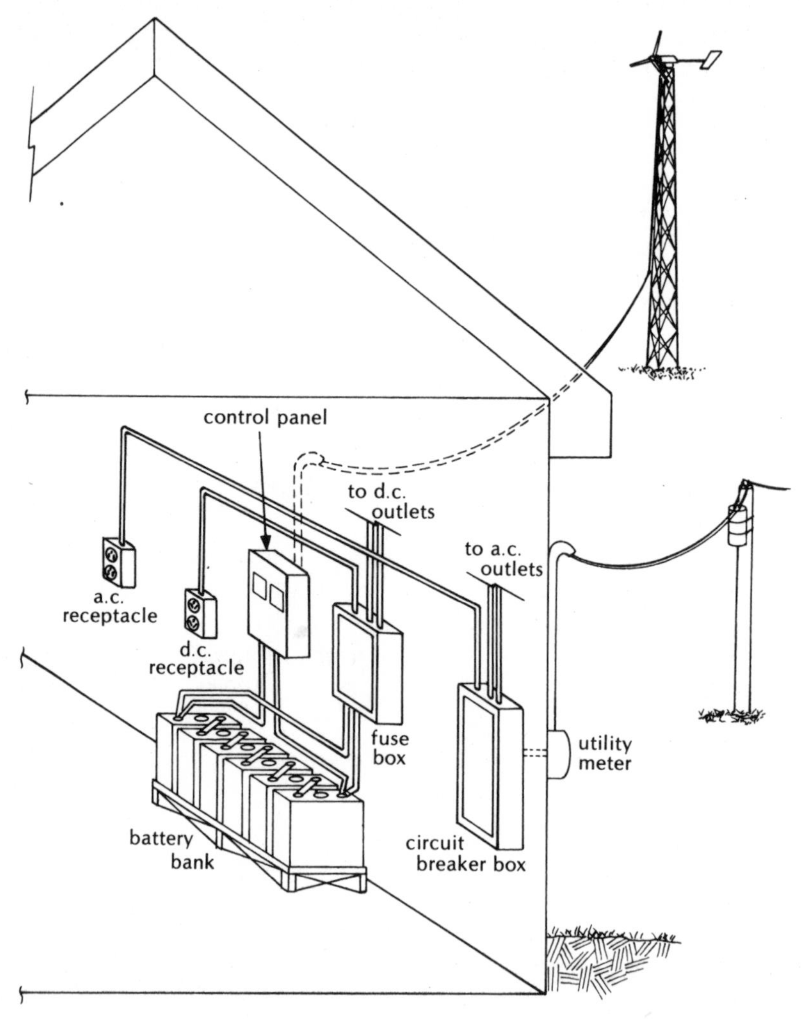

Illus. 6-4 Combination of D.C. and A.C. Outlets—The d.c. outlets are connected to battery-stored electricity, and the a.c. outlets are utility-connected.

verter is another option to be considered. The extra expense of using batteries in conjunction with the synchronous inverter may not be justifiable for the smaller home-size wind systems, but for larger systems it can reduce the cost of the system. For example,

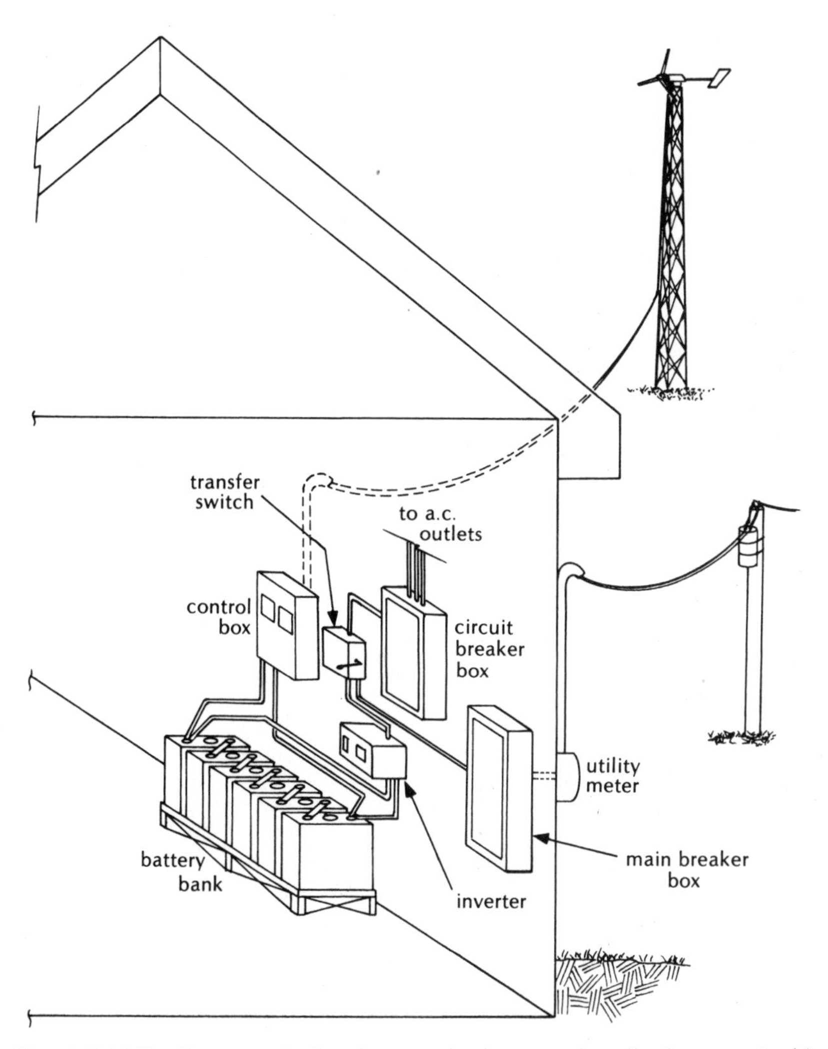

Illus. 6-5 Utility Power and a Synchronous Inverter as a Standby System—In this combination of utility-connected power and battery-stored power, a synchronous inverter changes the battery d.c. into a.c. so that only a.c. outlets are necessary.

Terry Mehrkam in Hamburg, Pennsylvania uses a 500-ampere-hour battery bank in conjunction with an 8-kilowatt Gemini synchronous inverter with his 40-kilowatt (at 25 mph) wind system. Mehrkam points out that a 40-kilowatt synchronous inverter would cost 3 times as much as the 8-kilowatt unit he is using. The difference in price enabled him to buy a set of batteries.

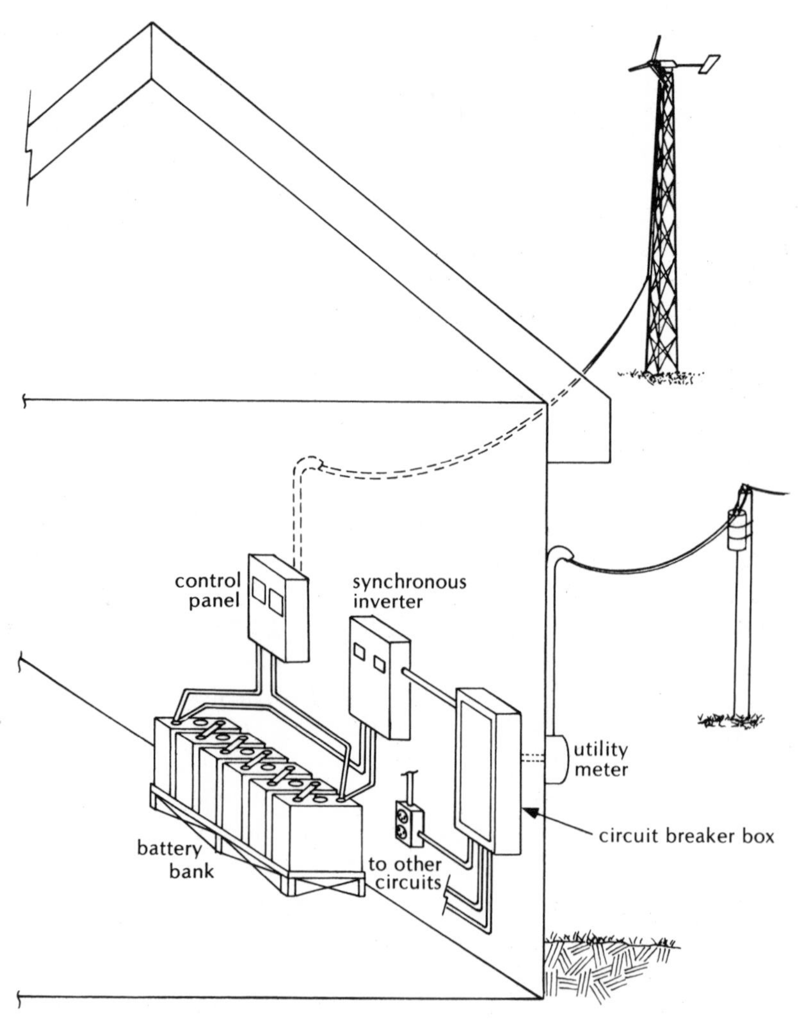

Illus. 6-6 Batteries in Combination with an A.C. Synchronous Inverter—This hybrid set-up may be advantageous with larger systems.

A further benefit for Mehrkam is that the batteries act as an electrical "load leveler." By adding the batteries between the generator and the inverter, the amount of power that the inverter has to handle is leveled out over a period of time. Mehrkam's system is also arranged so that when the batteries are drained too low the synchronous inverter is automatically shut off. The first power that comes from the wind generator starts refilling the batteries. As the batteries are filling, the power also operates appliances and lights.

Although battery storage can be used with an induction generator, there is no apparent advantage in doing so. Batteries cannot be placed between the generator and the utility connection, since both utilize a.c. power. However, a battery bank could be charged with a battery charger and used for d.c. power in emergencies.

Load Control

A load-control unit can increase the amount of wind energy used in any type of system by connecting to different electrical "loads," such as connecting a hot water heater to the wind system whenever excess energy is generated during periods of high wind. For battery-storage systems, a load-control switch may be a simple relay that is set to close its contacts at a specified voltage when the battery voltage exceeds a certain limit. The switch contacts of the relay can turn on a water heating element to preheat water going into the household hot water tank, or it can operate a space heater. Other possible applications include turning on a refrigeration unit or charging the batteries of an electric vehicle.

Of course, a hot water tank that is heated by wind power should be fitted with the usual pressure-relief valve to avoid overheating the water. Also, a "high limit switch," available from heating or electrical supply outlets, should be installed. The switch can be set for the desired temperature range, and, when the temperature is reached, the wind system is disconnected from the hot water heating element. In addition, the switch can be used to connect the wind system to another load, such as a space heater. Load-control switches have additional features, such as a low-voltage sensor that can be used to light an indicator or start a standby power source if the battery bank voltage runs too low.

As I've said before, in utility-connected applications, wind en-

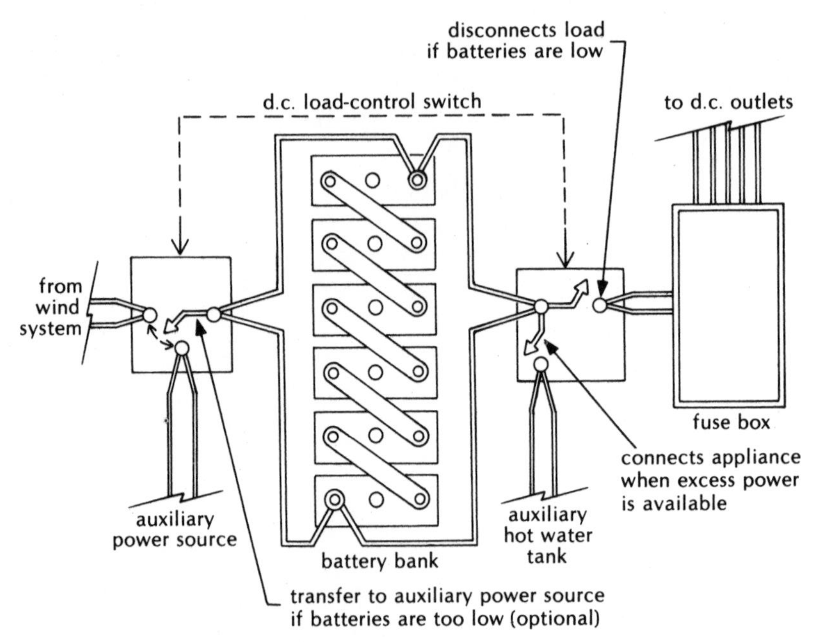

Illus. 6-7 A Load-Control Switch for a D.C. Battery-Storage System—The electronic circuit transfers charging to the auxiliary power generator if the batteries need to be charged and the wind speed is low. If the batteries are fully charged and the system generates excess energy, the load-control switch connects the battery bank to another load, such as a water heater, to use the extra electricity.

ergy is fed back to the utility whenever the household use is less than the output of the wind system. If considerable amounts of power are fed back on the grid to the utility company and the credit for that power is little or nothing, a load control device can increase efficiencies. In this use, a wattmeter senses whether the power is coming from the utility grid or from the wind system. If power is flowing back to the utility from the wind system, the power sensor actives a switch that connects a load such as a heating element in a hot water tank. Such a device is used in a home operated by the New Mexico State University Livestock Research Center, discussed in Chapter 11.

A load control unit can play an important role in utilizing any energy generated by your system beyond the amount you need to run your appliances. Most of your electricity will go toward running the appliances, however, and after finding out what kind of system suits your location and average wind speeds best, you'll want to know just what kind of appliances you can use with that system, particularly if you choose a d.c. wind system. The next step, then, is to learn about the appliances that you can operate, and where you can obtain them.

Chapter 7

Appliances You Can Use

Concerning the use of appliances with your wind system, if yours is utility-connected, you'll have no problem with compatability. Whatever appliances you use at present will operate equally well powered by the wind system as they do connected directly to utility power. This, of course, is one of the advantages of a utility-connected system. Likewise, if you install a battery-storage system, you can still use your present appliances by powering them through an electronic inverter (see Ch. 5).

Some people choose to operate their appliances on direct current from storage batteries. If you live in an area remote from utility power or if your goal is to be energy independent, this is the option to consider. You should be aware, though, that a wide range of appliances will work on d.c. from batteries, but such items as automatic dishwashers and microwave ovens are not among them. People who have cabin-size wind systems or smaller home-size systems usually choose the d.c. appliance option because they consider the cost of an inverter an unnecessary expense. Also, in areas of moderate to low wind speeds, the power drain of an inverter may represent too large a percentage of the energy output of a smaller wind system.

Before deciding to use a d.c. system, sit down and make a list of the appliances you intend to use that operate on d.c. electricity.

By making this list, all members of the household can discuss and learn of any limitations in their use of the energy supplied by the wind, which should prevent many frustrations in the future use of the wind system. If you or the members of your family are not prepared to leave the all-electric world that most of us come from, then a d.c. electrical system is not for you, and the other options should be considered. Be assured, though, that many people, including myself, do use d.c. systems successfully.

There are literally thousands of electrical appliance models on the market, and it is impossible to consider them all individually here. However, we can consider types of appliances, the basic features to look for in them, and how compatible they are with d.c. wind systems. Appliances specifically made to be used with battery power are usually marked as such on the label. Some appliances that are made for operation from 120-volt a.c. also operate from 120-volt d.c. To determine what appliances can operate from 120-volt d.c. or can be modified to work with other voltages, you will have to know if the appliance has a heating element, what type of switch or thermostat it uses, what type of motor it has, if any, and if it contains any solid-state electronic components.

Appliance Voltage and Characteristics

The availability of appliances in different voltages will affect the choice of what voltage you choose for your d.c. wind system. In pre-R.E.A. days, the most commonly used voltage was 32 v. Many wind systems with 32-volt generators manufactured in that period have been refurbished by wind power enthusiasts and are being used now. These systems require 32-volt appliances which are generally available only as restored "antiques," and as such are not available to most people.

Because of the recreational vehicle (RV) boom, there is a wide range of appliances available that operate on 12-volt d.c. These are, of course, convenient for the owner of a small, 12-volt d.c. wind system. Many appliances, including pumps, TV's, refrigerators, power tools, lights, and stereo systems, are available in 12 v.

through RV or marine (boating) supply outlets. Popular sources for 12-volt equipment are the J.C. Whitney automotive parts catalog or the Sears, Roebuck and Company RV catalog. One note of caution should be made about purchasing small 12-volt appliances, however. Those made for the RV market are sometimes of low quality construction. This may make them adequate for the intermittent use they receive in an RV but they may not stand up if you use them daily. Often, appliances made for the marine market are of better construction because greater reliability is required on boats to withstand wind and salt water. Also, the quality of marine equipment is superior in order to meet the demands of boat owners, who generally are able and willing to pay more for such reliability.

Many 120-volt a.c. appliances will also operate on 120-volt d.c., even though they are not so marked. Thus, it is possible to install a 120-volt d.c. battery-storage system and operate appliances directly from it. Such systems possess an advantage in that the house wiring need not be modified to work with 120-volt d.c. (explained in Ch. 9). A disadvantage is that a greater number of battery cells are required for a 120-volt system and that the battery bank must be carefully protected to prevent an accidental shorting of the terminals which could result in dangerous arcing of electric current. Also, since most 120-volt a.c. appliances are not labeled as to whether they will operate on d.c. or not, you must become somewhat knowledgeable about the inner workings of the appliances before using them. Also, warranties may be voided on new appliances if they are used with d.c. electricity and are not labeled for such use, or if the appliances are modified by the user.

Heating Elements and Light Bulbs

The most important component of many appliances such as toasters, ranges, or electric space heaters is the heating element. Generally made from a nickel-chromium alloy, the heating element gives off heat when an electrical current flows through it. It makes no difference whether the electricity is a.c. or d.c.; a 120-volt a.c. heating element will give off exactly the same amount of heat if it is operated from 120-volt d.c. This little fact of nature solves a lot of problems, especially since incandescent light bulbs are actually heating elements that convert some of their heat (5 percent) to light.

This means that the same incandescent light bulbs used with a.c. electricity can be used with d.c. wind systems.

If a lower voltage is used with a heating element than the element was designed for, no harm will be done, but the amount of heat will be less. For example, if a heating element designed to give off 1,000 w. of heat when used with 120 v. is instead connected to a 12-volt battery, only 10 w. of heat will be given off. This is because both the voltage and current are decreased by a factor of ten and the power is equal to the voltage multiplied by the current.

Switches and Thermostats

On-off switches for appliances are rated as "a.c. only" or "a.c.-d.c." If the rating is "a.c. only," it means just that; the switch contacts may burn up or be damaged if they are used with d.c. On the other hand, if the rating is "a.c.-d.c.," the switch will work with either type of electricity. The difference between the two types of switches is really one of quality. To be used with d.c., the switch must be made a little sturdier, and the switch contact points must open a little wider to prevent electricity arcing across the contacts when the switch is opened or turned off. This means that the "a.c.-d.c." switch will last longer when used with a.c. too.

In some cases it is possible to replace an "a.c. only" switch with an equivalent "a.c.-d.c." switch. In other cases, it is possible to leave the switch on permanently and install another switch (a.c.-d.c.) elsewhere on the appliance or in the cord. With very small appliances, you can tape the switch permanently "on" and turn the appliance on and off by connecting or disconnecting the power cord. This is inadvisable with appliances that use substantial power since arcing would occur when the cord is disconnected, possibly damaging the electrical outlet.

Most appliances with heating elements also include a thermostat, or thermally operated switch, used to control the temperature of the heating element. A thermostat consists of a bimetal strip that moves up at its free end when it is heated and thus opens a set of switch contacts. Modifying the thermostat in any way will change either the temperature at which the thermostat operates or the time it takes to operate. Very often it is the thermostat that gives an appliance an "a.c.-only" rating, even though the heating element or motor it uses may itself run on d.c. Unfortunately, thermostats are usually designed specifically for each appliance and no equivalent "a.c.-d.c." thermostat is readily available.

One solution, which requires you or someone else to be electrically handy, is to use what I call an "outboard relay" in conjunction with the thermostat. As shown in Illustration 7–1, a relay (electrically operated switch) is mounted outside of the appliance, an iron for example, and inside a small electrical box. The appliance is rewired so that the thermostat turns on the relay instead of the heating element. In turn, the switch contact of the relay turns the heating element on. Only a small amount of electric current is required to operate the relay so the thermostat is not harmed by the d.c. electricity. The switch contact of the relay should be of a sufficient rating to handle the current that flows through the heating element. Relays that operate on d.c. are available from most electrical supply outlets. (See d.c. appliances in the Source List following the Appendices.)

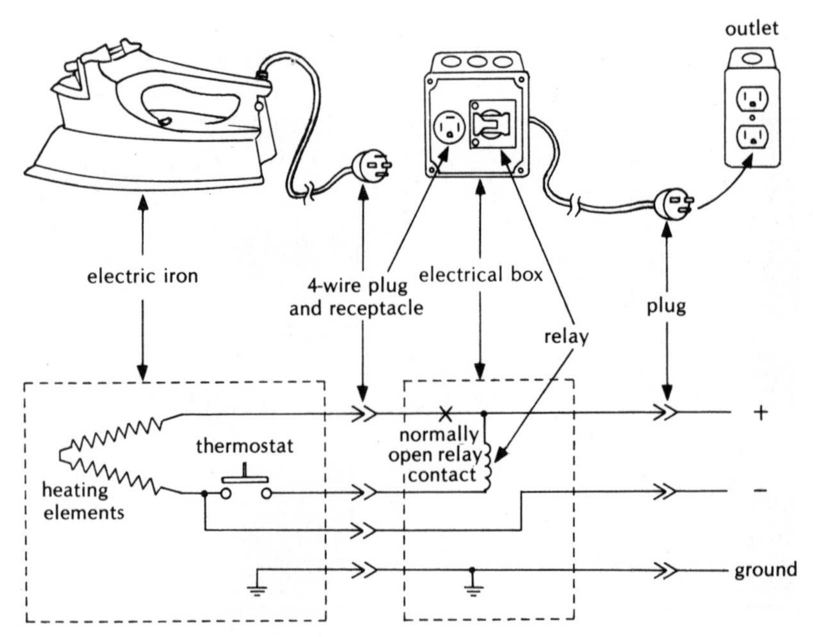

Illus. 7-1 Modifying a Thermostat for D.C. Power—By connecting a relay to the appliance, and rewiring the appliance so that the thermostat turns on the relay instead of the heating element, d.c. power will not harm the appliance.

Motors

Electric motors run many appliances and tools in the home, and it's not always possible to know what kind of motor is used in the appliance. However, most appliances use the same type of motor with only a few exceptions.

There are various types of a.c. motors: capacitor-start, shaded pole, split phase, and repulsion-induction, all variations of the basic induction motor. To the uninitiated, they look fairly similar in their outside appearance. Any a.c. induction motor will operate on a.c. only—either from utility power or from an inverter. An a.c. motor that is accidentally plugged into a d.c. outlet will heat up and burn out its electrical windings fairly quickly.

Universal motors are so called because they will run on either a.c. or d.c. electricity of the same voltage rating. They have carbon brushes and commutators just as a d.c. motor or generator does (see Ch. 4) but they are constructed in such a way that they also operate on a.c. Universal motors are easily recognizable by the brush caps that hold the carbon brushes in place. Examine your power drill or saw, for example, and you'll see that the small, round brush caps

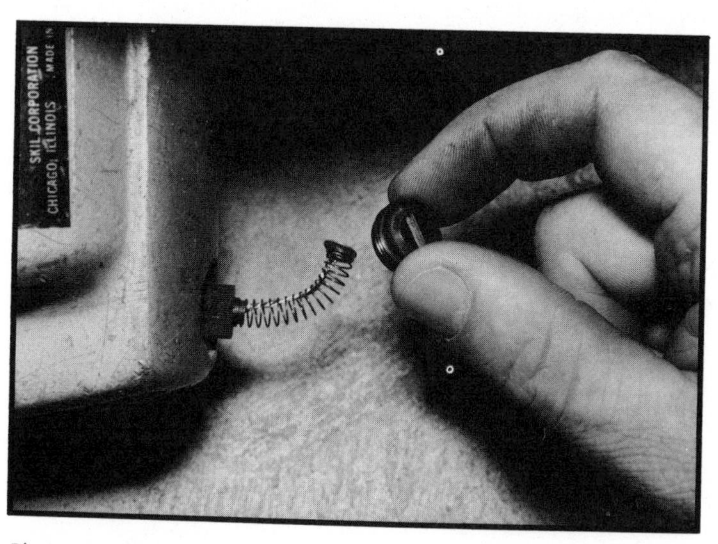

Photo. 7-1 Brush Caps on a Universal Motor Power Tool—Brush caps make universal motors easy to recognize. The advantage of universal motor tools is that they operate on both a.c. and d.c. electricity.

will either be visible on the outside body of the tool for easy replacement or covered by ornamental metal or plastic. Universal motors are usually designed to run at high speeds such as 5,000 to 10,000 rpm compared to 1,800 or 3,600 rpm for the typical a.c. motor. Wind power users sometimes overlook this fact when they seek to replace an a.c. motor with a low cost d.c. motor from a surplus outlet. A motor must not only be replaced by one with an equivalent power rating, but also with one of the same operating speed.

One type of a.c. motor, the repulsion-induction motor, has carbon brushes and could possibly be confused with a d.c. or universal motor. Few of these are manufactured now, however, and they are almost always used in large power systems, such as on farms and in shops, but seldom with household appliances.

Usually, d.c. motors are not found in the average home (except for universal or a.c.-d.c. motors). A few manufacturers make them available, but because so few are made, they can cost 3 times as much as an equivalent size of a.c. motor. Generally, d.c. motors are made for commercial or industrial use. Motors designed for 12-volt operation are relatively easy to obtain because of their use with RV appliances.

Permanent magnet d.c. motors, called "adjustable speed/-torque drive" motors, are used widely in industry because their speed can be changed by varying the voltage applied to them. Such control is useful in regulating the speed of conveyors and in other industrial processes. In general, though, you cannot adapt such d.c. motors for home use because their armatures operate on 90-volt d.c. instead of 120 volt.

Electronics

Home entertainment appliances such as radios and TV's contain electronic circuits and, as a rule, these kinds of appliances are not made to be plugged into the outlet of a d.c. system. However, many portable appliances exist that operate on dry cell batteries or on 12-volt battery systems.

Interestingly enough, most electronic circuits are powered by d.c. electricity. The 120-volt a.c. is converted to d.c. voltage with a "rectifier" circuit (see Ch. 5). Unfortunately, there is no standard voltage that electronic circuits are designed to operate on. The appliance might operate on 18.4 v. or any other voltage, and might even use several different voltages for different circuits inside the appliance.

Other electronics to be aware of in home appliances are the speed control devices on power drills, and dimmers on light switches. Since they use solid-state electronic components called silicon-controlled rectifiers (S.C.R.'s), variable-speed drills and light dimmer switches will not work with a d.c. system.

Household Appliances

Let us consider all of the appliances typically found in the home and see which ones are available or adaptable for use with d.c. wind systems. Household appliances can be divided into lighting, small appliances, home entertainment appliances, and major appliances.

Lighting

Incandescent bulbs with tungsten filaments are available in a wide variety of sizes and voltages. As discussed previously, regular 120-volt a.c. bulbs will work equally well on 120-volt d.c. The only problem is that a 120-volt d.c. battery bank will vary from 110 v. when it is discharged to 140 v. when it is being heavily charged by the wind system. The higher voltage tends to cause the bulbs to burn out faster than they normally would.

A regular 120-volt bulb should last from 750 to 1,000 hours, which amounts to a year or more with moderate use. With my own 120-volt d.c. wind system, I estimate that I have to replace an extra two to four bulbs per year because of the varying voltage. You can purchase special 130-volt bulbs (sometimes called long-life bulbs) from lighting supply houses. Since these bulbs cost more than regular ones and are not available in every hardware store, the extra expense and effort to find them is not worth their added durability.

A 100-watt, 120-volt bulb will give off the same amount of light whether operated on a.c. or d.c. However, as the voltage fluctuates, so will the amount of light. In fact, a 5 percent reduction in the voltage results in a 15 percent reduction in the amount of light produced. Although this reduction is not particularly annoying, the d.c. wind system user will notice a definite change in the brightness of the lights on different days depending upon the batteries' state of charge. In fact, the lights actually become an indirect indicator of the charge of the batteries.

Automotive bulbs also work with 12-volt wind systems. They have bayonet-type bases, however, and a special bayonet-to-screw base adapter from marine supply houses must be purchased to adapt the bulbs to regular lamps.

You can also obtain low-voltage 12- or 32-volt bulbs from lighting supply centers. An important advantage of low-voltage light bulbs is that they put out more light for the same wattage rating as 120-volt bulbs. For example, take a 25-watt, 12-volt bulb such as that in a service station mechanic's trouble light. Compare it with the brightness of a 100-watt, 120-volt bulb. You'll see that you can use a 50-watt bulb with a 32-volt wind system, or a 25-watt bulb with a 12-volt system to cast the same brightness of light received from a 100-watt bulb powered by a 120-volt system.

The reason for this difference in brightness is that low-voltage bulbs have thicker tungsten filaments to handle the heavier currents. (The current through a 50-watt, 12-volt bulb is 4 amp. while that of a 50-watt, 120-volt bulb is 0.4 amp.) This extra thickness allows the bulb to reach higher temperatures without melting the filament, and higher temperatures mean a brighter light for the same amount of power.

Many people use fluorescent lights because they are 4 times as efficient as incandescent bulbs. Although it is possible to operate fluorescent bulbs directly from a 120-volt d.c. system, it is not recommended. To operate on d.c., the ballast or starter (an electric circuit in the tube that excites the gases to produce the light) must be replaced with an electrical resistor or an incandescent light bulb, which in itself will use about as much power as the fluorescent bulb does. The efficiency advantage over an incandescent bulb is thus lost. Also, a special on-off switch must be used with the fluorescent light to switch the polarity of the connections with each use.

Fluorescent light fixtures are available that can operate on 12-, 24-, 32-, or 120-volt d.c. They use small, built-in electronic inverters to convert the d.c. to 120-volt a.c. to operate the fluorescent light tubes.

Small Appliances

Small home appliances for cooking, cleaning, or for shop use can be divided into those with heating elements and those with universal motors.

Appliances with heating elements include toasters, irons, waffle irons, slow cookers, roasters, and automatic coffee makers. Slow cookers usually do not have thermostats and can operate on 120-volt d.c. All of the other appliances mentioned above have thermostats and are unsuitable for use with 120-volt d.c., although it is possible to install an outboard relay on most irons, waffle irons, roasters, and automatic coffee makers. Also, coffee makers and travel irons are available in 12-volt models from RV and marine supply houses.

Universal motors are common in such appliances as vacuum cleaners, floor polishers, sewing machines, mixers, blenders, hair blowers, electric toothbrushes, electric shavers, power saws, drills, routers, and sanders. If the appliance is marked as "a.c. only" on the label, it's often because the switch is an a.c. switch, even though the motor itself will run on a.c. or d.c. In some cases you can still use the appliance if you either replace the switch with an a.c.-d.c. switch, or simply leave the switch on and unplug the cord to turn it off. Of course, this is not advisable with small power tools, because the switch is necessary for reasons of safety.

Most vacuum cleaners will run on 120-volt a.c. or d.c. although the on-off switch may need to be replaced on some models. An exception is the Electrolux vacuum cleaner which has an automatic shut-off device that stops the machine when the cleaner bag is full. The switch on this device is not meant for d.c. operation, however, and will burn up every time the bag gets full. I found this out myself after burning up several switches. Also, small versions of vacuum cleaners for cleaning automobiles are available in 12-volt d.c. models.

Except for those that have electronic circuits to power complicated stitching, most sewing machines have universal (a.c.-d.c.) motors. Electric shavers, electric toothbrushes, and hair blowers work fine on d.c. as well, once you're past the problem of an incompatible on-off switch. The only hair clipper that works on d.c. is the Oster brand, which is used by many professional barbers.

Most electric mixers operate on d.c., although they too may have "a.c. only" switches. Blenders, on the other hand, will not work on d.c. if they have built-in timers.

Hand power tools such as saws and drills should not be used on d.c. unless they are specifically marked as "a.c.-d.c." on the label. Most power tools of commercial or tradesman quality are marked "a.c.-d.c." The W. W. Grainger catalog (see your local Yellow Pages) has a good range of a.c.-d.c. tools under the Dayton Label. Radial arm saws usually have capacitor start motors that work on a.c. only, but the DeWalt Model 7700, 8-inch radial arm saw has universal motors. Power drills that operate on 12 v. are available from RV and marine supply houses.

Appliances that rely on the speed of the motor for timing, such as electric clocks, typewriters, phonographs, and small fans almost always use motors that run only on a.c. Electronic clocks and small fans that run on 12-volt d.c. are widely available, however. You can install d.c. motors on I.B.M. Standard and Executive model typewriters by special order (except for the Selectric). They are available for 65-, 115-, 130-, 200- and 230-volt d.c. Oscillating or desk fans that function on 32- or 120-volt d.c. are made by the Merrin Electric Company, in Hoboken, New Jersey.

Home Entertainment Appliances

Radios, televisions, and stereos usually operate only on a.c., except for portable units that work on 4.5-, 6.9-, or 12-volt batteries. Home entertainment appliances can easily be powered by using a small inverter, since their power drain is relatively small. Televisions and stereo equipment designed to operate on 12-v. are particularly plentiful. Portable 12-volt TV's operate on as low as 33 w. of power for a black-and-white set, less power than for most light bulbs. Automotive stereo radios and tape decks work fine in the home, too.

If you appreciate antique-hunting, it is possible to locate old radios and TV's that operate on 120-volt a.c. or d.c., or on other d.c. voltages. These radios and TV's contain vacuum tubes instead of modern solid-state electronics. (One source of information on old tube radios is Vintage Radio.) Although not all vacuum tube TV's will run on a.c. or d.c., many will. To find out if a TV will run on 120-volt d.c., you will have to look at its circuit diagram. Most TV repairmen maintain a complete file of circuits for TV sets and can look up any model that you ask about. Tell the repairman that you are looking for a TV constructed with tubes, a series-string filament, and no input transformer. I have used early model TV's and radios, but in the long run I feel they aren't worth the effort since repair parts are difficult to find. It's easier to use 12-volt home entertainment equipment with a d.c. wind system, or to use a small electronic inverter to operate a.c. appliances.

With a little ingenuity, users of 32- or 120-volt d.c. wind systems can operate 12-volt TV's or stereos, too. A 12-volt TV can be operated from 32 v. by using an integrated circuit regulator of the appropriate size to limit the current, and thus the voltage, to the TV. Available from electronic supply stores, regulators will limit the output voltage to 12 v. when the input voltage from the batteries is 32 v. These regulators are limited typically to handling a 1-amp

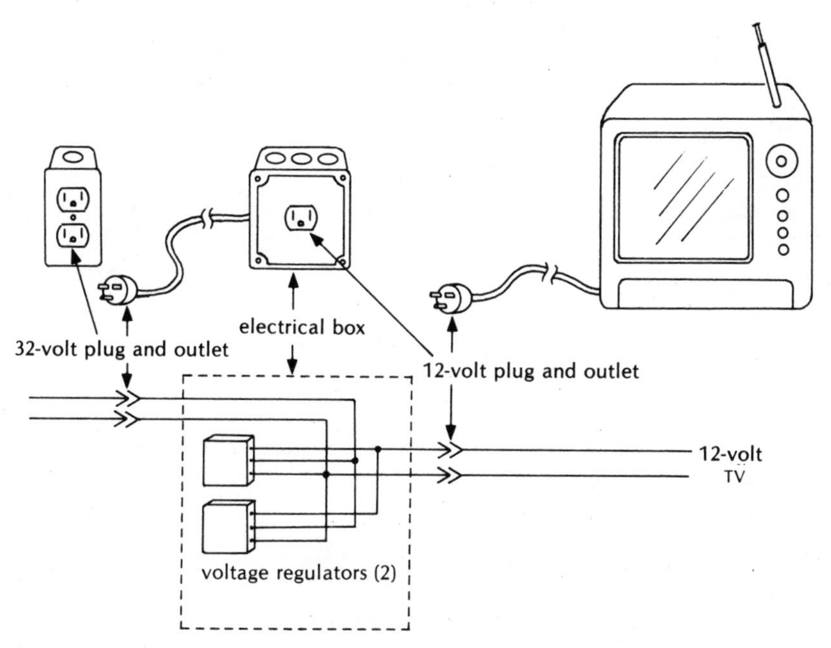

Illus. 7-2 Connecting a Voltage Regulator—If a voltage regulator is connected in series with the 12-volt TV to limit the voltage, a 12-volt TV can run off of a 32-volt system.

load, but they can be connected in parallel to provide more current (see Illus. 7–2). For example, a 12-volt, 36-watt television requires 2 amp. (36 w. ÷ 12 v.) to operate, and thus requires three 1-amp, 12-volt to 32-volt regulators.

To operate a 12-volt TV or radio from a 120-volt d.c. system requires another arrangement because of the large differences in voltages. One way is to use a "volt box," such as that developed by John McGeorge of Norwalk, Connecticut. My family has used one of these for over a year without experiencing any problems.

The volt box shown in Illustration 7–3 consists of a 12-volt battery mounted in a box with a cigarette lighter socket into which the 12-volt TV plugs. The TV operates off the 12-volt battery. Since the battery will discharge after several hours of use, it's necessary to hook up a "battery charger" consisting of a light bulb connected between the 120-volt d.c. outlet and the 12-volt battery. The electricity going through the light bulb also goes through the battery to keep it charged. Although somewhat reduced in intensity, the light from the bulb can be used for reading. For my own 15-watt set, a 100-watt lamp keeps the 12-volt car battery charged at all times whenever the TV is being watched.

Again, be sure to mount any electrical components in an electrical box. In the box, the battery should be inside a plastic case, such as those used in RV's or boats, to prevent acid from spilling in the house. Locate the battery where it cannot be tampered with and away from any electrical fixtures or mechanical devices that might give off sparks. The sparks could ignite small amounts of hydrogen gas emitted by the batteries during charging. Although this arrangement may sound complicated, anyone skilled in electricity can understand and build it and thereby eliminate the need to transport a messy battery around to be recharged. A small, 12-volt fan and a vent mounted in the box should adequately vent the hydrogen.

Major Appliances

With few exceptions, the major home appliances for cooking, clothes washing, water pumping, air conditioning, heating, and refrigerating are designed for operating from a.c. electricity. Since these appliances also use the largest amounts of energy, the small d.c. wind system owner has to be selective in the number of appliances used.

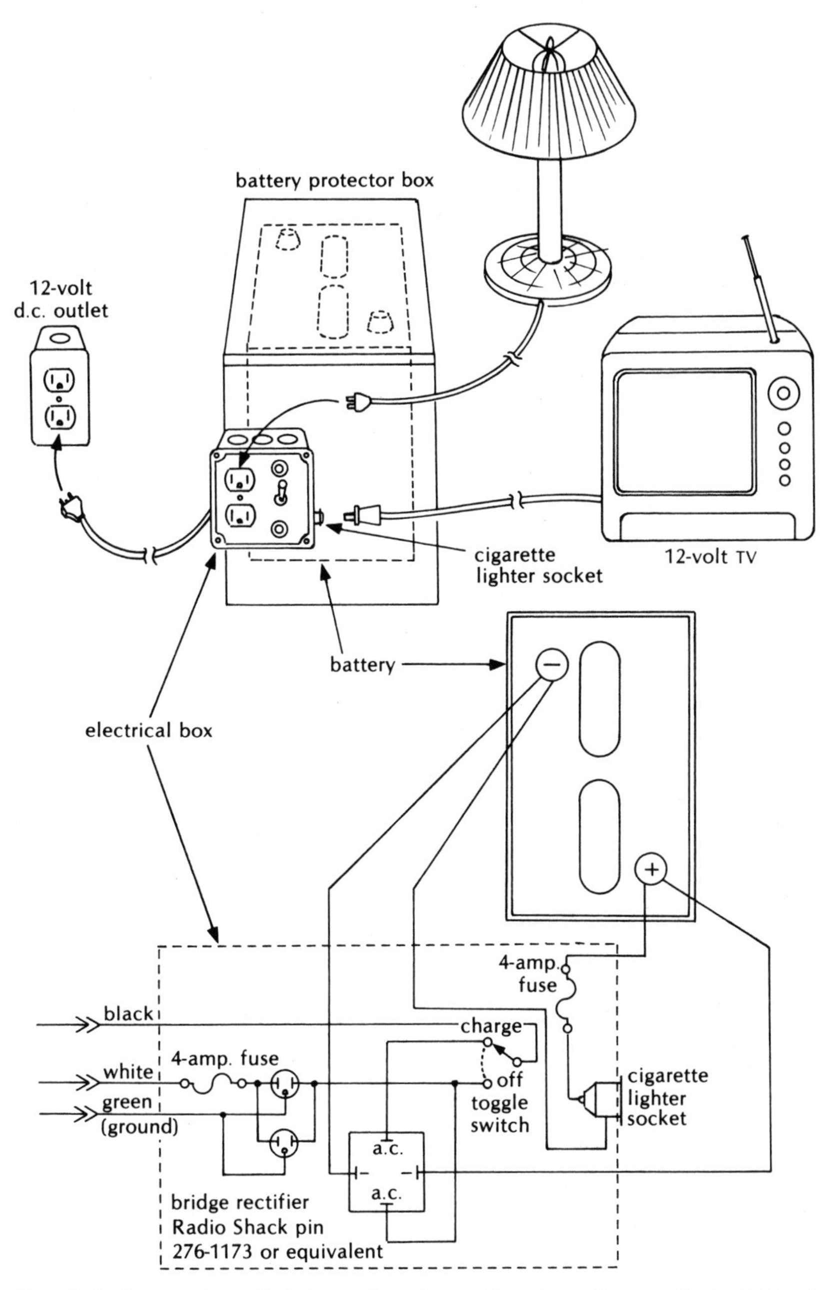

Illus. 7-3 Connecting a Volt Box—Running a 12-volt appliance off of a 120-volt system requires the use of a volt box to limit the voltage. The light bulb keeps the 12-volt battery charged, while also providing sufficient light for reading.

Cooking

Electric ranges cannot be operated on a d.c. wind system because of the large amounts of power they require and because electric ranges are designed for 220-volt a.c. instead of 120 volts. Therefore, most d.c. wind power system owners employ a gas or wood cookstove. It should be noted that gas ranges made since 1980 all have automatic igniters that require electricity to operate; you cannot bypass them and light the stove with a match.

Microwave ovens require about ¼ as much energy as electric ranges. They can be operated by using a sine-wave inverter that does not use S.C.R. components (see Ch. 5). Check with the inverter manufacturer, however, to see if there are other problems in operating a microwave oven with his inverter.

The most unique electric stove I have seen is that used by Douglas and Flo Brooks of central Michigan. The Brookes bought old-style range burners from an appliance dealer and replaced the heating elements with 110-volt elements from a dryer. According to the Brookses, the burners give just the right amount of heat from their 32-volt system.

Refrigeration

Electric refrigerators use a.c. motors that, along with the compressor unit, are sealed in a metal box. Any inverter capable of starting induction motors can be used to power an electric refrigerator. Also, it is possible to remove the sealed compressor unit from the refrigerator and replace it with a belt-driven piston compressor such as that available from Tecumseh Corporation, which can then be run with a d.c. motor. This conversion requires $150 to $200 in parts and should be done by someone familiar with refrigeration systems.

Douglas Brooks removed his sealed compressor from his refrigerator, replaced it with a belt-driven compressor, and used a 32-volt d.c. motor. He disconnected the evaporator (cooling) coils that are inside the refrigerator box on modern units and replaced them with a freezer compartment that also served as the evaporator on an older refrigerator (see Illus. 7–4). For quieter operation, Brooks placed the compressor and motor in the basement underneath the

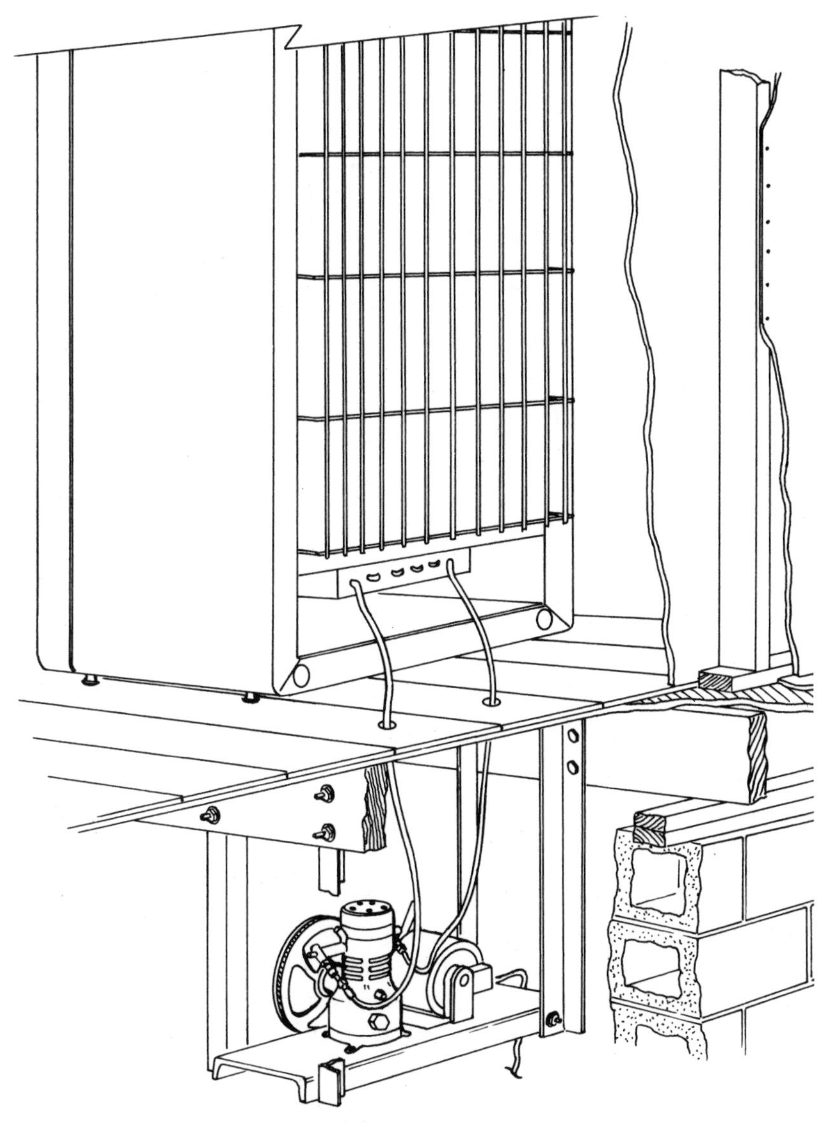

Illus. 7-4 Douglas Brooks's Refrigerator Compressor—To cut down on noise, Brooks suspended his belt-driven compressor from underneath the refrigerator floor/basement ceiling.

refrigerator. He had earlier tried using the air conditioner from a car as the refrigeration compressor, but found that it was very inefficient to operate.

An acquaintance of mine who repairs refrigerators converted an older type of refrigerator to run on d.c. for me. He replaced the sealed compressor with a belt-driven piston compressor and a 110-volt d.c. motor (¼ h.p. rating). Next, he reloaded the refrigeration lines with freon. He also installed a line filter dryer in the refrigeration lines to remove any trapped moisture—a standard procedure when replacing freon. An outboard relay operated by the refrigerator's temperature-control switch was added to start the d.c. motor, since otherwise the d.c. electricity might harm the switch contacts. The refrigerator has operated well for over two years.

Refrigerators designed for RV use are widely available and usually operate either on 12-volt d.c. electricity or on liquid propane (L.P.) gas. These are called absorption refrigerators and use heat from a gas flame or from an electric heating element to circulate the refrigerant through the system. The old Servel gas refrigerators that were made in the 1930's and 1940's operate on this principle. Many of these are in use today in different regions of the country. They are, however, very inefficient and most people (including myself and my family) usually end up looking for a better refrigerator after using one for two years. These old Servels are also notorious for stopping once or twice a year, and they have to be tipped on their side or turned upside down to get the system going again.

Most RV refrigerators are small, about half the size of a regular home refrigerator. The largest is made by the Dometic Sales Corporation of Elkhart, Indiana. The Dometic is the same size as a regular home refrigerator and runs on L.P. gas, natural gas, or 120-volt a.c. or d.c. It is one of the most efficient gas refrigerators made, but it costs somewhat more than a regular compressor refrigerator.

One of the most efficient of the small RV refrigerators is made by the Norcold Company of Sidney, Ohio; they are also sold by the Montgomery Ward Company. The Norcold AC/DC model operates on either 120-volt a.c. or 12-volt d.c. It is a compressor refrigerator with a small built-in inverter that allows it to operate on 12-volt d.c. To operate an absorption refrigerator on d.c., a small heating element is used to power the refrigerator instead of a gas flame. This is an inefficient process and the drain on the 12-volt battery is 2 to 3 times as much as that on an equivalent size Norcold compressor refrigerator (Norcold also makes absorption refrigerators).

Water Pumps and Other Water Appliances

Pumping water with a d.c. wind system does not present too great a problem. However, a jet pump or a submersible pump will not operate on d.c. electricity. Jet pumps are the most widely used pumps and work with either shallow wells (less than 25 feet deep) or deep wells (over 25 feet deep). An a.c. motor spins a plastic or brass impeller past a venturi tube (shaped like a nozzle with a curved tip), creating a partial vacuum to lift the water from the well. If a d.c. motor were used with such a pump, the pump would not work as efficiently because the motor speed would vary with the battery voltage. If its exact design speed is not maintained, the efficiency of a jet pump drops off quickly. Likewise, submersible deep well pumps are sealed and work only with a.c., so modification is not possible.

For shallow wells of less than 25 feet deep, a piston pump (such as those available from the Sears, Roebuck and Company) is the best solution. Any type of motor can be installed and connected to a piston pump using a "V"-belt, since the speed of the motor is not as important as with jet pumps. Small 12-volt pumps for shallow well use are also available from RV supply stores.

A pump jack, appropriate for use with deep wells, is basically a gear box that changes the circular motion of the pulley wheel to a linear up-and-down motion. The pump jack raises and lowers a pump rod attached to a brass pump cylinder that goes to the bottom of the well. The cylinder fills with water on the downstroke of the pump rod and lifts the water up the pipe on the upstroke. Leather packing in the cylinder needs to be replaced every few years, so the pump rod and well pipe must be pulled up at that time, a considerable job if you have a 100-foot or deeper well. Pump jacks, cylinders, and pump rods, including models that operate on an electric motor, are available from the companies that make water-pumping windmills. (See pump jacks in the Source List following the appendices. For more information, see Ch. 10.)

Water softeners are almost a necessity in many parts of the country and should be considered along with the water system. Most water softeners use a.c. motors and timers. A manually operated water softener that is simple to run is available through the Montgomery Ward Company; however, it takes one hour to regenerate the chemicals each week for this model.

Automatic clothes washers and dryers, with their maze of a.c. motors, switches, and timers, definitely cannot be operated on d.c. The alternative is to use a wringer washer, such as the popular Maytag, in which a d.c. motor replaces the usual a.c. motor.

Electric hot water heaters require 220-volt a.c. and use more energy than the output of cabin-size and smaller home-size wind systems. An alternative is to use a gas or oil-fired hot water heater. Many d.c. wind system owners also have a water preheater connected to a wood stove or a solar water heating system.

Heating and Cooling

Gas and oil-fired furnaces do not operate on d.c. electricity. Although the fan motor can be replaced with a d.c. motor, the furnace controls operate only on a.c. and thus need to operate either from utility power or through an inverter. Most d.c. wind system owners rely on wood heat or a combination of wood heat and passive or active solar energy for space heating, thus eliminating the need for expensive controls.

All in all, most users of d.c. wind systems are pioneering people who are more interested in an energy-independent lifestyle than in being able to use every possible modern appliance. If you are one of these people, then the d.c. wind system is a good option for you. Otherwise, a utility-connected or inverter wind power system should be used.

Chapter 8

Installing the Tower, Generator, and Rotor

At this point, if you have measured your wind speeds, obtained any necessary permits or approvals, and picked the system that will fit your needs, you are now ready to install the system. You will have to make careful plans to install the tower, raise the wind power generator on top of it, wire the system, and if you intend to have a stand-alone system, install your storage batteries.

The first decision for you to make is how much of the work you will do yourself, if any. If you have a trained wind system installer do the work, you pay to have the work done, but the contractor is then responsible for the safety of himself and his crew as well as for the quality of the work. Some companies guarantee their system only if it is installed by trained people, while some allow the owner to install it but require their own inspection to assure proper installation before honoring a guarantee. The larger the system is, the more likely it is that the dealer will install the system with a trained crew. Even if the dealer or contractor does all of the work, however, it is worthwhile for you to become familiar with what is involved to be able to judge the quality of their work.

Many people have installed their own cabin-size or smaller home-size wind systems, realizing a savings in cost and also the satisfaction of knowing that they did the work themselves. If you have this in mind, you should learn as much as possible about what is involved before tackling the job. Also, remember, as mentioned in Chapter 3, that friends who help you with the installation are not covered under your homeowner's insurance policy. Even after learning the basics, the best rule is to have the help of one person who has experience installing wind systems or in similar work to direct the job and check the equipment.

Rigging

The basic rigging or lifting tools for installing a wind system are nylon rope, wire rope, tackle blocks, winches, and gin poles, which we'll consider in turn.

Both nylon rope and wire rope, or cable, are used to lift the tower parts and the aboveground components into place. You'll need 250-foot lengths of cable to meet the requirements of most situations when installing a wind system. Nylon rope is made from braided strands of nylon fibers, and is the modern replacement for manila rope. It's very flexible and easy to handle but, as it ages, its strength is difficult to predict, compared to that of cable.

Because of its flexibility nylon rope is well-suited for use as a hand line or as a tag line. Hand lines are ropes used to raise and lower tools and small parts up and down the tower. A canvas tool bag or pouch attached to the hand line is a convenient item for carrying small items. A tag or guide line is a rope that is used to control the generator (or any other component) as it is being raised to the top of the tower to keep the generator from bumping the tower and thus causing damage.

Cable is made from braided strands of steel wire and is preferable for supporting heavy loads since its strength characteristics are more predictable than nylon. You can find out the strength of specific sizes of cable from the manufacturer. Cable should be treated with respect, because if it does break, it snaps and lashes back with lightning speed. Anyone in its path can be seriously hurt. Also, cable cannot be coiled loosely like nylon rope. Cable must be coiled onto a spool when not in use to prevent any kinks or knots from forming. It is best to wear gloves when handling rope to prevent rope burns, but it's almost imperative to wear gloves when handling cable because small metal burrs on the wire can cut your hands otherwise.

Breaking Strength of Rope and Cable

Rope and cable are rated by the diameter of their cross section and by their breaking strength. The breaking strength, also called the tensile strength, is simply the weight in pounds that the rope or cable can support before breaking. Some typical breaking strengths of different size rope and cable are shown in Table 8–1. The figures shown are for illustration purposes only. You should confirm the actual breaking strength of a particular rope or cable you intend to use from the manufacturer's data.

Obviously, you do not want to raise a 5,600-pound object with a certain diameter cable if that weight is the cable's breaking strength. A good rule of thumb is to allow a safety factor of at least

TABLE 8-1
Representative Breaking Strengths of Nylon Rope and Steel Cable*

Nylon Rope		Steel Cable	
Diameter, Inches	Breaking Strength, Lbs.	Diameter, Inches	Breaking Strength, Lbs.
1/4	1,750	1/16	250
5/16	2,400	1/8	1,760
3/8	3,200	5/32	2,800
7/16	4,700	1/4	4,780
1/2	6,600	7/32	5,600
		5/16	7,420
		3/8	10,620
		7/16	14,380
		1/2	18,700
		9/16	23,600
		5/8	29,000
		3/4	41,400
		7/8	56,000

*Actual breaking strengths will vary with the manufacturer.

5 when raising wind equipment. In other words, you would use a rope or cable with a breaking strength of 5,000 lbs. to lift something weighing 1,000 lbs. However, in a situation in which the danger of accident could risk someone's life, a safety factor of 10 is advisable. The safety factor allows for unsuspected weaknesses in the rope or cable and for sudden stresses due to jerking of the load. Sudden acceleration of the rope or cable can increase the tension on it by 100 percent or more. For this reason, you should use a relatively slow and steady rate of ascent—10 to 20 feet per minute, for example—when raising heavy tower and generator parts.

Any rope or cable that you intend to use for raising towers and generators should not have been used previously for other jobs. If the rope or cable has been nicked or jerked violently from pulling a car out of a ditch or some other use, this rope is not one to rely on for raising wind equipment.

Joining Rope or Cable

To join wire cable to a clevis or a hook, first put a cable thimble through the eye of the hook and then wrap the end of the cable around the thimble. The cable is then fastened by using cable clamps. The "U" bolt side of the cable clamps should bear on the short or "dead" end of the cable so that the long end is not crimped or cut by the "U" bolt. The number and spacing of the cable clamps

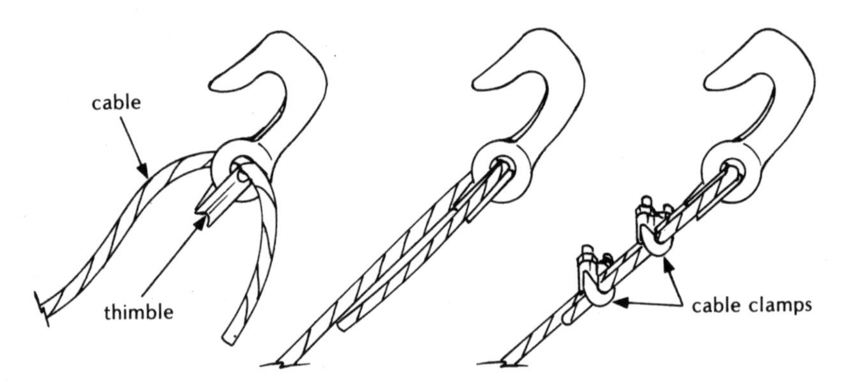

Illus. 8-1 Joining Wire Cable to a Hook—(1) Insert a cable thimble through the hook's eye. (2) Wrap the cable end around the thimble. (3) Fasten the cable with clamps, assuring that the "U" of the clamps are on the short end of the cable.

is important and is usually specified by the manufacturer. For cable ⅜ inch or less in diameter, a need for two cable clamps spaced six cable diameters apart is typical (see Illus. 8–1).

Of course, joining rope is a matter of learning to tie knots. Some basic knots useful for tower work are shown in Illustration 8–2. Although it's not impossible to learn knot-tying from pictures, my experience is that it's easier to learn by watching a skilled person tie knots, and then to tie some yourself under expert guidance. So, if you know someone experienced in tying knots, ask to be taught how. An example is permanently joining a rope to itself. The best knot to use is an eye splice, and since this knot is somewhat complicated, you should learn it from a person who knows how to tie it. Even then, the knot will still demand practice on your part.

For a temporary connection to an eye or ring, two half hitches

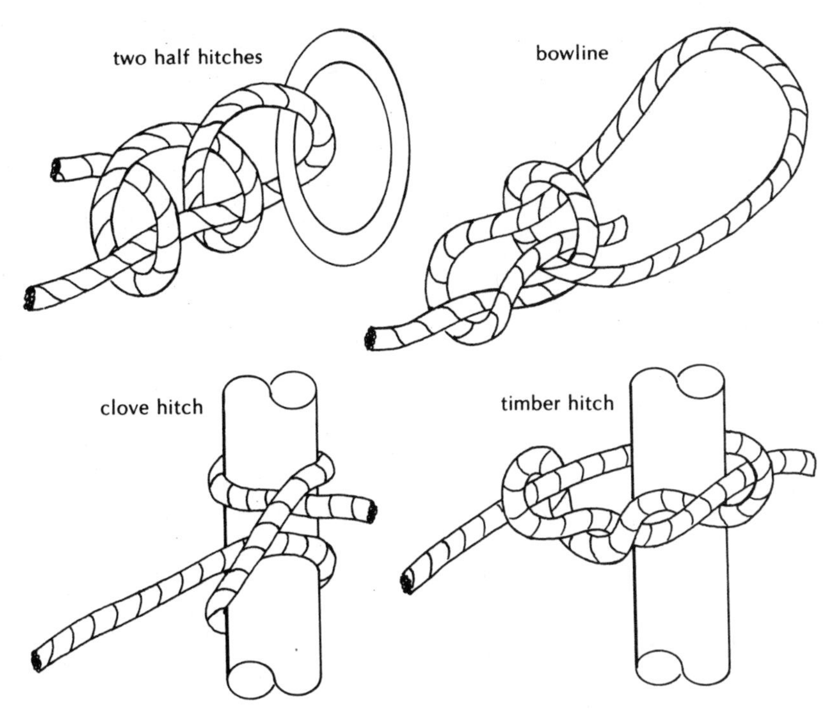

Illus. 8-2 Tying Knots for Tower Work—Half hitches, bowlines, clove hitches, and timber hitches are some of the useful knots for securing equipment when working on your wind system tower.

can be used, although a bowline or a bowline-on-a-bight is probably less prone to slip. Gin poles and tower sections can be hoisted up using either a well pipe hitch, or a timber hitch with a half hitch.

When raising pipe or tubing, a hook on the end of the rope and a simple half hitch near the upper end of the pipe is the most convenient way of hoisting. The half hitch is easy to release. Either the clove hitch or the timber hitch is useful for fastening rope to a pole. When raising or lowering a fairly heavy part by hand, such as a gin pole or tail vane, using a snubber—that is, simply wrapping the rope around a post or tree—greatly reduces the amount of pull needed, and you can suspend the load at any time simply by applying a bit of extra tension to the rope.

Tackle Blocks

Tackle blocks are used with ropes or cables to lift loads, to change the direction of the pull on a line, or to reduce the effective pull on a line. A tackle block consists of a frame and one or more pulley wheels, called sheaves. Usually a hook or an eyebolt in the top of the frame connects the block either to the load being lifted or to a fixed point to lift from, such as the top of a tower.

A tackle block with a frame that opens at the top to allow the rope or cable to be strung over the sheave is called a snatch block. These are handy if the rope or cable you're using has a hook or other device at the end of it that makes it difficult to thread the rope through the body of a regular block.

Tackle blocks, sometimes called pulleys, are rated according to the load they can lift and according to the size of rope or cable that will fit on the sheave. A heavier weight can be lifted than that possible using a block with just one sheave if you use tackle blocks with several sheaves to reduce the tension on the rope or cable. This reduction in the effective pull on the rope is called the mechanical advantage and is similar to the mechanical advantage gained when a lever is used to move a heavy object. Illustration 8–3 shows an upper tackle block connected to a fixed point to where the load is to be lifted. The lower tackle block is connected to the object to be lifted. The tension or pull on the rope is then approximately the weight of the load divided by the number of parts of rope supporting the lower tackle block. The number of parts of rope depends upon the configuration. Usually there's one more pulley wheel in the upper tackle than in the lower one. A three-part arrangement is shown in Illustration 8–3. The pull on a rope lifting a 1,000-pound generator with a three-part arrangement is only 333 lbs. With a safety factor of 5 this means that a cable with a breaking strength of at least 1,667 lbs. can be used with the three-part arrangement instead of a cable with a 5,000-pound breaking strength for a one-part arrangement.

Actually, in this example a slightly greater pull is required because of friction between the pulley and the rope. For each part of the rope, friction usually requires about 3 percent more effort than that necessary for the pull. For the example given, then, friction adds about 30 lbs. to the pull (3 parts × .03 × 333 lbs.), for a total of 363 lbs. of pull to lift the 1,000-pound weight instead of the 333 lbs. calculated originally.

Winches

You can raise light tools and parts up to the tower by putting together a simple hand line and a snatch block hooked at the top.

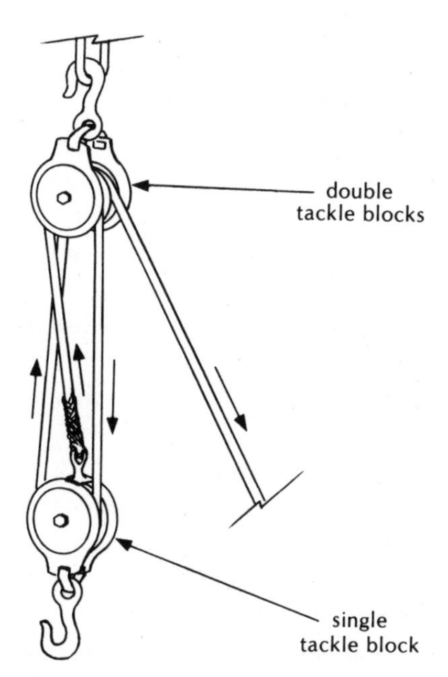

Illus. 8-3 Tackle Blocks—A three-part arrangement reduces the pull required on the rope to lift the load by a factor of nearly 3.

It may have occurred to you that you can use a vehicle as the pulling force to raise a generator or lift a section of tower. A vehicle, though, is relatively difficult to control. There tends to be a lot of jerking of the line that can add to the acceleration stress on the cable, and as I mentioned before, you do not want to test the strength of the cable just when you are raising your new wind system.

A hand-operated or electric winch with the right capacity for the job can lift a load with a steady pull and little effort. Again, a safety factor of at least 5 is desirable, so to lift a 300-pound generator, for example, a winch with a minimum capacity of 1,500 lbs. would be used.

Any winch that you use should have a mechanical or electrical brake that allows you to suspend the load at any point while it is being raised. With an electric winch, the brake should activate if power is accidently disconnected from the winch motor. This means that you should use a commercial or industrial-quality winch designed for lifting loads vertically and equipped with a brake rather than typical boat or vehicle-mounted winches, which are generally designed for pulling instead of lifting and do not usually have brakes.

Gin Poles

Gin poles, also known as davits, are used with various rigging to raise tower parts and wind system components above their installation point so that the parts can be lowered into place. Gin poles are really small cranes that attach to the top of the tower.

A simple gin pole consists of a length of seamless steel or thick wall aluminum pipe. (Water pipe should never be used because of its insufficient strength.) The gin pole is clamped to the tower. The pole should extend high enough above the tower for you to be able to place the component properly. Whatever the design, the pole should be clamped so that it cannot sway sideways or downward when it's used to lift components. The main consideration is to use for your gin pole a pipe that is strong enough to take the side-to-side pulling on it that will occur during lifting.

Gin poles often bear a horizontal crossarm that allows the tackle block to hang over the center of the tower. This allows for more flexibility in use than that of a straight piece of pipe. The crossarm should be braced on the gin pole from above or below to prevent the arm from bending. Flimsy gin poles will bend over like a pretzel under stress, so don't stint on yours.

Some manufacturers sell gin poles that are designed for lifting their model of wind system. Others may recommend a gin pole design. Gin poles have been used successfuly for systems weighing up to about 600 lbs. For heavier systems, a crane is often used to lift the generator in place.

Safety

When working on a tower, safety is not only a matter of using the proper equipment but also of common sense. Basic safety equipment for tower work includes hard hats, leather gloves, and lineman's safety belts or rock climber's sit-harnesses.

You should wear a hard hat when you're on the tower to protect yourself in case a spinning rotor blade hits your head or a part falls on you from above. Workers on the ground need hard hats, of course, in case a bolt or tool falls from the tower. But even with hard hats, no one should be working directly under the tower when work is going on at the top, and no one should ever climb a tower when the rotor is spinning. You can assume that at least one bolt or wrench will be dropped during the course of an installation. The fact that it will be traveling at least 60 mph when it hits the ground should be incentive enough to stay out from near the tower area. An area of at least 20 to 30 feet in diameter around the tower should be considered a target area for falling parts that might ricochet off the tower.

Besides a hard hat, a leather or nylon webbing safety belt that wraps around the tower is a must for working on towers. The belt should be sized to your waist and it should be comfortable. After your tower is installed, an additional safety feature you can put in for maintenance work is a safety belt that attaches to a cable or pipe that runs the length of the tower. If you slip while climbing, a special mechanism that connects the belt to the cable tightens immediately and stops your fall. (See Installation Equipment in Source List.)

Hoist Signals

The best safety rule is to have one person in charge of the work so that there are no conflicting instructions being given at critical times. This person should be the most experienced tower worker, of course, and the one who feels most comfortable and confident —as the "boss." Ideally, this person should not suffer from acrophobia (fear of heights).

The crew "boss" should check all rigging personally and give hand signals for raising and lowering equipment. Everyone should be familiar with these signals since verbal commands are difficult to hear on the ground from a distance of 60 to 80 feet in the air. Some suggested hand signals for raising, lowering, and stopping are shown in Illustration 8–4. Some installers use walkie-talkies for tower to ground communication, too.

Tower Installation

Of course, the first step in installing a tower is to check the manufacturer's specifications and instructions. These will include the dimensions and placement of footings as well as the number of reinforcing rods (re-bar) and amount of concrete required. Installation techniques vary with the type of tower, but the general princi-

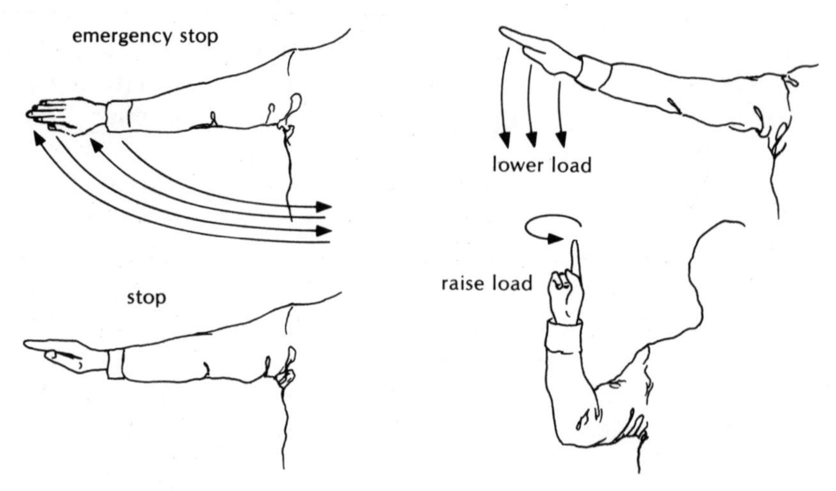

Illus. 8-4 Directional Hand Signals—Hand signals assure safety for work around a wind system tower where height and noise can make hearing spoken directions almost impossible.

ples of using gin poles and rigging are the same with all.

Installing a tower is like building with a big erector set. Each piece or section follows the last according to the instructions from the manufacturer or designer of the tower. The time it takes to install the tower will constitute at least half of the total time spent on the whole system. Actual erecting of the tower usually takes a full day for two people. This time will vary considerably depending upon the experience of the crew, the height of the tower, and on the weather conditions. It is difficult and hazardous to raise tower parts on a breezy day with wind speeds above 15 mph or so. Likewise, you do not want to be on a tower during stormy weather because of the threat of lightning.

Depending upon the type of soil, it usually takes part or all of a day to dig holes for concrete footings. A few hours are needed to pour the concrete base and footings and these should dry, or "cure," for at least two weeks before putting up the tower.

Guyed Towers

Digging the holes for the base footing and the guy wire footings is the first and most arduous part of installing a guyed tower, or any tower for that matter. The guyed tower base or pier is typically about 4 feet deep and 4 feet wide at the bottom. The correct spacing and size of the holes is spelled out in the manufacturer's drawings. A base plate that the first tower section sits on is anchored to the top of the pier. Total concrete required may be about 1½ cu. yds. weighing about 6,000 lbs. (1 cu. yd. = 3 × 3 × 3 feet of concrete).

Footings for the guy wire anchor rods are typically spaced 120° apart (with the tower as the circle center), and at a distance from the base of the tower of about 80 percent of the tower height. Laying out of the holes can either be done by using a surveyor's transit or simply by laying out stakes and measuring distances with a string or tape. The most important dimension is the distance between anchor rods. If each one is an equal distance from its neighbor, then the guy wires will be equally spaced. If not, the guy wires will be pulling unequally on the tower.

These footings are typically about 3 × 3 × 3 feet and require ⅓ to ½ cu. yds of concrete along with reinforcing rods in each of

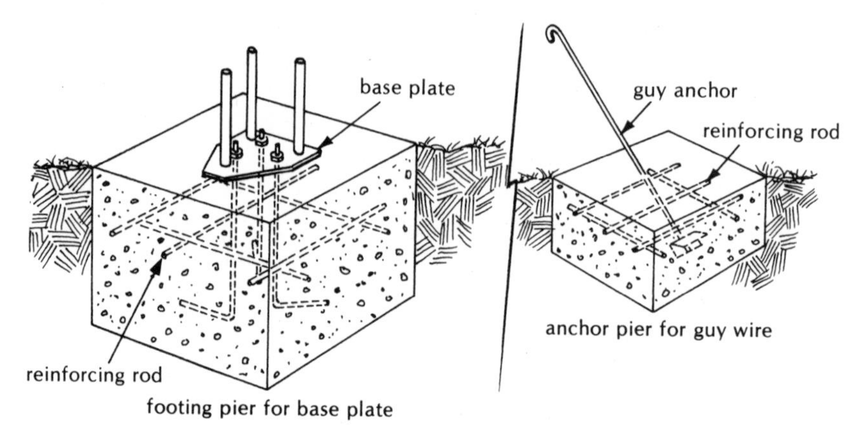

Illus. 8-5 Tower Base and Guy Wire Footing—The holes for these footings are about 4 feet deep by 4 feet wide. The tower base plate is then anchored in the concrete pier.

them. The anchor rod to which the guy wire attaches is set into the concrete at the specified angle, generally about 45°, according to the manufacturer's instructions. The rod can be propped up at the correct angle with a board until the concrete sets.

After the concrete has dried and cured, the next step is to erect the tower itself. The first section is bolted onto the baseplate. Then each succeeding section is lifted into place using a gin pole attached to the last section installed. The gin pole must extend high enough to be able to lift the section into place. One person can pull up the section on the ground and then it can be put into place by one, but preferably two, people on the tower.

Guyed towers between 40 and 100 feet high usually have 2 to 3 sets of guy wires specified for them. Each set is connected to the same anchor rod but connects to the tower at different heights. The first set of guy wires is attached after the specified number of sections are in place. A guy wire bracket is attached to the tower and the guy wires first attach to the bracket at one end, and then to the anchor rod at the other end by a turnbuckle. Each guy should not be tightened more than the other or the tower may bend to one side. Also, check the tower to make sure that it is level, using either a 6-foot level or a transit after each set of guys is put into place, and again when the tower is completed.

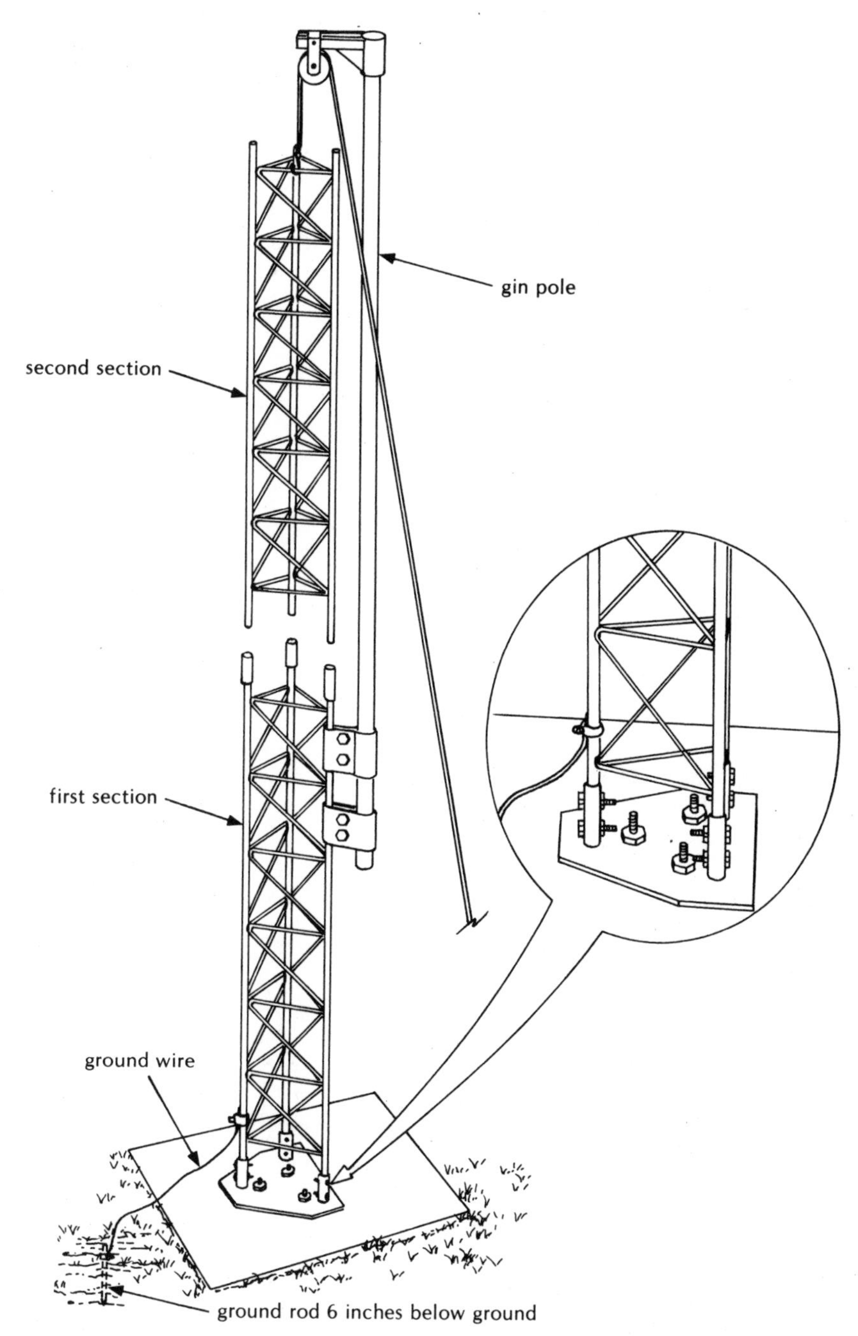

Illus. 8-6 A Gin Pole Attached to the First Tower Section—The gin pole is used to raise the next section of the tower. The pole is then secured to the newly raised tower section in order to bring up the next piece of tower.

Set the tension on the guy cables according to the manufacturer's specifications. Although it is possible to set the tension by "feel"—not too tight and not too loose—the correct way (particularly important to follow with towers 100 feet or more in height) is to read the tension in pounds of pull with a tensionmeter. The

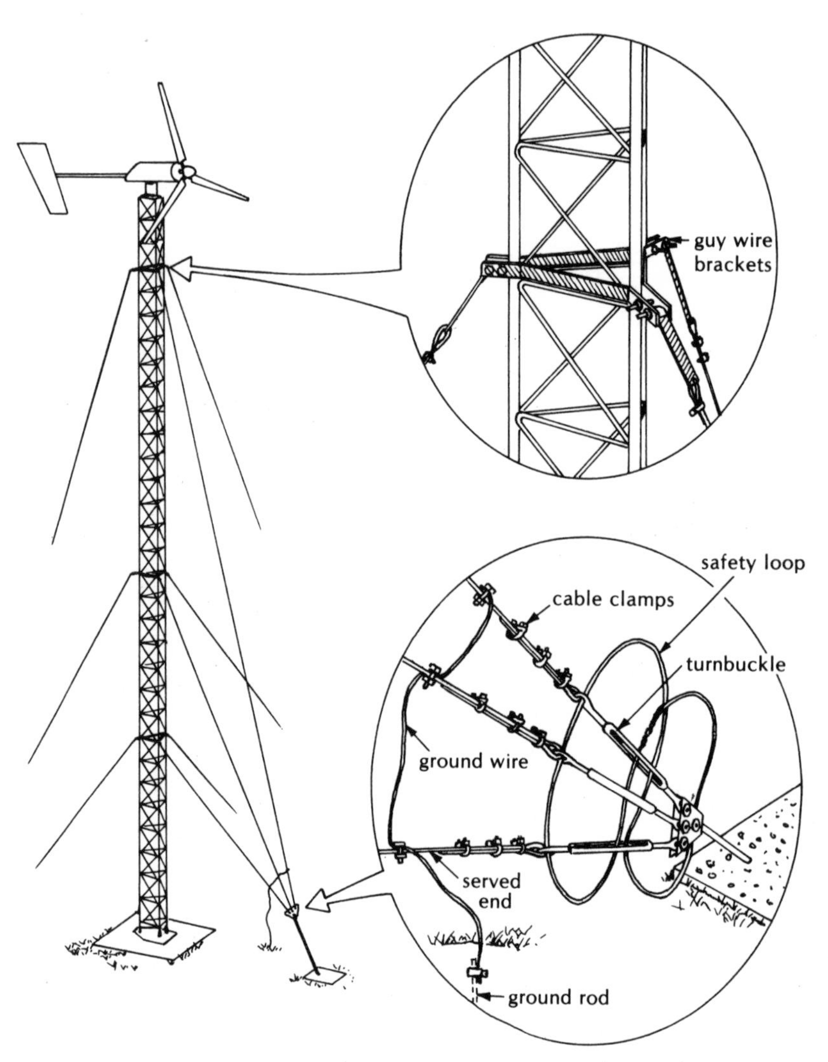

Illus. 8-7 Setting Guy Cables on a Tower—Guyed towers for wind systems usually have two to three sets of wires. Each set is anchored by one rod but connects to different points on the tower for an equal distribution of support.

tensionmeter registers tension by the amount of sag or deflection in the cable.

Continue this process until all sections and guy wires are in place, including the last section, the stub tower, upon which the generator is mounted. Be sure to check that the last guy wires set are low enough on the tower so that the rotor blades will not hit them. Blades can flex toward the tower several inches during a strong wind and if the cable is too high up, it can knock a nice chunk out of the blades.

After the last section is put into place, the tower should be checked again to be sure it's level. The tightness of bolts and guy wire tension should be checked as well. Finally, the ends of each guy wire are "served" to prevent them from unraveling. To "serve" a wire end, unwrap each strand of the free end of the wire and then rewrap it around the end of the wire under tension. Also, loop a safety wire through the turnbuckles and guy thimbles in a double "S" to prevent the turnbuckles from turning and loosening the cables.

In some cases, an obstruction prevents a guy wire anchor from being located where it should be. In that case, a stub anchor, usually a steel "I" beam set into a concrete footing of the right length to maintain the required angle on the guy wires, is used.

Self-Supporting Towers

As I mentioned in Chapter 4, several types of self-supporting towers are available for mounting a wind generator. The wide-based, water-pumping windmill type of tower has three or four angle-iron legs with angle-iron lattice work, or bracing. Another narrow-based type has round tubing legs (three) with angle-iron lattice work. The octahedron tower is usually constructed of round tubing with pinched ends for joining sections.

Self-supporting towers can be built either from the ground up or they can be assembled on the ground and then tipped up with a gin pole or hoisted into place with a crane. Building the tower from the ground up, piece by piece like an erector set, is the method used where the terrain doesn't permit on-the-ground assembly of the tower for tipping or hoisting it up. Tipping the tower up with a gin pole is convenient in the sense that the tower can be assembled more easily on the ground, but great stress is put on the tower when it is being raised into place. Since the stress increases with an increase in the tower's height, towers more than 50 feet high should not be assembled on the ground to be tipped up. There is no hard and fast rule on this, however, and I have seen an experienced installer, such as Curt Eggert of Garrison, Minnesota, tip up a 70-foot water-pumping type tower with no trouble. Using a crane to hoist the assembled tower up into place is quick and convenient, too, but it's not always physically possible to get a crane to the site.

Although self-supporting towers vary in construction details, most of them are put together in a similar manner. First, you dig holes for the footings for each tower leg, each about 5 to 8 feet deep. If you use a fencepost auger to dig the holes, they will probably have to be finished by hand since most of these augers go only 3 feet down. The base on these can be made level by assembling the first section of the tower, propping it up over the footing holes, leveling it with a plumb bob (when there is no wind), and then pouring the concrete.

After the concrete has cured for about two weeks, the complete tower can be assembled. The corner sections are first bolted into place and then the cross braces, or struts, are installed. On water-pumping windmill towers, bolts should be tightened, but not completely, until the tower construction is finished, so that any misalignments can be adjusted later. As with a guyed tower, a gin pole is used to raise each tower piece into place. For the inexperienced installer, towers with 10-foot leg sections are preferred over those with 20-foot legs, for easier handling.

Some towers attach to anchor bolts set in the pier footing. These anchor bolts are set at the correct distance using a template that extends to each corner. The bolts are held in place by the template until the concrete dries. The template is also kept level so that the tower itself will be level when it is set on the anchor bolts.

Assembling a Tower on the Ground

If you plan to tip up your tower intact or hoist it up with a crane, you should assemble it on a relatively flat piece of ground to prevent serious difficulties in its construction. Start from the top section and work toward the bottom. Assembling the tower with the sides propped up on concrete or wooden blocks allows you to keep the tower straight and enables you to reach to the underside to tighten bolts.

After all sections are assembled, tighten all of the bolts because any bolt failure during the tower-raising could be disastrous. Also, never raise your wind system along with the tower. This not only

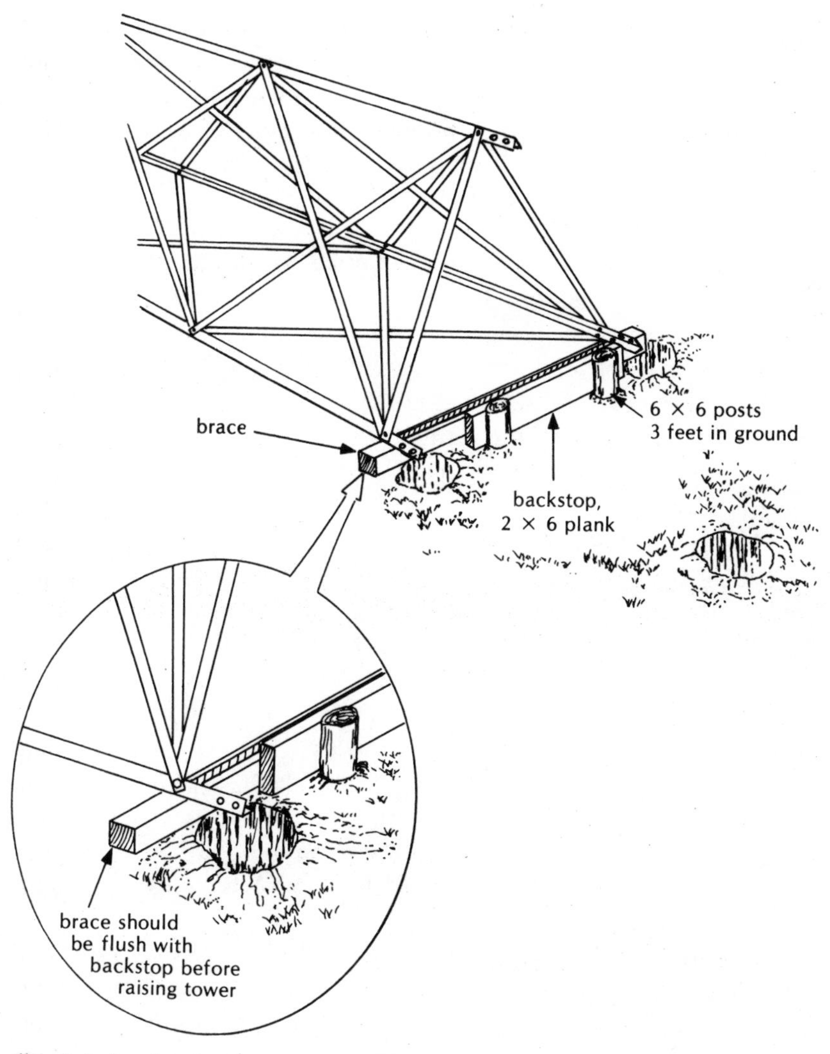

Illus. 8-8 Bracing the Base to Tip up a Tower—When the tower is tipped upright, a 4-× 4-inch beam secured between two of the tower legs hits up against a backstop supported by two vertical posts dug into the ground.

puts an additional strain on the tower but also presents an unnecessary risk to the generator.

You should assemble the tower where it will be raised in order to eliminate the need for moving it any farther than necessary. Since the bottom legs receive the greatest stress during raising, the very bottom section of the tower should be braced with a section of angle iron on the two vertical sides (while the tower is horizontal). Also, fix a strong wooden beam, such as a 4 × 4, between the bottom legs. This beam not only braces the legs, it will also bump up against two backstop posts (dug into the ground) to prevent the tower from sliding as it is pulled up (see Illus. 8-8).

Tipping Up a Tower

Before pulling the tower up, dig holes for the footings and set the anchoring leg sections loosely into the footing holes. Next, 6 × 6-inch beams are set at the corners so that the braces at the bottom of the tower can rest on them until the concrete footings are poured. The beams should be as close to the corners as possible so that the bottom braces on the tower do not bend, since the braces have to support the whole weight of the tower until the legs are set into the footings.

The key to tipping up a tower is to use a gin pole for an initial advantage when you first start to pull on the tower. This gin pole is not the same as the type used to raise parts to the top of the tower. It is simply a pole about 20 feet tall that is placed in such a way that you do not have to start pulling the rope or cable of the tower directly horizontal to the ground, but at a slight angle above the horizontal. This decreases the initial strain on the rope considerably. You can use as a gin pole, a 4 × 4-inch beam, a telephone pole, or even two sections of a guyed tower. The pole should be braced to each side so that it cannot tip sideways. Also, dig a small depression into the ground for the pole so that the pole can't kick out of place during its use.

You can determine the correct lengths of rope or cable required for raising the tower through trigonometry and arithmetic, but it's actually easier to build a scale model of the tower out of toothpicks or balsa wood and see what lengths of string are needed to raise it.

The rope or cable used to raise the tower should be strong

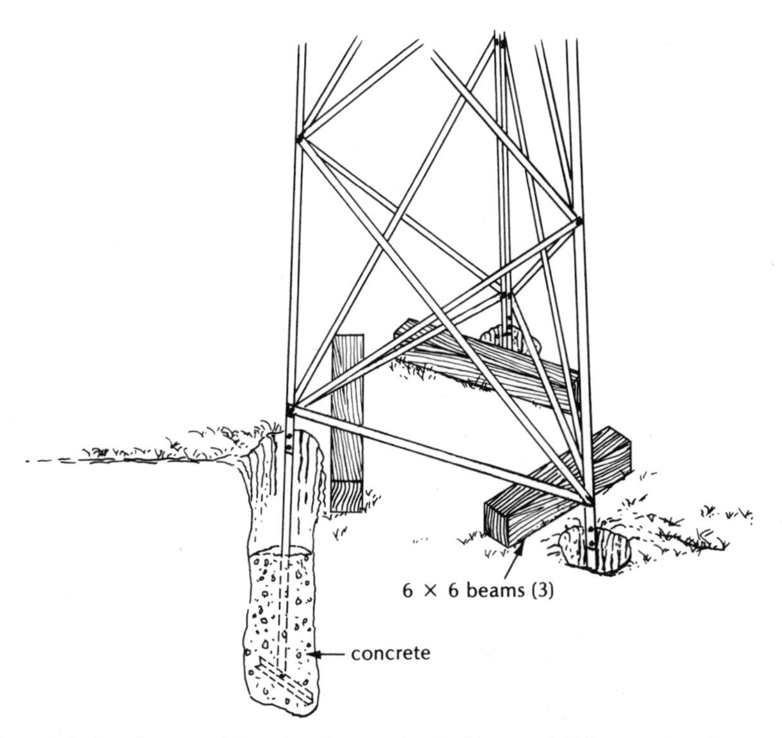

Illus. 8-9 Footings and Bracing Beams for Raising a Self-Supporting Tower—Beams 6 × 6 inches placed near the footing's holes act as support for the tower while the concrete footings are poured.

enough to take the initial strain when you first start to lift the tower. The tension on the rope is greatest when pulling first begins, and least when the tower is almost vertical. Consider the following example: a 60-foot tower weighing 2,000 lbs. is to be tilted up. Its center of gravity is at 25 feet. The gin pole is high enough so that the initial angle between the cable and the central axis of the tower, the lift angle, is 15°. The initial angle between the towers and the ground, the tower angle, is 0°. The initial tension on the cable can be figured out as follows:

$$2{,}000 \times \frac{25}{60} \times \frac{\text{cosine } 0^\circ}{\text{sine } 15^\circ} = 2{,}972 \text{ lbs.}$$

When the tower is almost up, at an angle of 80°, and the lift angle is 45°, the tension on the cable is only 204 lbs., as shown in the following computation:

$$2{,}000 \text{ lbs.} \times \frac{25}{60} \times \frac{\text{cosine } 80^\circ}{\text{sine } 45^\circ} = 204 \text{ lbs.}$$

To provide a safety factor of 5, however, the cable should have a breaking strength of at least 14,860 lbs. (5 × 2,972) to withstand the initial tension. Depending upon the rating of the cable, this might require a ½-inch cable, which is rather large and expensive. Someone using a ⅜-inch steel cable, then, would be testing it almost to its breaking limit on the initial pull. This is one reason that most installers prefer building the tower from the ground up or hoisting it up with a crane instead of tilting it up.

To pull the tower up, the cable is attached at or near the top of the tower, and it then runs over the top of the gin pole (starting with the gin pole in the vertical position). Then the cable continues to the winch or vehicle that will help in the pulling. The cable can be tied to the top of the gin pole, or it can be run through a tackle block attached at the pole's apex. Once the tower is raised about one-third of the way up, the gin pole is no longer necessary. At this point, if the rope is tied at the top, the gin pole will then begin to slide along the ground. If a tackle block is used, the gin pole will be lifted off the ground and will slide down the cable, which can be hazardous unless the pole is restrained.

To prevent the tower from tipping sideways during raising, guy wires should be attached from the top of the tower to an anchor or post located at least 50 to 80 percent of the tower height away from the tower, and in line with the final position of the center of the tower base.

As an additional safety precaution, a restraining line tied to the tower's top and drawn by a vehicle or winch in the opposite direction of the tipping will prevent the tower from toppling over to the other side when it's pulled all of the way up.

Once the tower has been assembled and the rigging is in place, the actual raising of the tower takes very little time. Of course, all of your helpers should be well out of the way in case the tower falls or a cable snaps.

I raised my own water-pumping type tower this way and found it to be quite a dramatic moment when the tower tipped into place and rocked back and forth for a few seconds. I used a tractor to pull the tower up and found it to be less than desirable for the task because releasing the clutch jerked the line unnecessarily. When the tower was about 80 percent up and its center of gravity was in line with its base, it began to fall into place by itself. After the tower was in place, we attached the anchor legs, leveled the tower by pulling on the guy wires, and anchored the legs with concrete.

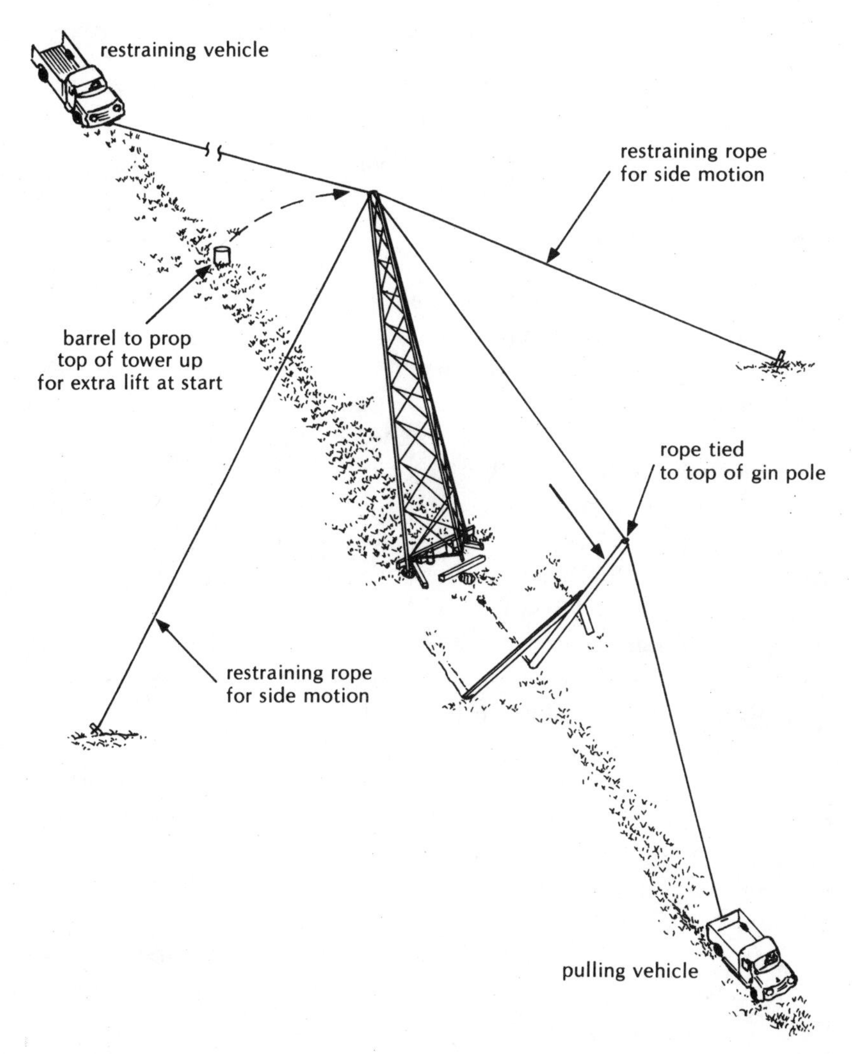

Illus. 8-10 Tipping a Tower into Place—Guy wires and support vehicles stabilize the tower as it is tipped into place.

Pole Towers

Self-supporting, steel pole towers rely greatly upon a proper foundation for their strength, so you should do this work with special care, according to the manufacturer's specifications. Since several yards or more of concrete are required for the base, you should use premixed concrete. You can raise a small steel pole tower into place section by section with a gin pole. For larger steel pole towers you'll need a crane to lift the sections.

Installing the Wind Generator and Rotor

You can raise wind generators weighing 500 pounds or less into place by using a gin pole attached to the top of the tower. For larger, heavier systems, a crane is often necessary. Another method is to use sections from another smaller guyed tower, which you can borrow or rent for the purpose, as the gin pole. The smaller tower is temporarily clamped to the larger one.

To use a gin pole, run the cable through the hook on top of the generator, up the tower, and over a tackle-block pulley on the pole, then back down the tower through a tackle block attached to the bottom of the tower. This tackle block insures that the pull on the gin pole is straight down so that the pole is not pulled to the side. This tackle block does not offer any mechanical advantage to reduce the effective tension on the cable, it simply directs the pull.

If you intend to use a hand winch, you can also clamp it at the base of the tower, although this will necessitate standing under the tower. It is safer to have the winch attached to a vehicle or some other heavy object away from the tower.

Although it's possible to use a gin pole to raise some wind systems complete with tail vane and rotor, I do not advise this practice. The unit must be kept far enough away from the tower to prevent damage to the blades, a complicated process if the whole system is already assembled. Usually, the generator is raised first, and the rotor and tail vane then follow. When raising the generator, use a tag line to keep the unit about 6 inches to a foot away from the tower.

Although the details of assembling different wind systems vary, there are common procedures for assembling them all. With horizontal-axis machines it is important to check the tracking of the rotor blades. This should be done on a calm day with the wind plant turned off. To check the tracking, clamp a stick or some other type of indicator to the tower at the level of the blade tips. Then slowly rotate the blades with your hand past the stick "indicator." If one

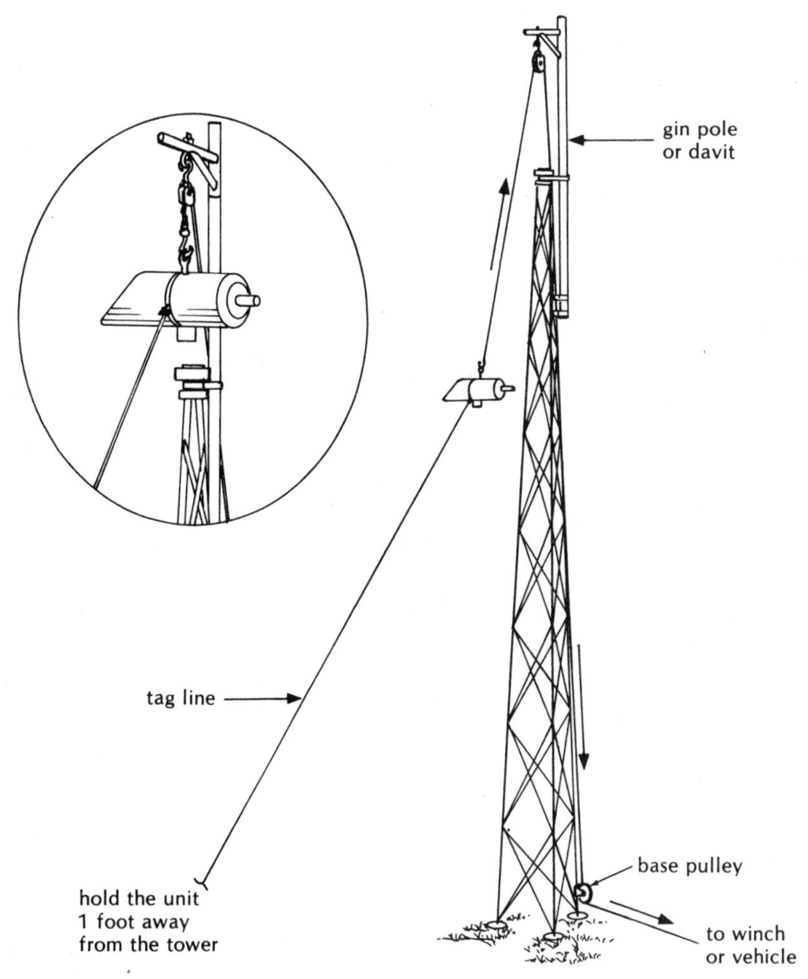

Illus. 8-11 Raising the Wind Generator with a Gin Pole—A pulley at the base of the tower redirects the pull outward to a winch or vehicle. The tag line is used to keep the generator from hitting the tower during the hoisting.

of the blades is more than a fraction of an inch in or out of the plane of rotation of the others, then it is not tracking properly. This can lead to vibrations, rotor noise, and loss of power. If your rotor blades are out of alignment, consult the manufacturer concerning adjustments.

Also be sure to check to see that the unit is level and turns freely on its turntable bearings. If the unit is out of balance, it will favor one position over others when the unit is swung around. Downwind units are particularly sensitive to imbalance since the rotor can get stuck in one direction and end up facing the wind 180° out of line.

With a gear-drive machine, check the gearbox to make sure no oil was accidently lost during installation. The gears may be stiff for several weeks until they wear in.

Chapter 9

Wiring a Wind System

Wiring the wind system to your house and making any necessary modifications to the house wiring are critical to the proper installation of your system. In some areas, the work may have to be done by a licensed electrician, and in all cases the local electrical inspector will have to approve the work.

Electrical Codes and General Wiring

With modifications made by state or local authorities, your area's electrical code is based upon the National Electrical Code (N.E.C.) published by the National Fire Protection Association. The N.E.C. is updated every three years and references here are based upon the 1978 code. The N.E.C. does not yet contain a section specifically drawn up for wind systems. General wiring provisions of the N.E.C. apply to wind systems, of course, but specifics peculiar to wind systems may be a matter of interpretation. The articles of the N.E.C. that may bear upon wind systems are summarized in Table 9-1, but for the exact wording of these articles, refer to the most recent N.E.C. Keep in mind particularly that when matters of interpretation are involved, your area's electrical inspector is the one who makes the interpretations and can approve or disapprove any exceptions. As far as you are concerned, the inspector is the code, so talk to the inspector first if there are any questions. You may save yourself the trouble of redoing work later on.

You may be aware that some easy-to-read books have been published that explain household wiring in simple terms and interpret the most commonly used provisions of the N.E.C. These books are available in many stores that sell electrical wiring to the general public, and some are available from Sears, Roebuck and Company as well. You'll find the titles of some of these electrical handbooks listed in the Bibliography.

As with any wiring task, the basic concerns in wiring a wind system are choosing the right type and size of wire for its intended use, and proper grounding. The proper tools and methods for wiring are the same for wind systems as for any other wiring job.

Types of Wire

In most home wiring only a few of the many available types of electrical wire are used. These types of wire are distinguished by their outer insulating cover. Local codes often specify which types of wire can or cannot be used, so be sure to check first before purchasing any wire. One of the most common types of wire is designated as either type T, TW, or THW. The T stands for thermoplastic, which describes the plastic insulation covering the wire. The W stands for weatherproof and indicates that the wire can be used in wet locations. The H stands for high temperature, with the temperature range specified in the N.E.C. or by the manufacturer. Another type of wire designation is type RH or RHW, which indicates a rubber insulation covering and, at present, is not used as commonly as is type T.

Types of wire are further designated by their application; that is, whether they are used outside, underground, or inside the house. The house can be wired from the wind system or elsewhere either by overhead or aerial cable, or through underground cable. Underground cable can be designated as type USE for Underground Service Entrance, or type UF for Underground Feeder. When you run underground or aerial cable into your house, either connect it to the service panel using SE cable (for Service Entrance), TW, THW, or RHW

TABLE 9-1
A Summary of the *National Electrical Code* (1978) Provisions That May Relate to Wind Systems

Article	Title	Summary of Contents*
110-17	Guarding of live parts	Exposed live parts of electric equipment operating at 50 v. or more must be enclosed in an approved cabinet or located in a room so that people or objects cannot come in contact with the exposed electrical parts.
210-7	Receptacles and cords: noninterchangeable types	Where mixed voltages or types of electricity (a.c. and d.c.) are used in the same building, noninterchangeable plugs and receptacles should be used for the different types of circuits.
250	Grounding	Systems that generate d.c. must be grounded except for those with voltages of 50 v. or less.
310-16 through 310-54	Allowable ampacities of insulated conductors	Lists ampacities for all sizes of wire, temperature ranges, types of insulation, and for aluminum or copper.
445	Generators	Lists wiring, overcurrent protection, and ampacity requirements for generators.
480	Storage batteries	Gives requirements for grounding, insulating, venting, and safety for storage batteries.
551	Recreational vehicles	Gives wiring and overcurrent requirements for low-voltage (24 v. or less) wiring in recreational vehicles. Conceivably, some of these requirements could be applied to low-voltage wind systems.

TABLE 9-1—Continued

Article	Title	Summary of Contents*
720	Circuits and equipment operating at less than 50 v.	States that No. 12 copper or equivalent wire (No. 10 for appliance circuits) is the smallest size wire to be used with a.c. or d.c. systems operating at 50 v. or less. Also states that lampholders should not be rated at less than 660.
750	Standby power generation systems	Discusses wiring consideration for standby generator systems.

*Refer to the *National Electrical Code* for exact wording and details.

wire that is inside a metal or polyvinyl chloride (PVC) conduit.

The service panel contains either fuses or circuit breakers to stop the flow of electrical current if any overload occurs. Circuit breakers are convenient because they can be reset by flipping a switch, whereas fuses need to be replaced each time they are burned out. Circuit breakers do not operate on d.c. electricity, so only a fuse panel should be used with d.c. wind systems. Circuit breakers that operate on d.c. are manufactured but they are not generally available.

At the service panel, the electrical current is divided into several branch circuits for distribution throughout the building. Each branch circuit has its own fuse or circuit breaker. Branch circuit wiring comes contained in either cable or conduit. The common types of branch cable are type NM or BX. Type NM (NMC for wet locations) is called nonmetallic sheathed cable and is sometimes called Romex, one of its brand names. Type BX is called armored cable and has flexible metallic sheathing. Thin-wall conduit, known as electrical metal conduit (EMT), is often used for branch circuit wiring. After the conduit is installed, type T or TW wiring is pulled through it.

Aluminum vs. Copper

Wire for conducting electricity is usually made of copper, although aluminum is often used in outside wiring. Aluminum wire is not recommended for inside wiring because apparently it has been the cause of many electrical fires. When the aluminum wire oxidizes, it creates an electrical resistance at the connection to the electrical outlets. This resistance leads to heating of the wire, which causes the aluminum to expand slightly beneath the copper screws on the outlet. If the screws become loose, electricity can arc across the gap between the aluminum wire and the screw, causing excessive heating and even fire. Aluminum is cheaper than copper but it is just not worth the risk to use it for inside wiring. For outside wiring, connectors have been designed for both aluminum and copper that greatly reduce the risk of loose connections. Also, antioxidizing grease compounds are used on both aluminum and copper service entrance connections to prevent oxidization and the resultant heating and loose connections.

Wire Size

Wire comes in various diameters. The greater the diameter of the wire, the more current it can carry without overheating. Wire size is indicated by American Wire Gauge (AWG) numbers, and they are written, for example, as No. 12 and No. 14. The higher the gauge number, the smaller the wire. The largest wire is No. 0000 (also written as No. 4/0 and pronounced as "four ought"); it's as big around as your little finger. The smallest wire is No. 44; it is smaller than a human hair.

No. 12 and No. 14 wire are the most common wires used in household branch circuits for general purpose lighting and small appliances. In many areas, No. 12 is the smallest gauge wire that is allowed. Heavier wiring is used for large appliances, such as electric ranges and hot water heaters.

Ampacity

The N.E.C. specifies the maximum allowable current-carrying capacity for each gauge of wire, which is called *ampacity.* The ampacities of all types of wire are specified in Article 310–16 through 310–54 of the N.E.C., but the ampacities of some common types of wire are summarized in Table 9-2. For example, the ampacity of No. 12 wire is 20 amp. when it is used in conduit or cable. This means that the largest size of fuse or circuit breaker that can be used with a circuit using No. 12 wire is 20 amp. Thus, if you look at the fuse panel in your home, you can assume that the circuit is probably wired with No. 12 wire if a 20-ampere fuse is used, No. 10 for a 30-ampere fuse, and so on.

Note that the ampacity table makes no reference to the voltage used or the length of the wire. The code is concerned only with safety, and so specifies the smallest wire that can safely carry the indicated electrical current regardless of the amount of voltage used or wire length.

Volt Drop

The length of the wire and the voltage used are of importance, however, when you consider the efficiency of the wire. Just as there

(*continued on page 224*)

TABLE 9-2
Ampacities of Insulated Copper Wires*
(*National Elecrical Code* [1978])

Wire Gauge	In Conduit Cable, or Underground		Single Conductors in Free Air	
	Type RUW, T, TW, UF	Type RH, RHN THW, USE	Type RUW, T, TW	Type RH, RHW, THW
14	15	15	20	20
12	20	20	25	25
10	30	30	40	40
8	40	45	55	65
6	55	65	80	95
4	70	85	105	125
2	95	115	140	170
0	125	150	195	230
00	145	175	225	265
000	165	200	260	310
0000	195	230	300	360

*For ambient (room) temperatures of 86°F. For other types of wire and other temperature ranges, refer to the *National Electrical Code,* Article 310-16 through 310-19 (as of 1978 code).

TABLE 9-3 Volt Drop Table*

Recommended Maximum (one way) Length, in Feet, of Copper Wire to Limit Volt Drop to 5%†

Circuit Voltage: 12 V.

Maximum Current, Amp.	Wire Gauge 12	10	8	6	4	2	0	00	000	0000
5	37	59	94	146	232	370	588	740	935	1,179
10	19	30	47	73	116	185	294	370	467	589
15	12	20	31	49	77	124	196	247	312	393
20	9	15	23	37	58	93	147	185	234	295
25	...	12	19	29	46	74	118	148	187	236
30	...	10	16	24	39	62	98	123	156	197
35	...	...	13	21	33	53	84	106	134	168
40	...	...	12	18	29	46	74	92	117	147
45	...	...	...	16	26	41	65	82	104	131
50	...	...	...	15	23	37	59	74	94	118
55	...	...	...	13	21	34	54	67	85	107
60	...	...	...	...	19	31	49	62	78	98
65	...	...	...	...	18	29	45	57	72	91
70	...	...	...	...	17	27	42	53	67	84
80	...	...	...	...	...	23	37	46	58	74
90	...	...	...	...	...	21	33	41	52	66
100	...	...	...	...	...	...	29	37	47	59
110	...	...	...	...	...	...	27	34	43	54
120	...	...	...	...	...	...	25	31	39	49
130	...	...	...	...	...	...	...	28	36	45
140	...	...	...	...	...	...	...	27	34	42
150	...	...	...	...	...	...	...	...	31	39
160	...	...	...	...	...	...	...	...	29	37
170	...	...	...	...	...	...	...	...	...	35
180	...	...	...	...	...	...	...	...	...	33
190	...	...	...	...	...	...	...	...	...	31

*Based on the formula $L = V/2RI$ where L = length, in feet, V = volt drop, R = d.c. resistance per foot of wire, I = maximum current in amp.

†For volt drops of 2%, divide table values by 2.5.

TABLE 9-3—Continued
Circuit Voltage: 24 V.

	Wire Gauge									
Maximum Current, Amp.	**12**	**10**	**8**	**6**	**4**	**2**	**0**	**00**	**000**	**0000**
5	74	118	187	293	463	741	1,177	1,480	1,869	2,358
10	37	59	94	146	232	370	588	740	935	1,179
15	25	39	63	98	154	247	392	493	623	786
20	19	30	47	73	116	185	294	370	467	589
25	...	24	38	59	93	148	235	296	374	472
30	...	20	31	49	77	124	196	247	312	393
35	...	...	27	42	66	108	179	211	267	337
40	...	...	23	37	58	93	147	185	234	295
45	...	...	...	33	52	82	131	164	208	262
50	...	...	...	30	46	74	118	148	187	236
55	...	...	...	27	42	67	107	135	170	214
60	...	...	...	...	39	62	98	123	156	197
65	...	...	...	...	36	57	91	114	144	181
70	...	...	...	...	33	53	84	106	134	168
80	...	...	...	...	...	46	74	92	117	147
90	...	...	...	...	...	41	65	82	104	131
100	...	...	...	...	...	...	59	74	94	118
110	...	...	...	...	...	...	54	67	85	107
120	...	...	...	...	...	...	49	62	78	98
130	...	...	...	...	...	...	...	57	72	91
140	...	...	...	...	...	...	...	53	67	85
150	...	...	...	...	...	...	...	...	62	79
160	...	...	...	...	...	...	...	...	58	74
170	...	...	...	...	...	...	...	...	...	69
180	...	...	...	...	...	...	...	...	...	66
190	...	...	...	...	...	...	...	...	...	62

TABLE 9-3—Continued
Circuit Voltage: 32 V.

Wire Gauge

Maximum Current, Amp.	12	10	8	6	4	2	0	00	000	0000
5	99	157	250	390	618	988	1,569	1,973	2,492	3,143
10	49	79	125	195	309	494	784	986	1,246	1,572
15	33	52	83	130	206	329	523	658	831	1,048
20	25	39	63	98	154	247	392	493	623	786
25	...	31	50	78	124	198	314	395	498	629
30	...	26	42	65	103	165	261	329	415	524
35	...	...	36	56	88	143	228	282	356	449
40	...	...	31	49	77	124	196	247	312	393
45	...	...	...	43	69	110	174	219	277	349
50	...	...	...	39	62	99	157	197	249	314
55	...	...	...	36	56	90	143	179	227	286
60	...	...	...	...	52	82	131	164	208	262
65	...	...	...	...	48	76	121	152	192	242
70	...	...	...	...	44	71	112	141	178	225
80	...	...	...	...	...	62	98	123	156	197
90	...	...	...	...	...	55	87	111	139	175
100	...	...	...	...	...	...	78	99	125	157
110	...	...	...	...	...	...	71	90	113	143
120	...	...	...	...	...	...	65	82	104	131
130	...	...	...	...	...	...	...	76	96	121
140	...	...	...	...	...	...	...	71	90	113
150	...	...	...	...	...	...	...	...	83	105
160	...	...	...	...	...	...	...	...	78	98
170	...	...	...	...	...	...	...	...	...	93
180	...	...	...	...	...	...	...	...	...	87
190	...	...	...	...	...	...	...	...	...	83

TABLE 9-3—Continued
Circuit Voltage: 48 V.

Maximum Current, Amp.	Wire Gauge 12	10	8	6	4	2	0	00	000	0000
5	148	236	375	585	927	1,482	2,353	2,959	3,738	4,715
10	74	118	187	293	463	741	1,176	1,480	1,869	2,368
15	49	79	125	195	309	494	784	986	1,246	1,572
20	37	59	94	146	232	370	588	740	935	1,179
25	...	47	75	117	185	296	471	592	748	943
30	...	39	63	98	154	247	392	493	623	786
35	...	...	54	84	132	215	342	423	534	674
40	...	...	47	73	116	185	294	370	467	589
45	...	...	...	65	103	165	261	329	415	524
50	...	...	...	59	93	148	235	296	374	472
55	...	...	...	53	84	135	214	269	340	429
60	...	...	...	...	77	124	196	247	312	393
65	...	...	...	...	71	114	181	228	288	363
70	...	...	...	...	66	106	168	211	267	337
80	...	...	...	...	...	93	147	185	234	295
90	...	...	...	...	...	82	131	164	208	262
100	...	...	...	...	...	...	118	148	187	236
110	...	...	...	...	...	...	107	135	170	214
120	...	...	...	...	...	...	98	123	156	197
130	...	...	...	...	...	...	...	114	144	181
140	...	...	...	...	...	...	...	107	135	170
150	...	...	...	...	...	...	...	...	125	157
160	...	...	...	...	...	...	...	...	117	147
170	...	...	...	...	...	...	...	...	...	139
180	...	...	...	...	...	...	...	...	...	131
190	...	...	...	...	...	...	...	...	...	124

TABLE 9-3—Continued
Circuit Voltage: 120 V.

Maximum Current, Amp.	Wire Gauge 14	12	10	8	6	4	2	0	00	000	0000
5	233	370	589	937	1,463	2,316	3,704	...	...	...	...
10	117	185	295	469	732	1,158	1,852	2,941	3,699	...	...
15	78	124	197	312	488	772	1,235	1,961	2,466	3,115	...
20	...	93	147	234	366	579	926	1,471	1,850	2,336	2,947
25	...	...	118	188	293	463	741	1,176	1,480	1,869	2,358
30	...	...	98	156	244	386	617	980	1,233	1,558	1,965
35	...	...	...	134	209	331	538	854	1,057	1,335	1,684
40	...	...	...	117	183	290	463	735	925	1,168	1,473
45	...	...	...	...	163	257	412	653	822	1,038	1,310
50	...	...	...	...	146	232	370	588	592	935	1,179
55	...	...	...	...	133	210	337	535	673	850	1,072
60	...	...	...	...	...	193	309	490	617	779	982
65	...	...	...	...	...	178	285	453	569	719	907
70	...	...	...	...	...	166	265	420	528	668	842
80	...	...	...	...	...	...	231	368	462	584	737
90	...	...	...	...	...	...	206	327	411	519	655
100	...	...	...	...	...	...	...	294	370	467	589
110	...	...	...	...	...	...	...	267	336	425	536
120	...	...	...	...	...	...	...	245	308	389	491
130	...	...	...	...	...	...	...	...	285	336	453
140	...	...	...	...	...	...	...	...	266	312	424
150	...	...	...	...	...	...	...	...	...	292	392
160	...	...	...	...	...	...	...	...	...	...	368
170	...	...	...	...	...	...	...	...	...	...	347
180	...	...	...	...	...	...	...	...	...	...	327
190	...	...	...	...	...	...	...	...	...	...	310

is pressure loss in a water pipe due to friction in the pipe, there is a loss of voltage or a voltage drop in wire due to electrical resistance. Therefore, in some cases, it is necessary to use a larger wire (smaller gauge number) than that designated by the ampacity rating in order to prevent an excessive voltage drop and thus, loss of power.

It's important to limit the voltage drop as much as you can, to below 5 percent and as low as 2 percent if possible. For example, if the voltage at the wind power generator is 120 v. and the wire to the house has a 5 percent volt drop, the voltage at the house will be 114 v. (120-.05 × 120).

Volt Drop Table 9-3 shows the longest (one-way) length of wire that can be used with a given gauge of wire, voltage, and fuse rating to limit the volt drop to 5 percent. For example, a No. 12 wire used in a 120-volt circuit with a 20-ampere fuse should not be over 93 feet long to limit the volt drop to less than 5 percent. In some cases a manufacturer of a.c. generator systems will specify shorter permissable lengths of wire than shown in Table 9-3, due to special power factor effects.

For low voltage battery-storage wind systems, (12- and 32-volt), the length of the wire becomes much more important than with 120-volt systems. A No. 12 wire with a 20-ampere fuse cannot be longer than 25 feet, for example, before the 5 percent volt drop is exceeded in a 32-volt system. The Volt Drop Table is particularly important in figuring the size of cable needed to carry the electricity from the wind system to the battery bank or service panel, as I will illustrate next.

Wind System Wiring for Various Voltages

The voltages commonly used in wind systems are 120 v. (a.c. or d.c.) and 48-, 32-, 24-, or 12-volt d.c. The branch circuit wiring for any house powered by a utility-connected wind system is the same as for any house wired for 120-volt a.c. The same applies to any house powered by an electronic inverter that changes the d.c. electricity from the wind system to 120-volt a.c. The size of the wire required to run from the wind system to the service entrance panel, or to the inverter, is determined by the ampacity and Volt Drop Table, or by the manufacturer's recommendations.

120-Volt Wiring

Wind systems that generate 120-volt a.c. or d.c. will use the same size wire in any house wired for 120-volt a.c. utility power. The wire size for 120 v. is No. 12 (and in some areas No. 14) for general purpose circuits, and heavier wire for large appliance circuits as specified by the code.

If 120-volt d.c. is to be used in the house without conversion to a.c., the same size of wire can be used, but you may have to make a few changes. These changes concern the type of overcurrent protection (i.e., fuses or breakers), the type of light switches that can be used, and the type of grounding. As I mentioned previously, circuit breakers are not made to operate with d.c., so a fuse panel must be used with a 120-volt d.c. system. In addition, light switches must be of the older snap-action type rather than the newer, quiet switches. Unlike the snap-action switches, the quiet switches are not designed to handle direct current. Snap-action switches are not always stamped as "a.c./d.c." but you can generally assume that they have that capability.

The general requirements for wiring a house are the same for 120-volt d.c. as for a.c. In Article 220 the N.E.C. dictates the minimum number of branch circuits that must be installed in a house. Although there are many additional requirements and exceptions, the basic rules declare that enough general purpose lighting branch circuits must exist in the house to provide 3 w. of lighting power per square foot of floor area, and that a minimum of two 20-ampere branch circuits must exist in the laundry area. The kitchen and laundry area requirements are based upon an assumed appliance load of 1,500 w. per circuit.

If you use No. 14 wire with 15-ampere fuses for the general purpose circuits, then the maximum power supplied by the circuit is 1,800 w. (15 amp. × 120 v.). Keeping in mind the 3-watt per square foot requirement, one 15-ampere branch circuit is required for each 600 sq. ft. of floor area (1,800 ÷ 3). This is a minimum, however, and the common recommendation is for one 15-ampere circuit per 375 sq. ft. of floor area (that is, 62.5 percent of 600 sq. ft.). Similarly, for a 20-ampere branch circuit using No. 12 wire, the maximum power is 2,400 w. and the minimum requirement is one such circuit per 800 sq. ft. of floor area (2,400 ÷ 3) with one circuit per 500 sq. ft. recommended by electricians (62.5 percent of 800).

As an example, consider a 2,000-square foot house that is to be powered by a 120-volt d.c., 3,000-watt generator, 150 feet away from the house. The maximum current of the generator is 25 amp. (3,000 w. ÷ 120 v.). Thus, according to the Volt Drop Table 9-3, No. 8 wire would be needed for the underground cable to the house as well as for the service entrance cable. If No. 12 wire is used for the general purpose branch circuits inside the house, a minimum of four such general purpose circuits are required (2,000 sq. ft. ÷ 500 sq. ft. per circuit). In addition, the required two 20-ampere small appliance circuits must be installed in the kitchen and one must be placed in the laundry area also. Additional circuits based upon the need to operate various major appliances and any other code requirements must be determined and installed as well.

32-Volt Wiring

A house wired to operate appliances and lights on 32-volt d.c. will either require more branch circuits or heavier gauge wire than for a 120-volt system. Article 720 of the N.E.C. covers requirements for circuits with voltage less than 50 v., and thus, this article pertains to low voltage wind systems. Basically, Article 720 states that No. 12 wire is the smallest gauge wire that can be used with a low voltage system, and that branch circuits for appliances must not use wire smaller than No. 10 wire. Thus, for a 32-volt d.c. system, No. 12 wire protected with a 15-ampere fuse for general purpose branch circuits and No. 10 wire protected with a 20-ampere fuse for the kitchen and laundry areas would be satisfactory as long as the wire lengths do not exceed 33 to 39 feet respectively (see Table 9-3).

The maximum power rating of appliances or lights you can connect to branch circuits using No. 12 wire fused at 15 amp. is 480w. (32 v. × 15 amp.). Thus the minimum requirement is one No. 12 general purpose branch circuit for each 160 sq. ft. of floor area (480 ÷ 3). Following the recommendations for 120-volt systems to reduce this area to 62.5 percent, one circuit is recommended for each 100 sq. ft. of floor area. In other words, about 5 times as many branch circuits necessary for 120-volt wiring are required for the 32-volt wiring.

Consider the following example: an 800-square foot home is to be wired for a 32-volt d.c. wind system. The wind system is rated at 2,800 w. and is located at a distance requiring 100 feet of wire from the generator to the batteries (see Illus. 9-1). The batteries are 20 feet from the service panel. A water pump with a power usage of 300 w. will require 100 feet of wire. Since the maximum current from the wind power generator is 87.5 amp. (2,800 w. ÷ 32 v.), and assuming that TW wire is being used, No. 2 wire will be required according to Table 9-2. However, as shown in Table 9-3, No. 2 wire will exceed the 5 percent volt limit if it is run over a length of 55 feet. Since you can run No. 2/0 wire over a length of 103 feet before it exceeds the volt drop limit, this is the wire size to use in this case.

Inside the house, a minimum of eight general purpose branch circuits should be installed when using No. 12 wire fused at 15 amp. (800 sq. ft. ÷ 100 sq. ft. per circuit). According to the Volt Drop Table, the wire size must be increased if the wires are more than 33 feet long. In the kitchen and laundry area, the prescribed No. 10 circuits fused at 20 amp. should be used if the distances it's run do not exceed 39 feet. The maximum current required for the water pump is 10 amp. (300 w. ÷ 32 v.), and according to the N.E.C., No. 12 wire would suffice. However, you would need to use No. 8 wire in this case in order to carry the 10 amp. for 100 feet to the pump with less than a 5 percent volt drop. If the main fuse in the service entrance panel is rated at 100 amp., then No. 2 wire (assuming it's THW) can be used for the 20-foot distance from the batteries to the panel.

12-Volt Wiring

The procedure for wiring a 12-volt d.c. system is similar to that for a 32-volt system. The wire size requirements are even greater, however, and it is therefore impractical to wire such a system any larger than one for powering a cabin-size building.

Your local inspector may rule that a 12-volt installation falls under the jurisdiction of Article 551 of the N.E.C. which relates to RV's. If so, the N.E.C. requires that wire designated as either type HDT, SGT, SGR, or SXL be used for such an installation. Although it is possible to use No. 12 and No. 10 wire for 12-volt installations and still remain in accordance with the code and ampacity requirements, No. 8 wire (with 30-ampere fuses) will probably be neces-

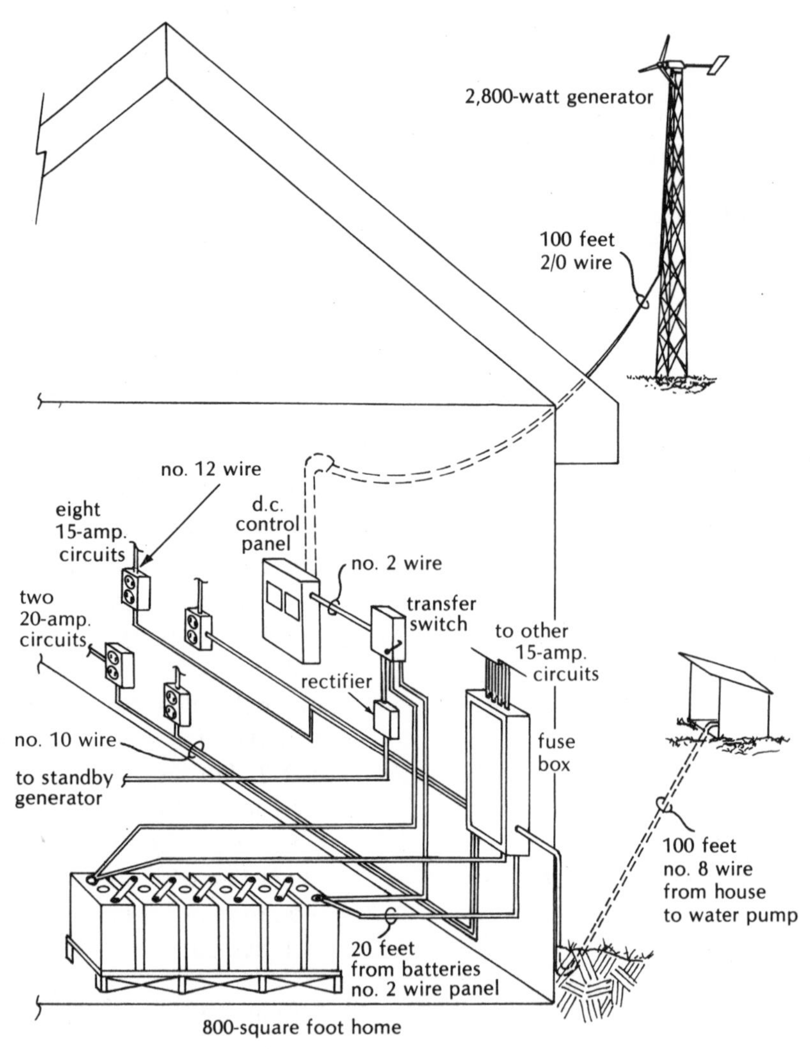

Illus. 9-1 Wiring a D.C. Stand-Alone Wind Power System with Storage Batteries —The standby generator maintains the charge in the battery bank if the wind doesn't blow for a considerable period of time.

sary at the minimum to limit volt drops to acceptable levels for most 12-volt installations. Using an approach similar to that for 120- and 32-volt systems, one general purpose branch circuit for each 75 sq. ft. of floor area is recommended, assuming that No. 8 wire fused at 30 amp. is run over lengths of no more than 15½ feet. If longer wires are required, the fuse size can be reduced to 20 amp. or the wire size can be increased to No. 6.

Consider a cabin with 300 sq. ft. of floor area that is to be powered by a 500-watt, 12-volt d.c. wind system located 100 feet from the batteries. If maximum current from the generator is 42 amp. the Volt Drop Table recommends No. 3/0 wire as the best for the 100-foot distance from the generator to the batteries. With No. 8 wire and 30-ampere fuses, count on installing four general purpose branch circuits in a 12-volt d.c. system (300 sq. ft. ÷ 75 sq. ft. per circuit).

Mixed Voltages

You may wish to wire your house using two different voltages with both a.c. and d.c. circuits. For example, a d.c. wind system can operate lights and small appliances, and a.c. utility power can run large kitchen appliances and a water pump. In combined systems such as these, the N.E.C. requires that receptacles (outlets) for each type of circuit must be noninterchangeable to prevent someone from plugging an appliance into an outlet powered by the wrong current (see summary of N.E.C. Articles 210-7f and 551-4e in Table 9-2). This means, for example, that all of the cord plugs on the appliances intended to operate on d.c. would have to be changed to fit only the nonstandard receptacles. Some of the plug and receptacle styles specified by the National Electrical Manufacturers' Association (NEMA) are shown in Illustration 9-2. For example, the standard "5-15R" configuration can be used for a.c. circuits and style "6-15R" or "7-15R" can be used for d.c. circuits. Before doing such an installation, however, you should check first with your local electrical inspector since, other than the requirement for noninterchangeable receptacles, the N.E.C. does not give specific requirements for mixed wiring in houses.

You must, of course, keep the different types of electrical circuits entirely separate, and you cannot use a common service entrance panel or joint electrical boxes for both circuits. Since the code requires a specific number of outlets per room—typically, an

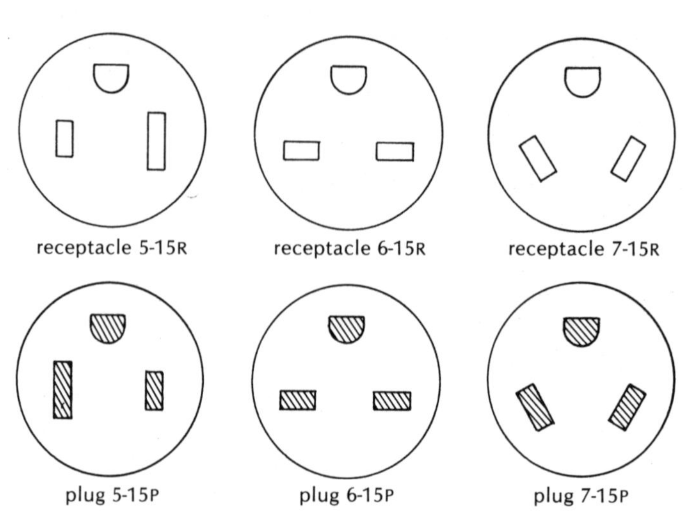

Illus. 9-2 Types of Plugs and Receptacles for Use with D.C. Power—When mixed voltages are used in a house, the plugs and receptacles for either type of current should not be interchangeable. Notice that the positioning of the slots on each receptacle match only their respective plug.

outlet must be within reach of 6 feet from any spot on any wall—the question arises as to whether or not the required number of receptacles have to be installed for *both* wiring systems. General purpose branch circuits, a.c. or d.c., would have to be duplicated by outlets for their counterpart current in each room. The inspector may argue that the reason for the code requirement for so many outlets in every room is to limit or eliminate the use of extension cords, a common cause of electrical fires. Installing only one or two receptacles for either d.c. or a.c. in a room might encourage the use of extension cords with them.

Another possibility is that the extra d.c. circuits from the wind systems may be considered a standby power system under Article 750 of the N.E.C. and dual wiring would not be necessary.

Grounding and Lightning Protection

In addition to using the right size wire for the job, proper grounding is an important aspect of any wiring project. An electrical grounding device consists of a copper rod or galvanized pipe driven at least 8 feet into the earth. This rod is connected to the house wiring as specified by the code. If the electric wires inside an appliance come in contact with the metal parts of the appliance, or if lightning strikes the wiring, the current is directed into the ground, thus preventing a fire, or an electric shock to someone touching the appliance.

Again, grounding for wind systems is not covered specifically in the N.E.C., although Article 250 addresses grounding in general. According to Article 250-3 of the N.E.C., all a.c. wiring and d.c. wiring above 50 v. should be properly grounded, although d.c. circuits with voltage of 50 v. or less do not require grounding. Whether this exception applies to wind systems or not depends upon interpretation by the electrical inspector.

Controversy exists on whether or not d.c. wind systems should be grounded. The N.E.C. requires that a d.c. circuit with 50 v. or more should be grounded. However, wind pioneer Marcellus Jacobs of the Jacobs Wind Electric Company reports that he noted during the pre-R.E.A. days that when wiring from the generator to the control panel was grounded, chemical electrolysis of the wiring occurred, particularly in the presence of moisture. When this condition existed, the wiring eventually disintegrated. Furthermore, if the generator developed a short circuit, electric current flowed from the generator to the tower and back to the grounded battery bank through the earth. This not only presented a shock hazard to anyone on the tower, but the current flowing through the tower legs eventually electrolyzed the legs and weakened them. This controversy of whether or not a d.c. system should be grounded does *not* apply to house wiring, however. House wiring circuits are isolated from the generator by the cut-out circuit in the control panel and are therefore not affected by electrolysis.

Shock from d.c. electricity is a concern, just as with a.c. As many people know from experience, a.c. tends to "freeze" your muscles, making it difficult to move or to let go of the wire or appliance, thereby subjecting you to a continuous shock. On the other hand, while d.c. will give you the same jolt as a.c., it does not tend to freeze your muscles. This does not mean that a d.c. system should be treated lightly, however. The damage from d.c. is more likely to come from burns. When d.c. electricity jumps from one metal conductor to another, it tends to hold an "arc," just as on a welder's torch, and can cause deep burns.

A couple of modern-day wind power users discovered this by accident. Having heard that farmers used their wind system batteries to weld, these people tried it by connecting welding cables to a 120-volt battery bank. What they did not realize is that the pre-R.E.A. farmer welded from a 32-volt battery bank (commercial welders reduce 120-volt a.c. to about 40 v. for welding). The attempt to weld from a 120-volt battery bank resulted in the arc disintegrating the first few inches of the welding cable. Luckily no one was hurt. However, welding from a 32-volt section of a 120-volt battery bank is not recommended either, because it drains the cells unequally. Also, a loose cable connection at the battery posts can cause hydrogen gas to ignite.

Lightning is definitely of concern to the wind power system owner because the system itself is a tall object located out in the open, the perfect situation for lightning to strike. But you can safeguard your system effectively by using proper grounding techniques and lightning arrestors. First, the tower itself should be grounded. Although you might think that the tower is grounded because its legs and guy wires go into the ground, usually this is not sufficient because the concrete footings provide poor electrical contact with the earth. (In one extreme case, wind system manufacturer Jim Sencenbaugh of Palo Alto, California reports that lightning struck the ungrounded guy wire of a tower with a direct hit that melted the wire and apparently caused the tower to topple.) For this reason the tower base and each guy wire should be grounded. For self-supporting lattice towers, each leg should be grounded.

Actually, the problems caused by lightning are rarely due to its directly striking equipment. Lightning is a concentration of high voltage electrical energy in the air, and the closer an electrical wire or metal object is to the center of that concentration (lightning bolts), the higher is the surge of voltage that will be induced in the metal or wire. Even at a distance of many yards from the center of concentration, this voltage can be several thousand volts, enough to damage equipment and wiring. Grounding the tower shunts any lightning surges in the tower metal to the earth. In problem areas your wind system manufacturer or a lightning protection installer may recommend the installation of a lightning rod on top of the

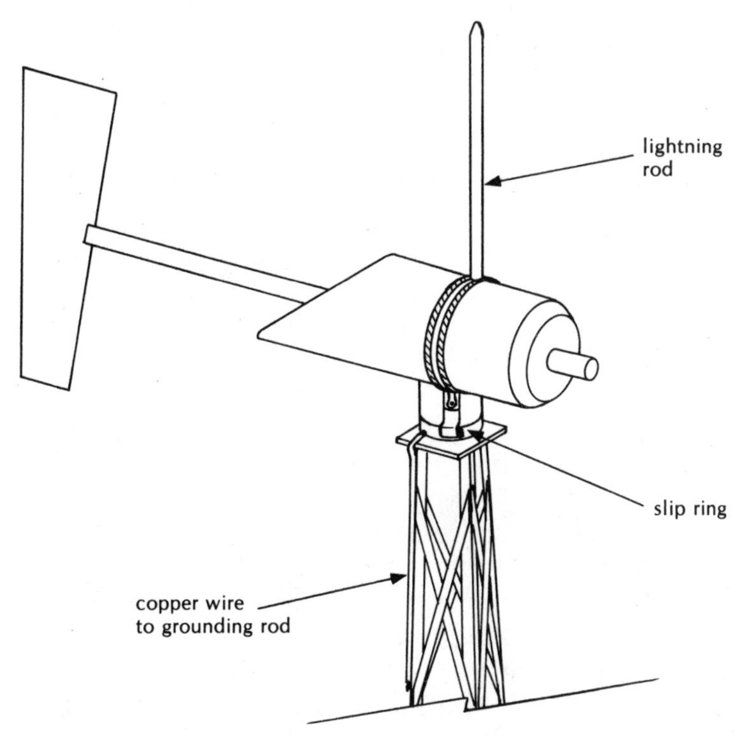

Illus. 9-3 Slip Ring on a Generator for Installing a Lightning Rod—In some cases the installer may recommend the mounting of a lightning rod on the generator. A slip ring allows the copper cable to attach to the generator without hindering the generator or rotor's rotation.

wind system itself, with a copper cable going from the lightning rod down the tower to connect to a ground rod at the base. Since the wind generator and rotor are rotating on a turntable, the copper cable going up the tower must connect with the lightning rod through a slip ring arrangement on top of the generator.

Lightning also induces voltage surges in the wiring between the tower and the house. I learned about this problem the hard way. On two occasions lightning surges were powerful enough to punch a pinhole into the insulation of my wind power generator's armature windings. I didn't realize it at the time, but the problem stemmed from the long length (over 400 feet) of aerial cable that carried the electricity from the generator to the batteries. By hanging in the open air, this long cable acted as an antenna ready to pick up any electrical discharges from nearby lightning strikes. To cope with this, I buried the cable from the tower to the building, using the earth itself as an insulator to protect the wires. Even though lightning does penetrate the ground, particularly at the points of direct strikes, the earth greatly reduces the lightning's effect. Also, many wind system installers bury the wiring in metal conduit for further protection.

Besides using buried cable, additional protection for the wiring is provided by connecting a lightning arrestor, often a surge-protection device, to the service entrance wiring. Lightning arrestors are designed to shunt any voltage surge exceeding a given limit (depending upon the type of arrestor) to the ground rod, whether the surge is due to lightning or to other causes. Various methods and components can arrest voltage surges, including spark gaps and choke coils, gas discharge tubes, and varistors. A spark gap consists of two pointed copper rods or bars, one of which is connected to the electrical wiring just before it enters the building, and the other is connected to a ground wire and rod. The gap between the two points is a fraction of an inch. If the voltage temporarily exceeds about 2,500 v., the electricity will jump across the gap and flow to the ground connection. Usually, a choke coil is used along with the spark gap to arrest high frequency surges caused by lightning. As the name implies, a choke coil is simply a coil of wire. A simple one can be made by coiling the lead-in wires to the house with two or three turns, each 8 to 12 inches in diameter.

Gas discharge tubes are gas-filled devices in which the gas "breaks down" and suddenly allows electricity to flow from the electrical wiring to the ground when the voltage exceeds a certain limit, typically about 600 v. Gas discharge tubes are common devices used in some of the small surge protectors that fit inside the service panel and connect to incoming wiring. Some question exists, however, as to how effective they work with d.c. electricity, and they are not designed to protect from direct lightning strikes. Varistors are similar to gas discharge tubes in operation. Varistors are solid-state electronic devices that allow electricity to flow within a fraction of a second whenever the voltage exceeds a certain limit —typically about 200 to 400 v.

Other lightning arrestors use a combination of these devices on the theory that each one will activate in succession. Gas discharge tubes and varistors provide the most protection for the insulation in motor windings or electronic circuits since it is the relatively low voltage surges that cause such damage.

Installing Batteries

When planning to install your batteries you should consider both safety and convenience. Common locations for the battery bank are in the house, in a garage or outbuilding, or in an insulated "battery shack." Locating batteries in a house offers the advantage that the batteries will be kept warm throughout the year. However, they will take up valuable space in your house. Also, proper venting and safety precautions become much more important in a house than in a garage or a shop.

If you're thinking of locating batteries in a garage or shop, you should give thought to what you can expect to experience as the lowest winter temperature. If temperatures are consistently below freezing, you would be better off putting the batteries in an insulated room. A small block-shelter built outside and partially banked with earth can serve as a good, relatively warm battery room in a temperate climate. However, if you have extreme temperatures in the winter or the summer, never locate batteries outside.

The batteries should be installed on a battery rack capable of supporting considerable weight. You can use a wooden rack if it's carefully constructed. Most battery racks are made of angle iron painted with acid-resistant paint, and the batteries are insulated from the metal with rubber or plastic.

According to Article 110-17 of the N.E.C., exposed or "live" electrical gear should be located either in a cabinet or, as in the case with batteries, in a locked room to prevent accidental damage. If the batteries are located in a garage, for example, someone could accidentally lay a tool across the battery posts. For this reason, the batteries should be enclosed in a screened or walled room.

Battery Room Ventilation

Ventilation is important because the batteries give off small amounts of hydrogen and oxygen during charging. Batteries are not as volatile as a tank of gasoline inside a building, but they should be respected. The gases given off by a battery may ignite if sparking occurs, or if a match is struck in close proximity to the battery vent caps. Sparking can occur from nearby equipment, such as welders, switches, or motors. Loose battery connections can also cause arcing, so always make your battery connections as tight as possible while being careful not to put too much pressure on the battery

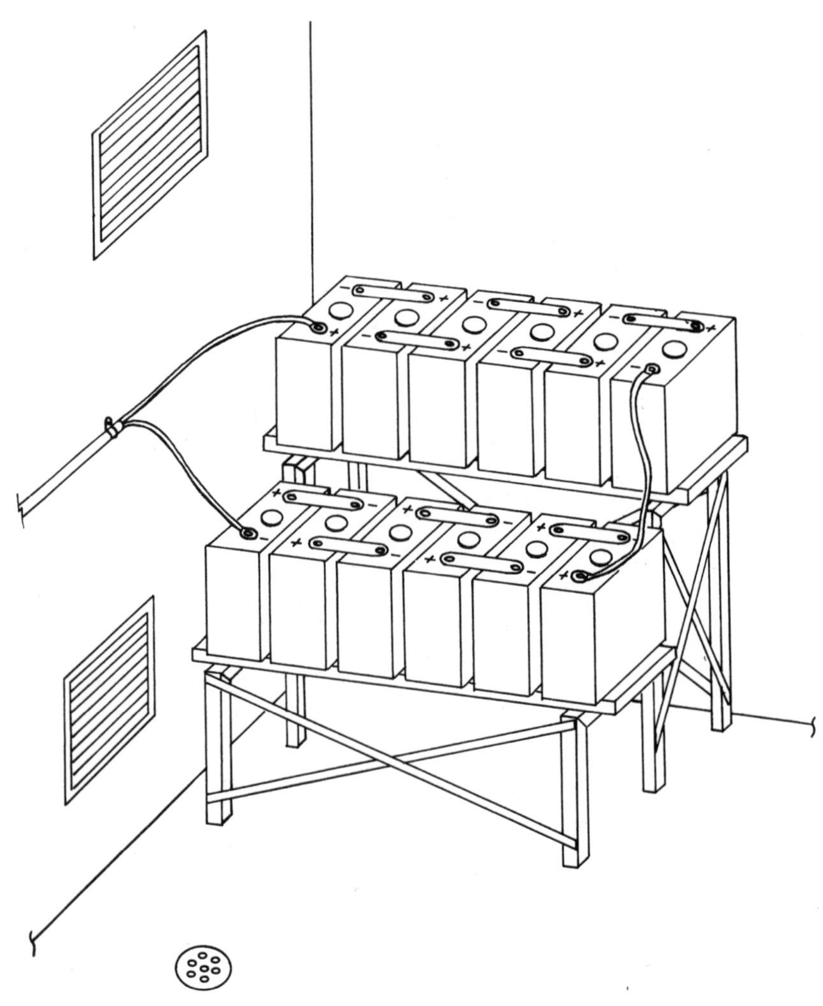

Illus. 9-4 A Battery Room—Venting, as well as temperature control, are important considerations in housing storage batteries. The battery rack should be sturdy and allow easy access for service to the batteries.

point while tightening it. Too much pressure in tightening could damage the battery's internal connections.

In a completely airtight room, gases from the battery bank eventually build up to an explosive mixture, so from three to four air changes per hour need to be provided, which is a moderate amount of ventilation. If your batteries are positioned in a small space, provide an inlet vent at the lower part of the space and an outlet vent at the top of the space.

Generally, normal ventilation in a room provides an adequate number of air changes per hour to remove hydrogen gases exuded when batteries are charging. However, in small, enclosed rooms, you may have to arrange for additional ventilation to ensure that explosive mixtures do not build up. If the volume of hydrogen gas reaches 3 percent of the room air volume, the mixture can be explosive. You can estimate the amount of hydrogen gas emitted if you know the volume of the room, the charging rate, and the number of battery cells being charged. The amount of hydrogen gas given off in cubic feet per hour is calculated by multiplying the number of cells by the charging amperage by .016. As an equation it is shown: cu. ft. hydrogen per hour $=$ .016 $\times$ charging amp. $\times$ number of cells

For example, if a battery bank of 20 cells is charged at a rate of 50 amp., the amount of hydrogen gas generated would be: .016 $\times$ 50 amp. $\times$ 20 cells, or 16 cu. ft. per hour. If this battery bank is located in a room with a volume of 800 cu. ft. (a 10-foot $\times$ 10-foot $\times$ 8-foot room), a volume of 24 cu. ft. of hydrogen would be explosive (3 percent of 800 cu. ft.). If the room is completely enclosed with no circulation, the volume of 24 cu. ft. of hydrogen would be reached in 1.5 hours ($24 \div 16 = 1.5$). To prevent hydrogen buildups in this case, the volume of the room air would have to be changed at least 1.5 times per hour.*

Situating the batteries in a well-ventilated place completes the installation of a stand-alone wind power system. The system is ready to begin generating electricity like its counterpart, the utility-connected system. All that remains for the wind power user to do is to become familiar with some fine-tuning and maintenance tips, detailed in the next chapter.

*Based on information from "Stationary Lead-Acid Battery Systems, Section 50.-00," Exide Power Systems Division, E.S.B., Inc., Philadelphia, PA 19120.

Chapter 10

Living with Wind Power

The coal or nuclear power plant that generates electricity for your area is probably many miles away from you and you may never have seen the plant, unless in passing. Most people's involvement with their source of electricity is minimal, and usually consists of paying the bill at the end of the month. However, like any other type of on-site power generation device, a wind system requires a certain amount of attention from the user. The amount of involvement you have with the workings of your wind system can simply be periodically checking the meters on a control panel of a utility-connected system. Or your involvement may be more complex for a battery-storage wind system, such as checking the state of charge of the storage batteries every few days and possibly operating a standby generator two or three times per month. Many wind power users also live in energy-efficient homes and use other renewable energy sources as part of their overall plan to become energy-independent.

Once the system is installed and operating, most people go through a period of learning what has to be done to operate and maintain the system as well as becoming aware of how much energy the system can produce. A frequent comment made by experienced wind power users is that you have to learn to accept the variability of the wind. The amount of electricity produced can vary considerably from month to month depending upon wind speeds, and if you expect the wind system to operate at its estimated energy-output level every month, you may be disappointed. Other than using a kilowatt-hour meter to measure the energy output, people generally monitor their wind system's performance by noting either the amount of reduction in their monthly utility bill, or for battery-storage systems they note the amount of time it is necessary to operate a standby unit.

Wind System Maintenance

Upkeep of most wind systems consists of a lubrication and inspection usually performed once every six months to a year. This may be done by the homeowner or by the dealer under a service contract. Malfunctions of components should, of course, be corrected by the manufacturer or the installer. Other difficulties are usually temporary ones caused by weather conditions.

Frank Hansen recalls that the only problem he has had with his battery-charging system in two years of operation was with the slip rings. During the winter, he noticed that the generator would stop putting out electricity every time the weather was cold. Perplexed as to what was happening, Hansen wrote to the well-known wind system expert, Martin Jopp. Jopp replied that even though the Dunlite was a "brushless alternator" type, it still had brushes—in the slip rings. He instructed Hansen to check to see if his slip rings were icing up. Hansen climbed the tower, and, sure enough, snow was packing up inside the slip ring assembly through a small crack. He patched the crack with tape used for fixing roof leaks in mobile homes, and the generator has been fine since. Similar problems have occurred with other systems when grease or oil from the generator or gearbox dripped onto the slip rings.

During ice storms, it is possible for ice to form on the rotor blades. This usually causes no problem as long as the snow is allowed to melt off by itself. Melting may take a day or two but the rotor will throw off any remaining ice in the first breeze. In this situation, trying to start the generator up manually by motoring it from the control panel can cause damage since the blades become very unbalanced under several pounds of ice. Wind system installer

Chris Chomiak of Montana devised a unique solution to the problem. He waxes his blades with cross-country ski wax to help keep snow and ice from building up.

Lightning can occasionally be a problem, even with a properly grounded and protected system. Wind power user Caleb Scott of Connecticut reports that he had to replace an electronic circuit board in the control panel three times due to lightning surges. Each time this occurred he simply sent the circuit board back to the manufacturer for repairs and the system was soon back in operation.

Utility-Connected Systems

Compared to battery-storage systems, utility-connected systems generally require much less attention from the user. Bob Howard uses a utility-connected, Enertech model "1800" wind system to power his rural Blair, Wisconsin home. Howard notes that his main involvement with the wind system is to watch the rotor spinning, simply out of fascination. He estimates that the system generates from 250 to 350 kwh. per month. The greatest amount of work he had to do came before installing the system, says Howard. His system was installed before the state of Wisconsin and his local utility had enacted a purchase agreement for power fed back by wind systems, and so there was considerable discussion over the interconnection issue. The parties resolved the issue with the utility allowing Howard to run his meter backwards until a formal agreement could be drafted. Also, although a wind system dealer installed the system, Howard became involved by digging an 800-foot-long trench for the wires from the wind system to the house. For the most part, the system operates unattended, but Howard points out that it still has affected his awareness of electrical use. Like other users of wind power, he too has become more conscientious in turning off unneeded lights, and he has also installed a timer on his hot water heater.

Thomas Zaborski, of Calumet City, Indiana found his utility-connected wind system to work well after he solved some initial problems he experienced with his synchronous inverter. A manufacturing defect in the generator led to a short circuit in the field windings. This short circuit damaged four of the S.C.R.'s in the

Photo. 10-1 Bob Howard's Utility-Connected Wind System—Situated on top of a hill in Wisconsin, Howard's Enertech system benefits from the increased winds at the higher ground. In periods of low wind he uses utility power to operate his household.

synchronous inverter (see Ch. 5), and had to be replaced after the generator was fixed.

Zaborski notes that his family has always been energy-conscious and that they have cut their heating bill by 60 percent in the last several years. Part of the house heating comes from a solar system in which heated water is stored in two 575-gallon tanks. Also, a steel jacket around Zaborski's wood-stove chimney heats water that circulates in pipes embedded in sand in the house's crawl space. The heat from the sand is stored, in turn, in 900 one-gallon plastic bottles filled with water.

Inverters

Problems with inverters arise when they are used in unsuitable applications. For example, Frank Hansen found that the small square-wave inverter that he used to operate his stereo system caused electrical hum on his ham radio set, and that a sine-wave inverter was more appropriate. Other wind power users have reported problems with components burning out in inverters. This is often due to loading the inverter beyond its rated capacity. It is best to stay within the limits of the power ratings and motor-starting capabilities of any inverter.

Gary Grotte and his family use a 48-volt, rebuilt pre-R.E.A. Jacobs battery-charging wind system in conjunction with a 5,000-watt electronic inverter on their North Dakota farm. The Grotte house remains connected to utility power so that the house circuits can be powered either from the inverter or from utility power by simply operating a transfer switch. Since the utility company charges for a minimum of 180 kwh. per month, Grotte tries to use this minimum every month and then allows the wind system to provide the rest. The inverter has an automatic feature that disconnects it from the house circuits whenever the battery voltage drops below 43 v. or exceeds 53 v. All members of the family know how to manually operate the transfer switch when the inverter disconnects itself, and Grotte points out that the disconnection, therefore, has not been an inconvenience for the family. The 5,000-watt inverter operates almost all of the appliances in the Grotte household, including a refrigerator and a freezer. The only nonelectrical appliances are the cooking stove and the water heater, which operate on gas.

Bob Bartlett and his family, in northwestern Pennsylvania use wind power five days per week and utility power two days per week. Two wind systems each supply half of the house's electricity. The Bartlett house is a modern rambler with all of the modern appliances, including an automatic dishwasher, a refrigerator, a color TV, and two freezers. Not wanting to change all of these appliances to d.c., Bartlett decided to store the wind energy in batteries, but to convert it to a.c. through large inverters that power the whole house. A transfer switch is used to change the house power source from the inverters to utility power (see Ch. 6).

Bartlett's first wind system is unique. It consists of two 1,800-watt rebuilt pre-R.E.A. Jacobs generators, but with one mounted on *top* of the other! The top generator is driven by a timing belt connected to the lower generator. The Bartletts installed the generator on the 42-foot tower all by themselves. The shaft on the upper generator was cut short so that the rotor blades would not hit it. Such a design is not recommended because of possible balance and installation problems. However, Bartlett's careful workmanship and ingenuity has led to a system that works well for him.

The "twin" generators put out 5,000 w. in high winds, Bartlett says. Electricity for the "twin" generators is stored in sixty Gould 100-ampere-hour batteries and converted to a.c. by a 3,000-watt Versa Count inverter. Power from this system is used for the kitchen and living room lights and appliances.

The second wind system consists of a 2,800-watt rebuilt pre-R.E.A. Jacobs wind generator mounted on a 44-foot tower. The electricity is stored in twenty 6-volt golf cart batteries and converted to a.c. by a 6,000-watt West Wind inverter. This wind unit powers the bedrooms and two freezers.

D.C. Wind Systems

Of the people who own wind power systems, those who use d.c. electricity from storage batteries are often most involved with the workings of their wind system. My own wind system is a rebuilt 120-volt pre-R.E.A. Jacobs unit, and my house is wired for 120-volt d.c. use. Like other storage-battery users, our family members tend to be conscious of the amount of electricity we are using—lights are not usually left on unnecessarily. In addition, we designed a house which in itself is energy-efficient. Our 1,600-square-foot, two story house has 240 sq. ft. of glass facing the south side of the house for maximum use of the sun's heat during winter days. At night the windows are covered with insulated curtains. I've calculated that the sun provides about 35 percent of the house's heating needs. The rest comes from a wood space-heater on the first level of the house as well as from other miscellaneous sources, such as cooking and heating water.

To eliminate the need for electrical fans or blowers, the wood stove is located in the center of the house, with a large grate above it so that heated air flows naturally up to the second floor and then cold air returns as it is cooled down along the outside wall. (In

Photo. 10-2 Don Marier's Energy-Efficient Home—Large, south-facing windows supply sufficient passive solar energy that, along with a wood-burning stove, completely heat the author's house. His wind system is used only to operate appliances and his water pump.

conventional structures, warm air is forced up along the outside walls using blowers and ductwork and cold-air returns are located on inside walls.) Natural air flow is aided by the lack of dividing walls in the second story, which includes the kitchen and living space. This type of heating eliminates a blower on the stove, which can use 50 to 200 kwh. of electricity per month, a large load for our wind system. Temperature variations of 10 to 15° F occur throughout the house, however, and some people find this uncomfortable. On the other hand, I find that houses with forced-air systems are uncomfortable and feel too warm to me, after having adjusted to our natural, or gravity type, of heating system.

I use a standby gasoline engine-driven generator to charge my batteries during periods of low wind. We need to run the standby generator, typically, for a couple of hours per day for several days in the low wind months of August and January. In many months we never need to use the standby at all. This is not because our wind speeds are so high, but because our electrical needs are relatively low, generally about 60 to 90 kwh. per month.

Other storage-battery users report similar patterns in their energy consumption. Frank Hansen and his family live miles away from utility power at their canoe-outfitting camp in the Superior National Forest of northern Minnesota. The Hansens scale their electrical use in their energy-efficient dome home to match the output of their 2-kilowatt Dunlite wind system. An attached hydroponic, solar-heated greenhouse also serves as a warm battery room. Even the Hansen's radiophone is powered by a photovoltaic panel. Hansen kept records during 1978 and found that he had to run his standby generator only twice during the period of February through August. In the nonwindy (for his area) period of December through February, he operated his standby generator for an average of only six hours per week.

Scott and Janet Nielsen of central Vermont began changing their home from all-electric power that used 2,750 kwh. per month to a home-size, battery-charging system, by first converting from electricity for heating and cooking to a wood-heat and cookstove setup. Scott also built an "ice-house" refrigeration system that uses no electricity. The ice-house is really an ice-silo, 14 feet in diameter, topped by a real silo cap that a neighbor was getting rid of. The first level of the ice-silo is an uninsulated 2,500-gallon water tank that

Photo. 10-3 and Photo. 10-4 Frank Hansen's Hydroponic Solar Greenhouse and Battery Room—The solar effect of Hansen's greenhouse helps to maintain warm temperatures in the nearby battery room.

is exposed to the outdoors all winter long. By spring, the tank is one block of ice. A summer kitchen is located on the second story of the ice-silo where food is cooled by lowering it down to the ice-tank on a dumbwaiter.

The Nielsen's first electrical change was connecting the house lighting to the rebuilt 32-volt pre-R.E.A. Jacobs wind system. They used lamps with 25- and 50-watt, 32-volt incandescent bulbs. To reduce lighting, and thus electricity requirements further, the Nielsens replaced frosted bulbs with clear bulbs. They also placed lamps near the work at hand rather than relying on overhead, direct lighting. This meant removing globes from some light fixtures. The Nielsens also found that some prismatic glass fixtures they discovered in a junk shop worked well at reflecting light. One of these is installed over the kitchen table.

Other appliances used by the Nielsens include a wringer washing machine with a 32-volt motor, a 32-volt mixer, and a 12-volt tape player that operates through a "dropping resistor." Nielsen uses a battery-powered electric shaver that is charged from the wind system batteries. For Christmas tree lights, he took a series string set of lights made for 120 v. and cut it into three equal lengths to operate off of the wind system.

In converting from an all-electric house to a low-energy house, the Nielsens have come up with a unique hot water system that uses no electricity at all, not even for pumps. Water is stored in the kitchen in an 80-gallon tank that sits inside a larger container filled with a mixture of antifreeze and water. This antifreeze mixture circulates through the heater box of the wood stove as well as through a 50-square-foot solar collector located on the south side of the house. The fluid flows through the solar collector by thermosiphon method (fluid expands as it is heated to use in the pipes to the storage tank), and so, requires no pump. The solar collector was made of copper pipe soldered to second-hand roof flashing material.

The antifreeze solution is in an open tank so that excessive pressures cannot build up. The top of the mixture is covered with a thin layer of mineral oil to prevent evaporation, and a clear plastic tube connected to the tank is used for adding fluid when the level drops due to evaporation. Also, the tube is positioned at a shallow

(continued on page 250)

Photo. 10-5 and Photo. 10-6 Scott Nielsen's Summer Kitchen—Located on the upper level of Nielsen's silo-shaped ice-house, the kitchen saves energy conveniently through natural refrigeration.

Photo. 10-7 and Photo. 10-8 Light-Reflective Fixtures—Nielsen uses these prismatic fixtures to reflect light from 25- and 50-watt bulbs.

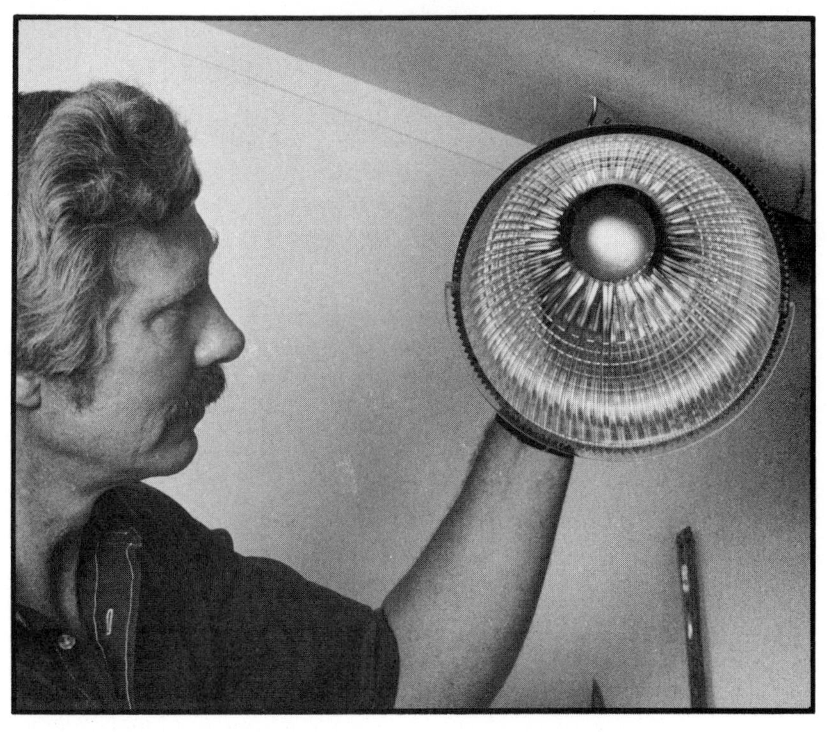

angle, nearly horizontal, causing the loss of a small amount of fluid in the tank to show up as a drop of several inches in the tube. Thus, a leak in the tank would be immediately noticeable. If the hot water tank does leak, the excess water leaks out the tube into the bathtub. The Nielsens estimate that the total cost of the hot water system was less than $500. It should be noted that some state codes require two metal walls between antifreeze solutions and a potable water supply, a requirement that the Nielsen system does not satisfy.

The Nielsens use about 50 kwh. of electricity per month, most of which is for lighting. Nielsen indicates that if he were to do it over again, he would test wind speeds before putting up the wind system. The average wind speed in the area is 8 mph, based upon local airport data. However, the Nielsens have had several periods of insufficient winds, and since their home is located in a mountainous region of Vermont where wind speeds can vary greatly, this seems to indicate that the average winds are less than 8 mph at the site.

Living with Storage Batteries

In a battery-storage system, the batteries themselves take the majority of the attention and maintenance required. With my own unit, I tended to be overly cautious about the state of charge of the batteries for the first six months of use. Several times I shut the wind system off during a windy period thinking that the batteries were being overcharged, when actually they were just beginning to get a good charge. I also learned that, in periods of low wind where charging is insignificant for several days, it is best to run the standby generator for only a short period of time, generally to store about one day's worth of electricity. That way the batteries will be able to take all of the charge when the next windy period occurs. With our electrical use, I found that the 5-kilowatt standby generator uses about one 2-gallon tank of gas for each two hours of operation and this provides one day's worth of electricity.

Our battery bank consists of 20 used, 6-volt golf cart batteries. The shop building in which the batteries are located is insulated but is often unheated. This has presented a problem on only one occasion, when the batteries were discharged to a very low level and the electrolyte started to ice over. Also, when it is very cold out, the reduced efficiency of the batteries is noticeable. However, it is generally windy enough during the cold part of winter to somewhat offset the batteries reduced efficiency.

Frank Hansen, who also uses golf cart batteries in his 114-volt,

Photo. 10-9 Nielsen Surveys his Water Heater—The Nielsens heat the water over their wood stove and pump it by the thermosiphon method, another way to conserve electric energy.

d.c. system, started out using 20 batteries in series but found that 19 were sufficient. During windy periods the voltage of each fully charged battery rises to 6.6 v., so that the total battery bank voltage with 20 batteries was 132 v., which Hansen felt to be too high. Hansen originally designed his wind system with two battery banks, hoping for extra capacity to carry him through windless periods. He found, however, that it takes quite a long time to get both banks back up to full charge when they are nearly empty. For this reason he recommends using only one battery bank that is sized to the output of the wind system.

Battery Maintenance

The basic step in maintaining your batteries is to try to keep them clean and dry. Just as with car batteries, you can prevent corrosion on the terminals with grease, petroleum jelly, or special battery post compounds. Dust and acid on the tops of the batteries can actually conduct small amounts of electrical current and somewhat discharge the battery. You should maintain about a ½-inch space between batteries to prevent electrical contact. If possible, install a drain in the floor of the battery room to flush away water after cleaning the batteries or in case of an accidental acid spill.

Batteries require periodic checking and maintenance, and this can be done most effectively by establishing a schedule, posting it in the battery room, and noting the date of the last maintenance. Maintaining batteries is like keeping up an automobile—if the oil changing and lubrication are done on schedule with periodic checks in between, problems are minimized and unnecessary worries are eliminated.

You should check the state of charge in your batteries every day or every few days. As discussed in Chapter 5, you can connect a voltmeter to the battery bank and mark the face of the meter with small dots or red lines to indicate critical voltages. If located conveniently and marked clearly, all members of the household can learn the procedures for checking the state of the batteries. Checking the batteries should be no more difficult than checking the fuel gauge in a car.

About once a month, check one or two of the cells with a hydrometer. This is also a convenient time to check the acid levels in the batteries. If the acid levels are below the marker on the battery case (or if there is no marker below ⅜ of an inch above the plates), add water. The water should be free of minerals that can degrade the battery performance. For this reason, it is safest to use distilled water. Sometimes you can request a battery supplier to test your water supply to see if it is suitable for battery use. You should have to add water only two or three times a year.

Your batteries should also get an "equalizing charge" each month. This means simply giving the batteries a slight overcharge to equalize the hydrometer readings of all the cells. Normally, the wind generator will do this on its own during the periodic windy storms that occur in most areas. If no storms of this sort occur in your area, or if the ones that do occur fail to equalize the cells, use your battery charger instead. Some control panels are designed so that an equalizing charge can be given to the batteries by simply operating a switch that allows the wind power generator to charge the batteries at a higher voltage.

Of course, you will want to be able to discern the difference between a slight overcharge and a damaging overcharge. During heavy charging, batteries often bubble and hiss, a normal condition and no cause for alarm. In even heavier charging, the battery solution will turn milky, which is caused by the hydrogen bubbles breaking up into very fine bubbles. This effect is similar to the cloudiness of water coming from the tap when it has great amounts of air trapped in it. Again, this cloudiness is no cause for alarm. A battery that is being severely overcharged, however, will feel warm to the touch, and reddish brown specks of material will begin to float in the battery acid, which indicates that the battery has reached a temperature of 125°F or more. The reddish specks are pieces of the positive plates that break off at such temperatures, and charging should be stopped to prevent further damage to the plates.

You should perform a complete maintenance checkup on your batteries about once a year. Give the batteries an equalizing charge and see that their battery acid is up to the correct level. If you add the distilled water right before giving the equalizing charge, the bubbling action of the charge will mix the water in nicely. After several hours, or a day later, check all the cells with a hydrometer and note the reading on a chart. If any cell registers a hydrometer reading of more than about .050 (50 points) lower than the others, then that cell should be checked further for possible damage. Listed below is a typical battery maintenance schedule for a week, a month, and a year:

- Weekly
  - Check battery bank voltage every one to three days.
- Monthly
  - Give equalizing charge.
  - Check electrolyte level and hydrometer reading in a few cells.
  - Add water as needed (two to three times per year).
- Yearly
  - Give equalizing charge.
  - Check hydrometer reading of all cells.
  - Check all battery connections.
  - Clean battery tops.

In order to keep track of your own battery bank maintenance, it might be worthwhile to put your schedule on paper, leaving space to mark dates after you've performed a certain chore.

Appliances

The wind power user who opts for d.c. electricity should be a bit handy in making adaptations and in learning to identify those appliances that will work on d.c. The experience of the Hansen family is typical. As with other wind energy pioneers, they had to learn which appliances could be used on d.c. electricity. Frank Hansen has found that most store clerks are not very helpful in describing the inner workings of an appliance. Once he learned that appliances with universal motors work on d.c., though, he made a practice of opening up an appliance right in the store before buying it, to see if it had a motor with brushes.

Another d.c. appliance user, Larry Elliott of central Pennsylvania uses a laboratory d.c. power supply (which converts 120-volt a.c. to d.c.) to test appliances before using them with batteries. The power supply automatically limits the electrical current to the appliance and thereby protects the appliance if it is not made to operate with d.c.

The Elliotts originally installed a rebuilt 32-volt pre-R.E.A. Jacobs wind system, but were not satisfied because of the limited number of 32-volt appliances they were able to find. For that reason they are in the process of converting the system to 110-volt d.c. by rewinding a generator for 110-volt operation. The Elliotts borrowed a d.c. power supply and tried out various appliances, such as a sewing machine, vacuum cleaner, and blender, to see if they would work on 110-volt d.c. when the conversion was made. So far no problems have been found.

Purchasing small appliances and light fixtures compatible with d.c. electricity is relatively easy. However, obtaining suitable refrigerators and water pumps often present special problems and usually require the user to modify some existing equipment.

The refrigerator that I use is an older model that I took to a refrigeration repairman for conversion to d.c. operation. He first took off the sealed compressor, common to all modern refrigerators. Next, he installed a belt drive, cylinder-type of compressor, and a 110-volt ⅓-horsepower d.c. motor to turn the compressor. The compressor and d.c. motor are positioned with rubber motor mounts on a board underneath the refrigeration box. Compressors of this kind are usually mounted on springs to minimize noise due to vibrations. If placed on just the rubber motor mounts, the refrigerator does make more noise than one with a sealed compressor.

As an experiment, I modified the refrigerator box to let in outside air during the cold winter months. I installed dryer vent hoses in the wall right behind the refrigerator. These hoses pipe cold air from outdoors inside to the refrigerator box. One outside air vent is located in the upper part and the other is in the lower part of the refrigerator box to let cold air circulate in, and then out. With this arrangement the refrigerator operates as seldom as three minutes in 1½ hours. It has the disadvantage, however, of letting in moist air during moderate weather (spring and fall), and the freezer coils tend to frost up. In the summer, insulating foam plugs are put into the vents, and the unit is operated in its regular way. A further modification was to add a small fan and a thermostat in order to control temperatures inside the refrigerator more closely to prevent such condensation.

To provide water for my house, I installed a homemade system that has worked well for several years. A regular jet pump will not work on d.c. electricity, so I constructed a pump jack by using an old water-pumping windmill for the gearbox at Martin Jopp's suggestion. After I removed the blades from the windmill head, I replaced them with a 2-foot diameter wheel that came from an old cultivator. This wheel serves as a pulley turned by a d.c. electric motor. Not having any 110-volt d.c. motors when the system was first installed, I rewound a car generator to work as the motor. When the windmill gears are turned, the pump rod moves up and down, which in turn moves two pump cylinders up and down. The first

Photo. 10-10 Don Marier beside his Water Pump Jack—At the suggestion of wind expert Martin Jopp, the author constructed his water pump jack out of an old cultivator wheel and a water-pumping windmill gearbox.

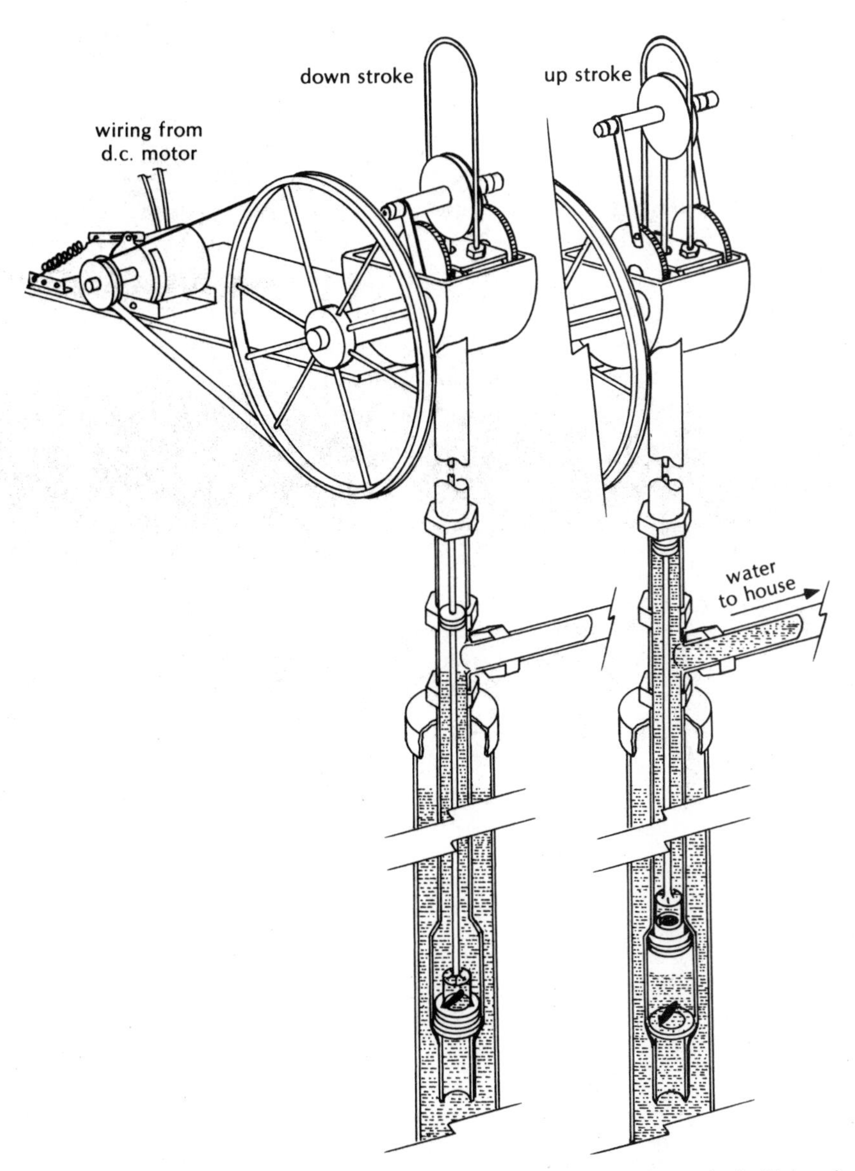

Illus. 10-1 Don Marier's Pump Jack—An old water-pumping windmill head serves as the gearbox, and an old cultivator wheel is used as the pulley. The lower brass cylinder fills with water on the downstroke, and on the upstroke the upper cylinder forces the water into the pipe leading to the pressure tank for the house.

cylinder is a regular brass pump cylinder located at water level in the well casing. The second cylinder is half the size of the first and allows the excess water on each stroke to be pumped to the hose through a plumbing "T" connection. This second cylinder is made of 1½-inch stainless steel pipe and uses three packing leathers to pressurize the pump system, just like any other water system.

Also faced with the need to devise his own water-pumping system, Scott Nielsen modified his existing jet pump to pump water from a deep well. Since the original jet pump motor would not run on d.c., Nielsen rigged up a belt-driven arrangement to power the jet pump with a ½-horsepower, 32-volt d.c. motor. To do this, he removed the electrical windings from the original a.c. motor and cut a hole in the case of the motor large enough to let a "v" belt go around the rotor of the motor. The rotor then acts as a pulley that is connected to the d.c. motor. This arrangement is inefficient and uses almost 1,000 w. of power to operate. However, the total amount of time in use is small, and Nielsen estimates that the pump really uses only about 5 kwh. per month.

No matter what type of wind system you choose for your use, either a battery-storage or a utility-connected system, you'll find yourself involved with its workings in some way or another. It may be as simple as periodically checking to be sure that the system is functioning properly, or as involved as the daily checking of storage batteries. Whatever the level of involvement, living with wind power gives many people the satisfaction of knowing that they are saving energy while becoming more energy-independent.

Chapter 11

The Future of Wind Power

The wind power system of the future may heat your house with a water churn, and the rotor may be enclosed in an aerodynamic shroud that concentrates the wind. These and other ideas are being investigated and tested by government and private researchers in many different fields. Developments in aerospace, helicopter, electric vehicle, and boat building technologies are being used to test new rotors, new storage systems, and new ways of using wind directly for mechanical power.

Mechanical Applications

The Persians used wind power for grinding grain and pumping water 4,000 years ago, quite a contrast to the history of the first wind powered electric generator systems, which were not devised until the turn of this century. Now there is renewed interest in using wind for mechanical power applications such as pumping water, operating refrigeration compressors, and heating water with mechanical friction. About one-third of the power required in this country can be directly supplied by mechanical means, and many researchers feel that the economics of mechanical wind power suit it quite well for use in these applications.

Pumping

The multi-bladed water-pumping windmill was an integral part of agricultural development in America and even now the traditional water pumper is seeing a revival in many parts of the western United States. Modern water-pumping experiments are being conducted by the U.S. Department of Agriculture (U.S.D.A.) under its Rural and Remote Areas Wind Energy Research Program. Large areas of the Plains states require irrigation for agriculture and since these same areas experience high average wind speeds, using wind power for irrigation appears to be a strong possibility.

Dr. Nolan Clark of the Southwest Great Plains Research Center in Bushland, Texas connected a 40-kilowatt vertical-axis turbine to a 56-kilowatt irrigation pump. This wind system supplied 40 percent of the pump's power and an electric motor provided the rest. Further research is needed, however, since for economic reasons the turbine should supply power for other applications during the six months of the year that irrigation is not necessary.

Cooling

Other wind power research under the U.S.D.A. program is being conducted toward the possibility of using wind for cooling. Although this research is carried on with farm and industrial use in mind, it is possible to imagine a home or apartment building being cooled with a similarly designed wind power system.

Researchers at Virginia Polytechnic Institute in Blacksburg, Virginia have developed an experimental wind power cooling system that keeps 2,000 bushels of apples in a storage room at a constant temperature of 30°F. In this system, a 6-kilowatt Elektro (Swiss) wind generator produces electricity that is stored in batteries. A 5-horsepower d.c. motor operating from the batteries runs a refrigeration compressor. Air that is cooled by being blown over the refrigeration coils then flows over the apples to cool them directly. Further storage is provided by blowing the cooled air over 6-inch steel pipes filled with a salt brine solution.

In a similar experiment at Colorado State University's Experimental Dairy Farm in Fort Collins, Colorado a wind power cooling system is used to cool milk. Rather than using a salt brine solution, cooling storage is provided by making ice with an ice-making refrigeration machine. The compressor is powered by a 20-foot-diameter, vertical-axis turbine that drives an induction generator. Further experiments using a heat pump instead of an ice-maker are planned; since a heat pump can be used for both heating and cooling, it could be used not only for cooling milk but also for heating water.

Heating

Because 40 percent of a typical home's energy use is for space heating, using wind power directly for heating space is now being seriously investigated. Some of the options are to heat water somewhat indirectly, with either a wind power electricity-generating system, or directly, with a mechanical friction device such as a water churn.

To prove the feasibility of heating a home with wind power, the University of Massachusetts at Amherst is heating their Solar Habitat I house with a wind system called the Wind Furnace. Originally the brainchild of Professor William Heronemus, an innovator of many wind power concepts, the work is now being carried on by Dr. Duane Cromack and others.

Solar Habitat is a modest-looking, 1,500-square-foot ranch house located on a hill overlooking the Amherst campus. Heat for the house is provided by a 32½-foot-diameter wind system and a 200-square-foot solar collector panel. Heating elements connected to the wind generator are immersed in a 1,000-gallon concrete water tank located in the basement of the house. The water in the tank is also heated by the solar collector.

The entire house is designed to be totally energy-efficient, and it's therefore well-insulated and sealed. Since the air volume in the house changes less than one time per hour, air is vented out of the house through a heat exchanger that heats incoming cool air with outgoing warm air. The heating requirements of the house are equivalent to about 25,000 kwh. per year. At first much of the stored heat was lost because the water storage tank was not well insulated. Later the tank was insulated with 4 inches of urethane foam. The original polyvinyl chloride plastic (PVC) waterproof liner in the tank also leaked and was replaced with a 6 millimeter (mil) thick polyethylene lining that has not leaked despite the tank reaching temperatures of 190°F.

The 16-foot fiberglass blades of the Wind Furnace have a pitch control mechanism that automatically changes the angle of the blades depending upon wind conditions. The generator itself is a 32-kilowatt synchronous unit. The gearbox between the rotor shaft and the generator consists of a one-ton truck rear axle with a chain drive to give a speed increase of 11 to 1 from the blades to the generator. Since the wind rotor is a down-wind model, a small motor, also with a chain-drive, rotates the generator into the wind whenever necessary. If the rotor is stopped and the wind starts blowing from the wrong direction, the rotor will actually go backwards for a time. However, the small motor then yaws the rotor into the correct position.

In the 1977-78 heating season, the Solar Habitat's Wind Furnace wind system provided 7,396 kwh. of energy out of 15,317 total kwh., or 64 percent of the house's total heat load. The relatively small solar system added 2,464 kwh., or another 16 percent. The goal of the project is to have the Wind Furnace supply 75 percent of the house's heat. Because the average wind speed in the Amherst area is only about 9 mph, Cromack feels that the 75 percent goal could be better met with a 40-foot-diameter rotor. Now that the feasibility of using a wind furnace has been demonstrated, whether or not they become widely used depends largely upon costs. According to Heronemus, the Wind Furnace concept becomes economical for widespread use if the systems can be mass-produced for about $6,000 each and are used in areas where electricity costs over $.10 per kwh.

One possible way to reduce such costs is to heat water directly by using a water churn, a device that relies on mechanically produced friction to raise the water temperature. This simple method for using wind power is based upon the principle that all of the mechanical energy employed in stirring water can be converted into heat energy. In other words, if you stir water or any other liquid hard enough, it gets warm. The principle was first confirmed in a famous experiment carried out by physicist James Joule in 1843. Joule put a simple paddle wheel in a water beaker and used a spool of string with a weight at the end, which turned the paddle as it dropped. Joule found that as the paddle stirred the water, the rise in the temperature of the water was directly proportional to the distance the weight had fallen.

Water churns designed by present-day wind researchers are

not much different in principle from Joule's original experiment except that the paddle wheel shapes vary and a wind turbine is used for power instead of a falling weight. One such device, called the Water Twister, shows promise as a heating unit for wind applications. Made by the All-American Engineering Company of Wilming-

Photo. 11-1 Wind Power for Heating Space—The Wind Furnace system heats a 1,000-gallon tank of water that acts as thermal mass in an energy-efficient house called Solar Habitat I. Solar panels supplement the heating of the house at the University of Massachusetts at Amherst.

ton, Delaware, the Water Twister was actually built to help slow down planes landing on aircraft carriers. Upon landing, the plane hooks onto a cable that is attached to a spooled tape on the Twister. The cable pulls the tape, which turns the paddle wheels inside the Twister. The result is that landing a 50,000-pound aircraft also raises the temperature of 90 gallons of water by 50°F in 15 seconds.

Engineers became interested in using such a device with a wind turbine when tests showed that the heating power of the churn is proportional to the cube of the shaft speed, just as the power of a wind rotor is proportional to the cube of the wind speed. This means that a water churn is well matched to the power output of a wind system.

Researchers at the University of Massachusetts are investigating the feasibility of using such water churns with a Wind Furnace. Two versions are possible—one in which the churn is on the ground with a shaft coming down the tower from the rotor to the churn, and the other in which the churn is on the tower directly behind the rotor or gearbox. With the churn on the ground, it is quite accessible for maintenance, but the long shaft requires bearing supports at regular intervals to keep it from wobbling. Mounting the churn on the tower eliminates the long shaft, but the subsequently long piping

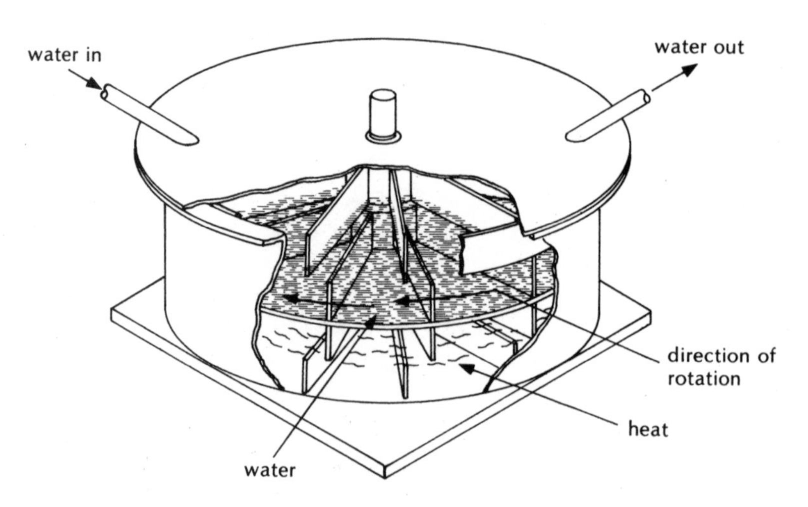

Illus. 11-1 Churn for Heating Water—Powered directly by a mechanical windmill, the paddle heats the water by friction.

necessary to the churn must be well insulated. Also, a fluid coupling joint, the equivalent of slip rings for an electrical system, must be used to prevent the piping from twisting every time the wind system rotates into the wind. Neither system possesses great advantages over the other. The question that these experiments will answer is if it is better and more economical to use either an electrical generator or a mechanical churn to heat a home with wind power. Interest in these direct heating systems is increasing because the storage system can be as simple as a water tank instead of a battery bank, and also because storage can be tied in with other sources of heat, such as a wood stove or solar heat system.

Other researchers are also investigating the workability of mechanical friction heat systems. At Cornell University's Agriculture Engineering Department in Ithaca, New York, researchers Dr. Wes Gunkel and Randy Lacey have installed a Pinson Energy Corporation "Cycloturbine C2E3" with a water churn connected at the base of the tower inside a small building. The diameter of the paddle wheel is only about 1⁄10 that of the wind system's rotor diameter. The hot water, which reaches 110°F, has potential for sanitation use in a dairy barn, among other applications.

Of course, water is not the only fluid that can be heated. Anyone who has worked with hydraulic power systems is familiar with the heat that can be generated—unwanted for the most part—in operating these systems. However, if your objective is to provide heat, an "inefficient" hydraulic turbine may just be the answer. With this goal in mind, Canada's Dominion Aluminum Fabricating Indal, Limited, has installed one of its Darrieus rotors on Prince Edward Island in order to power a hydraulic turbine. The heat will be used to heat the Ark, a self-sufficient home and laboratory originally designed by the New Alchemy Institute of Cape Cod, Massachusetts and now operated by the Institute of Man and Resources of Prince Edward Island.

Energy Storage

The introduction of utility-connected wind systems using either a synchronous inverter or an induction generator signified an advance in savings on costs compared to battery-storage systems.

There are indications, though, that the technologies for both types of systems will be improved even further in the future with the use of innovative "dump circuits" and new types of storage batteries.

Although federal regulations now guarantee that utilities must pay for power fed back to them by renewable energy systems, such

as wind systems, the payback rate depends upon many complex factors. As I stated earlier, it may be more advantageous for the homeowner to use any excess power in the home as it is generated rather than feed it back to the power grid at the utility's rate of credit. To use this power, you must install a special electronic control circuit that will sense when power is being fed back to the utility and then turn on a hot water heater or other device that makes use of the excess electricity.

Such a device was designed and tested by researchers Dr. Ken Barnett and Robert Wagner at New Mexico State University in Las Cruces, New Mexico. Called a "power dump circuit," the device senses whenever power begins to flow back to the utility and then sends that excess electricity to an auxiliary load (such as a water heater) in proportion to the power that is available. The researchers connected a Pinson "Cycloturbine C2E3" to utility power through a synchronous inverter in order to supply electricity to the test home at the Livestock Research Center of New Mexico State. Barnett and

Photo. 11-2 and Photo. 11-3 Water Churn Powered by Wind—A Pinson "Cycloturbine C2E3" at Cornell University turns the churn paddle mechanically, which raises the temperature of the water up to 110°F.

Wagner decided upon a 2,000-gallon tank of water to act as the auxiliary load in the house. They chose the hot water tank as the storage medium because it can store even small amounts of excess energy that become available. Solar collectors on the house also heat the water, and if the water rises above 190°F the excess power is then sent to a 4-kilowatt electric space heater.

Excess electricity often comes in short bursts as the wind changes speed or appliances are used in the house. Thus, if an appliance such as a refrigerator or freezer were connected to the

Photo. 11-4 A "Power Dump Circuit"—Dr. Ken Barnett and Robert Wagner look over the components of their "power dump circuit" at New Mexico State University. The device uses any excess energy generated by the wind system to heat a 2,000-gallon tank of water.

dump circuit as the auxiliary load, the compressor might never receive enough power to begin operating before the dump circuit would turn it off again. This might not be true of a wind system with a large output relative to the needs of the house, but such a situation would bring into question the cost-effectiveness of the size of such a wind system in comparison to its regular demand from the house. Thus the concept of a dump circuit works well, but the economic benefits of the device have yet to be proven. It is safe to assume, though, that enterprising engineers will soon develop a low-cost dump circuit that will be standard on many wind systems.

Dump circuits could easily be used in conjunction with a home computer system. Computerized remote control circuits that can turn any appliance on or off will plug into any outlet. A special

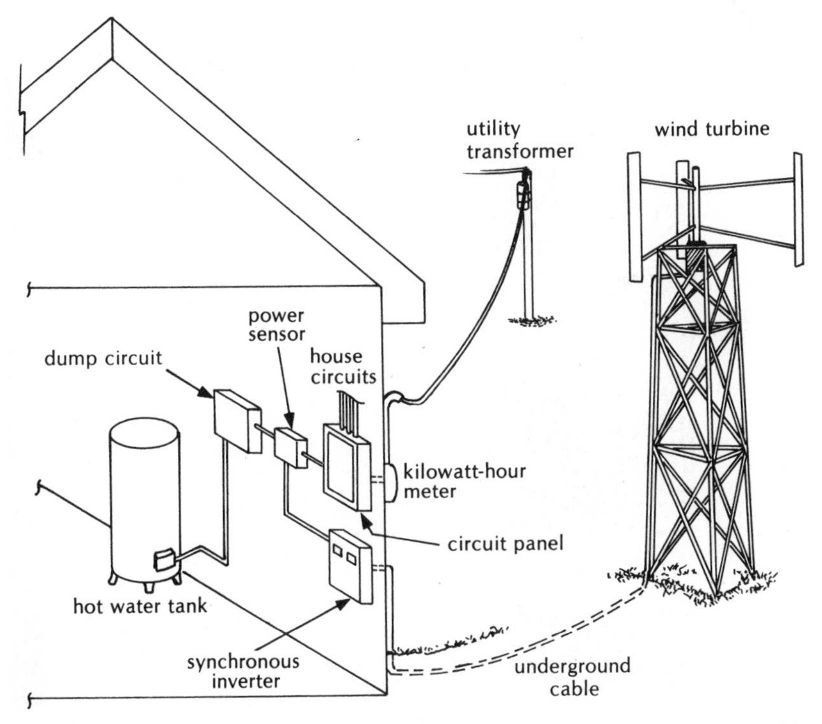

Illus. 11-2 New Mexico State University "Power Dump Circuit"—Designed by Dr. Ken Barnett and Robert Wagner, this utility-connected system uses a Pinson "Cycloturbine C2E3," an inverter, and a "power dump circuit" to heat a 2,000-gallon tank of water when the turbine generates excess energy.

power-sensing circuit connected to the wind system and to the household kilowatt-hour meter signals to the computer whenever excess power is available, and the computer then operates the desired appliances. Signals from the computer to the remote control device are sent through the house wiring circuits, so new wiring need not be installed in the house.

Research also continues toward improving storage batteries. Most developmental work in battery technology is being conducted for electric vehicles and for large-scale batteries designed for utilities to store power during off-peak periods. Researchers for electric vehicle batteries are intent upon developing the battery with the greatest energy density, or amount of storage capacity, per lb. of battery. Such light batteries will give a greater travel range to electric vehicles, but this characteristic is not really an advantage for a wind system. Many of these exotic batteries being developed have improved overall performances compared to lead-acid batteries, but most use expensive metals such as silver or cadmium. Also, batteries such as the lithium-sulphur or sodium-sulphur batteries operate at high temperatures of 400° to 800°F, a feature that makes them questionable for home or vehicle use from a safety standpoint.

The zinc-chloride battery, recently developed by Energy Development Associates, a subsidiary of Gulf & Western Industries, is one of the most promising innovations in battery technology in recent years. These batteries are so different in concept and appearance from the familiar lead-acid battery that Gulf & Western does not call them batteries. Instead, they are termed Electric Engine Units for electric vehicle use, and Electric Load Levelers for utility off-peak storage use. The advantage of the zinc-chloride battery is that it operates at low temperatures; it cannot be harmed by overcharging or overdischarging. It keeps full power until 95 percent of the stored energy is used up and only 2 percent of the power is lost if one cell malfunctions. This is quite a contrast to a lead-acid battery, which is disabled by one bad cell. Indications are that the new battery will have a lifetime of 2,500 charge and discharge cycles—comparable to or longer than a lead-acid battery lifetime. Also, although chlorine gas presents a potential hazard because it is poisonous, only small amounts of the gas can escape if the plastic case is damaged because the bulk of the chlorine is stored as solid chlorine hydrate.

The zinc-chloride battery is further unlike a lead-acid battery in that the electrolyte must be circulated through the cells by the use of a small pump. Any excess electrolyte is stored in a chamber separate from the cells. The positive plates of the cells are made of porous graphite, the negative plates are made of dense graphite, and the electrolyte is a solution of zinc chloride in water. During charging, a pump circulates the electrolyte through the cells and the charging current causes zinc metal to be deposited upon the negative plates while chlorine gas collects on the positive plates. The gas is carried away by the flowing electrolyte to a separate chamber called the "store." Here the gas is cooled by a small refrigeration unit until it condenses into a solid form—chlorine hydrate. During discharging, the electrolyte deposits chlorine on the positive plates and zinc leaves the negative plates to join with the chlorine to form zinc-chloride ions in the electrolyte. The electrolyte is warmed by the chemical reaction as it circulates back through the store chamber and melts from chlorine hydrate back into chlorine gas. When all of the chlorine hydrate is melted, the battery is completely discharged.

If batteries such as the zinc-chloride battery can be used successfully with intermittent renewable energy sources like wind and solar energy, new options will open up. It may be possible for the homeowner to generate electricity by using wind power and other renewable energy sources, store it in a super-battery such as the zinc-chloride type, and use utility-generated electricity only when necessary during off-peak periods.

Innovative Wind Systems

Research centers and inventors around the country are continuing to develop new and more efficient rotors for wind systems. The horizontal-axis (or propellerlike) rotor remains as the most advanced type of rotor, and further progress on new materials and lower cost production techniques is anticipated. Vertical-axis rotor design is really two or three generations behind horizontal-axis development, although efforts continue to reduce costs and increase efficiencies. This work includes development of more efficient Darrieus (or "eggbeater") rotors, straight-blade H-rotors, new blade designs, and novel, augmented systems that speed up the wind as it goes past the rotor.

Darrieus Rotors

The Darrieus rotor is commercially available at present through companies such as Dynergy Corporation and Alcoa, Incorporated (see App. E). However, research on the Darrieus also continues toward developing lower-cost blades and improving the system's overall performance. Rapi Rangi and Peter South of the National Research Council of Ottawa, Ontario revived the Darrieus concept in 1966, not knowing at the time that the idea had been patented previously by its French namesake in 1926. Later on, researchers at Sandia Laboratories in Albuquerque, New Mexico also began building a series of Darrieus rotors.

Although the Darrieus is about as efficient as a horizontal-axis rotor, the pitch of the blades cannot be varied. The rotor does not easily start up in low winds unless other means are provided (for example, using a small motor to start the rotor up or by motoring the generator). Thus, most researchers assumed that a Darrieus rotor would never start up on its own. This theory was unintentionally disproved in 1978 when the National Research Council's test rotor was left unattended with the brake off. The rotor began turning, apparently under just the right wind conditions, and eventually destroyed itself when one of the tower guy wires failed. The accident turned out to be a learning experience for future designers of Darrieus rotors.

In 1979 the Wind Engineering Company of Lubbock, Texas installed an innovative Darrieus rotor system at the Times Publishing building in St. Petersburg, Florida. Wind Engineering installed three Dynergy Corporation turbines, but with the rotors stacked one on top of another on a special 100-foot tower. The triple-tier wind system is now used to charge the emergency power batteries for the *St. Petersburg Times'* computer, as well as to light the building.

H-Rotors

Another vertical-axis design being developed is the H-rotor, which has straight blades instead of curved blades like the Darrieus.

The advantage of this rotor is that the blade pitch can be varied to make the rotor self-starting and also to control the speed, just as with horizontal-axis rotors.

The Pinson Energy Corporation's "Cycloturbine C2E3" is one commercially available H-rotor. Designed by Herman Drees while he was a student at the Massachusetts Institute of Technology, the "Cycloturbine C2E3" has been used for a variety of mechanical and electrical power applications. Its versatility is one of the attractive features common to all of the vertical-axis rotors. With the power shaft coming down to the ground it is easy to connect different devices to the turbine, depending upon the need. The "Cycloturbine C2E3" relies on a cam-actuated linkage to change the pitch of the blades during increases in wind speeds. The blades stay in the vertical position and change only their angle of attack in relation to the wind.

Other designs under testing use different methods of controlling the rotor speed. At Reading University in England, Dr. Peter Musgrove has developed an H-rotor that he calls a variable geometry vertical-axis wind turbine. In Musgrove's design, the crossarms are attached to the center of the blade. The blades are hinged at the crossarm so that the lower portion of the blade flies outward and the upper portion moves inward toward the center of rotation as the speed and, thus, the centrifugal force increases. Springs limit the amount of inclination of the blades. The increased inclination slows the rotor down because the amount of swept area presented to the wind is reduced. This concept is similar to the tilt-up or variable-axis rotor control system used on the North Wind Power Company model "HR2" (see App. E), which is a horizontal-axis machine.

Researchers at West Virginia University are testing a variation of Musgrove's design, a version of the H-rotor that they call a "tilting wing rotor." In this design, the blades tilt forward in their rotation plane instead of tilting outward, as with Musgrove's H-rotor. The tilting wing is supposed to be less subject to vibrations than the Musgrove design. The Musgrove blades may experience unequal centrifugal forces if the movement of each blade is not precisely controlled.

Another experimental vertical-axis rotor is the Giromill designed by the McDonnell Aircraft Corporation. The first Giromill was built for McDonnell by Valley Industries of Conway, Arkansas.

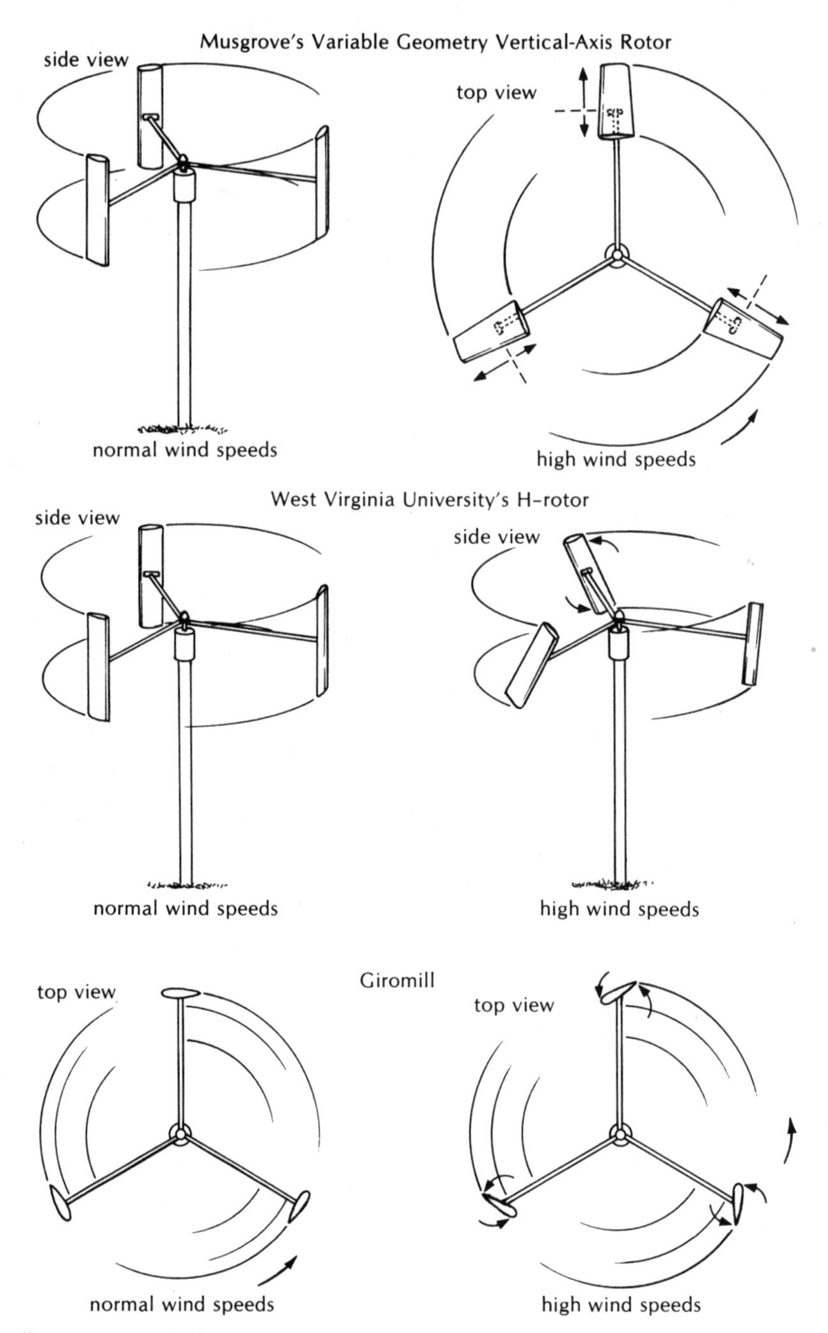

Illus. 11-3 Vertical-Axis Rotor Speed Controls—The hinged blades on Dr. Peter Musgrove's variable geometry vertical-axis wind turbine lean out at the bottom and in at the top during high winds. West Virginia University's "tilting wing H-rotor" is so called because the blades tilt forward in the rotational plane for speed control while the Giromill's blades change pitch.

Valley Industries also manufactures the familiar Aermotor water-pumping windmill and hence is interested in the pumping and irrigating capabilities of the Giromill. A small motor manages blade-pitch control for the 42-foot-high, 54-foot-diameter prototype Giromill.

Whatever the design, the vertical-axis rotor has come into its own as a versatile alternative to the horizontal-axis rotor. With a vertical-axis rotor, it's possible to switch from a friction heating device to an electrical generator, or to operate the two simultaneously. In the future it may well be a vertical-axis rotor wind system that heats or cools, or produces electricity for many of our homes.

Blade Design

The most widely used wind systems of the future will ultimately be those that use the lowest cost material with the longest lifetime to provide the most power. This depends in large part on advances in blade fabrication techniques and in the use of new materials. Although most present-day home-size wind systems depend upon wood or aluminum blades, new research is conducted constantly to develop improved types of blades. This research extends from testing various composites of glass fibers, graphite fibers, epoxy glues, urethane foam, and aircraft honeycomb material to evaluating new combinations of wood, steel, and aluminum.

One of the latest innovative designs is the Composite Bearingless Rotor (C.B.R.) that is being developed by the United Technologies Corporation of East Hartford, Connecticut. Simply put, the root of the blade, called the "flexbeam," is so flexible that the blade pitch can be changed simply by a twisting of the blade. The "flexstrip," a simple mechanical linkage connected from the center hub to the flexbeam, controls the blade pitch. Attached to the flexbeam are fiberglass blades that, instead of being formed in a mold, are manufactured by forcing fiberglass resin through a die, in a method called the "pultrusion process."

The flexbeam itself is a composite of long graphite fibers embedded in an epoxy resin, a combination that will not bend easily but will twist. Similar construction was employed for the tail rotor of a Sikorsky helicopter, and engineers saw the potential of the materials for wind turbines. The flexbeam eliminates the need for bearings and other mechanical parts, a development that may lead to the production of a lower cost rotor. The unique C.B.R. concept is being tested on an 8-kilowatt wind system that United Technologies built and had installed at the Rocky Flats test site in Golden, Colorado.

Wooden blades have always been reliable in wind system duty. The problem of locating enough wood of good quality for the production of large number of blades does exist, particularly for those system designs that incorporate longer and longer blades. Fiber and resin compositions attempt to imitate the properties of wood that enable blades to bend and flex without breaking. The technology for building such large wooden blades may come from the boat building industry. When one of the large 200-kilowatt wind systems being tested by the U.S. Department of Energy (described later in this chapter) developed cracks in its aluminum blades, other types of materials were investigated. The solution came from the Gougeon Brothers Company of Bay City, Michigan, producers of custom-made wooden sailboats.

Gougeon Brothers developed a special epoxy resin system called the Wood Epoxy Saturation Technique (WEST) that not only glues and laminates wood together but also gives it a protective coating that keeps the wood absolutely dry. The Gougeon Brothers Company replaced the 60-foot aluminum blades with laminated wooden blades built using the special WEST process. The blades were attached to the rotor hub with 24 steel bolts, each 15 inches long and an inch thick. These bolts were embedded and glued into the end of the wood. In one test, the bolts supported a 40-ton load, and in another, a stress test, a force of 10,000 lbs. was applied cyclically nearly a million times, causing the bolts to fail but without fatiguing the wood. If these composite wooden blades solve the problems associated with very large wind turbine blades, they certainly indicate a good possibility for the production of lower cost, longer-lasting blades for home-size wind systems in the future.

Augmented Systems

In an effort to increase the power output of wind systems to reduce costs, researchers are investigating various wind-augmenting designs. Just as a lens or parabolic reflector can be used to concentrate sunlight for solar energy applications and water power is concentrated by a dam, wind energy can be concentrated with various mechanical contrivances, too. It is known that an aerodynamically shaped duct or diffuser placed around a rotor can reduce the wind pressure behind the rotor and thus augment or increase the speed of the wind past the rotor. If the diffuser or duct can be built for less cost than a larger rotor, then the augmented wind turbine is more cost-effective than an unducted rotor.

Inventor Barry Lebost's augmented turbine looks like a silo cap with tail fins, which, in fact, is just what it is. The dome-shaped shell has an opening in it that is guided to face into the wind by two tail fins on the opposite side of the shell. The dome covers a 21-foot-diameter vertical-axis rotor with four curved blades. A curved deflector plate on the lower front portion of the dome serves to decrease the air pressure immediately underneath and thus further augments the dome's scooping action that speeds the wind passing over the rotor.

Tests have been conducted on a prototype Lebost turbine installed on top of New York University's Barney Building in New York City. The initial results encouraged researchers from the University's Department of Applied Science. The efficiency of the turbine actually approached the limits established by Betz (see Ch. 1). Since this high efficiency occurred at very low rotor speeds, the researchers decided that this turbine might best be adapted for use with mechanical power applications rather than with electrical generators, which require higher speeds. Further testing is being conducted with a Water Twister connected to the Lebost turbine, and Lebost is optimistic that even at such a low speed a domed rotor could be used in densely populated areas on building tops to provide heat and perform other tasks as well.

Another unique turbine is the tornado-vortex or Yen wind turbine invented by James Yen of Grumman Aerospace in Bethpage, New York. The Yen turbine uses a cylindrical tower with numerous shutterlike vanes in it that admit wind on the up-wind side. This creates a vortex turbulence like a tornado inside the structure. The vortex causes a low pressure area to form at the center of the tower that helps to pull in air at the bottom of a bell-shaped structure positioned at the base of the tower. A vertical-axis wind turbine sits inside the bell-shaped structure. Although as yet no full-scale prototype Yen turbine has been built, the inventor envisions augmenting the wind speeds even more with heated air by locating the towers over a heat source such, as a trash-burning plant or solar collectors.

Still another augmented turbine, the Torroidal Accelerator Rotor Platform, or TARP, has been designed by inventor Alfred

Photo. 11-5 The Lebost Augmented Turbine—The dome-shaped Lebost cap above the system reduces wind pressure behind the rotor while increasing the pressure in front as the wind passes the blades, thus raising the power available to the system from the wind.

Weisbrich of Connecticut. The TARP is a torroidal (doughnut-shaped) platform that can be located on the top of buildings, towers, or silos. The wind speeds up as it passes around the TARP, and its smooth shape provides a clean flow to the rotor, which can be mounted on either side of the platform. Although no TARP's have been built yet, wind tunnel tests at Rensselaer Polytechnic Institute in Troy, New York indicate that the wind power could be concentrated by a factor of 4 or 5 compared to the power of an unaugmented rotor.

Wind energy can be concentrated naturally by a mountain pass. For example, in the Pacheco Pass in California, the wind flows around the mountain and speeds up as it goes through the pass. Such sites are few and far between, however, and mechanical

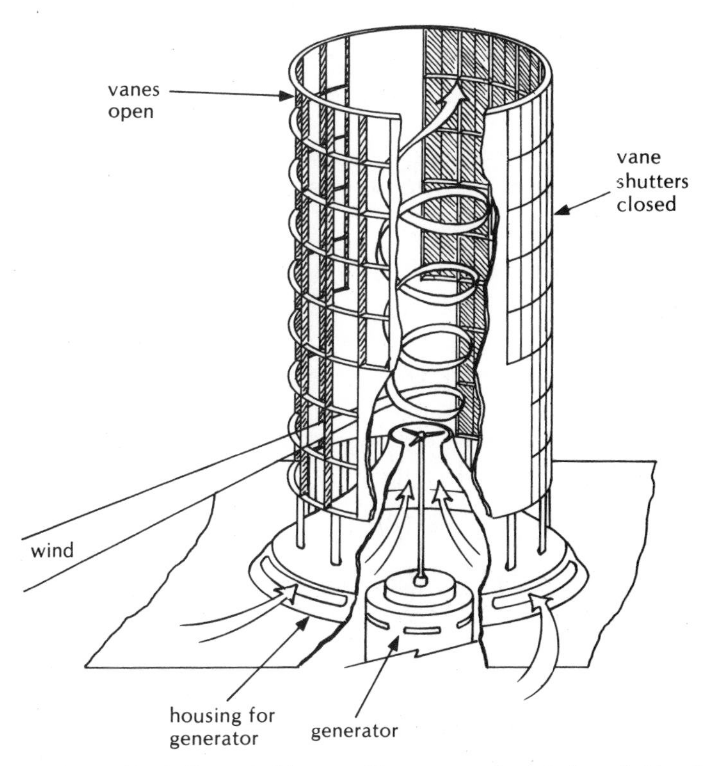

Illus. 11-4 James Yen's Tornado Vortex—Shutters on the windward side of Yen's vortex allow entrance of the wind, thereby increasing the draw of the wind below, past the vertical-axis wind turbine in the bell-shaped base.

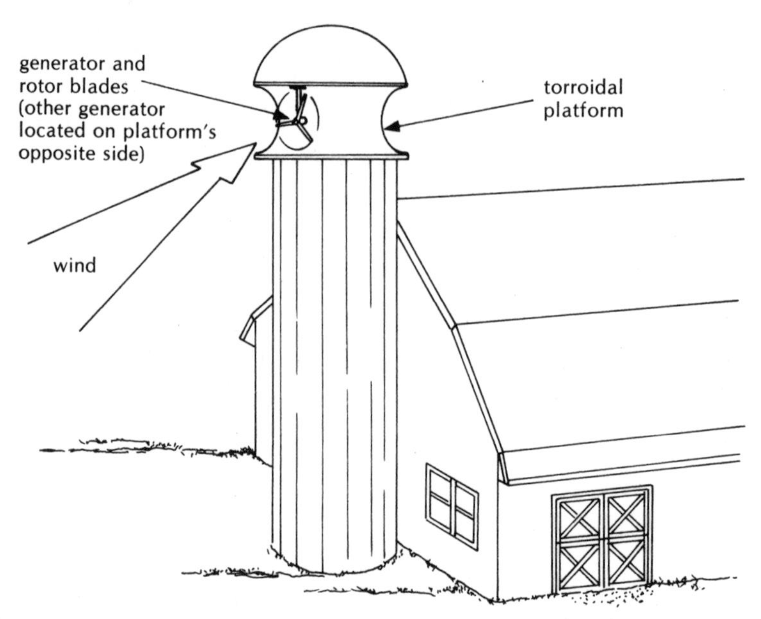

Illus. 11-5 Alfred Weisbrich's Torroidal Accelerator Rotor Platform—Weisbrich's TARP augments the wind by its smooth shape, which guides the wind to the rotors mounted on both sides of the structure. (Only one rotor can be seen from the view in this illustration.)

innovations for wind power augmentation seem to hold promise for wider-spread use. Undoubtedly more experiments with augmented turbine designs will take place in the future, and some of these turbines will become commercially available for use on the rooftops of homes, apartment buildings, and farm silos. Whether or not they can be made as economically as regular vertical- and horizontal-axis turbines for the same applications remains to be seen.

Large-Scale Wind Systems

While wind systems continue to be developed for home, farm, and commercial use, the Department of Energy (DOE), aerospace companies, and utilities are concentrating on large-scale wind projects as well. These projects involve both very large wind systems, such as those being built for the DOE under the Federal wind energy program, and large numbers of smaller wind turbines assembled into wind farms, being built by private industry. The idea is the same with both types of large-scale projects—to use wind power to generate electricity directly into the utility grid and to "wheel" or distribute the power wherever the utility needs it.

The DOE's wind program got off the ground in 1974 with the installation of its Mod O wind system at Plum Brooke, Ohio. The 100-kilowatt generator with its 125-foot-diameter, two-bladed rotor has since been used as an experimental turbine for other larger systems. Subsequently, larger Mod OA systems with 125-foot rotors and 200-kilowatt generators (at 19-mph wind speeds) have been installed in Clayton, New Mexico; Block Island, Rhode Island; Culebra Island, Puerto Rico; and Oahu, Hawaii. Each unit is capable of supplying power for about 60 homes. The first Mod OA unit, installed at the Clayton site, generated 380,000 kwh. of electricity for the Clayton Municipal Electric System power grid during 3,900 actual operating hours over a 2½ year period. As discussed earlier in this chapter, the only major problem encountered was that cracks developed in the aluminum blades. Otherwise the unit has been operational 80 percent of the time when winds have been available.

An even larger unit, the Mod 1, was installed at Boone, North Carolina in 1979. The Boone system generates 2,000 kw. (in a 25-mph wind) with a 200-foot rotor—enough to supply 500 homes. Ironically, this system proved to be plagued with a noise problem soon after it was installed. As wind passed over the tower legs, turbulence was created that affected the blades as they in turn passed by the legs. The blades vibrated slightly due to the turbulence, and the vibrations produced a low pitch sound that bothered the local residents for several miles around. In order to deal with this problem, the designers lowered the operating speed of the turbine, which caused the noise to substantially subside.

The latest and largest units installed under the Federal wind program are a cluster of three Mod 2 wind turbines located near Goldendale, Washington. Each unit has 300-foot blades that can generate 2,500 kw. in a 20-mph wind, enough power for 2,000 to 3,000 homes.

Other large-scale wind projects involve wind farms, large numbers of wind systems, located at a favorable site in order to generate large amounts of electricity into the utility grid. These proposed wind farms will be funded both by private and government sources. Sites being considered for arrays of wind systems are the Pacheco Pass near San Francisco, California; the San Gorgonio Pass near Palm Springs, California; and one in Medicine Bow, Wyoming.

The California Department of Water Resources plans to buy

Photo. 11-6 The Department of Energy's (DOE) Large-Scale Mod 1 Wind Turbine—The DOE constructed the Mod 1 in Boone, North Carolina. This giant wind system is capable of producing 2,000 kw. at wind speeds of 25 mph. (Photograph for DOE by Dick Pebody.)

wind-generated electricity from hundreds of wind systems to be installed by private companies in the Pacheco Pass area. Wind speeds average as high as 20 mph in the Pass and it is anticipated that 100 megawatts (100 million w.) of wind power capacity can be installed at the site.

In a similar project, the California Energy Commission is considering a wind farm for the San Gorgonio Pass area where wind speeds average 22 mph. Wind-powered generators with a total capacity of 60 megawatts could be installed at this Pass. At Medicine Bow, Wyoming, the Water Power and Resources Service anticipates the installation of an array of wind systems with 100 megawatts of generating capacity. The turbines will work in conjunction with the Colorado River Storage Project's hydroelectric facilities at the Glen Canyon and Flaming Gorge dams to constitute the first large-scale, integrated wind and hydropower facility.

The future of wind, then, involves development of both large and small wind systems. Although it's sometimes suggested that large-scale wind systems may hinder the development of smaller systems or make them unnecessary, this is not expected to happen. Large-scale wind systems are limited by the number of favorable locations that are available for their operation. Also, the main difference between large-and small-scale is just that—size. Any developments in one size of system can be used in any other size system. Indications are that as fossil fuel supplies dwindle, all types of renewable energy sources will be utilized to their maximum potential, including both large utility-grid wind systems and smaller home and farm-size systems. Undoubtedly, wind power is not only here now as a workable source of energy, it will grow in importance as an integral part of our energy future by turning wind into watts to light our lights and into BTU's to heat our water.

APPENDIX A

Betz's Law: The Available Power in the Wind

$$P = (0.593)\tfrac{1}{2}\rho A V^3$$
$$= 0.0000030168 A V^3$$

Where:

(0.593) = the Betz Limit or the maximum percentage of power in the wind which a wind system can extract

ρ = Air density = 0.002378 slugs/ft.^{3} = 0.002378 lb. ft.$^{-4}$sec.^{2} at a standard air pressure of 760 millimeters and temperatures of 59.9°F.

A = Rotor area = 3.14159 $D^2/4$, in ft.^{2}

Available Power in the Wind in Kilowatts According to Betz's Law

Wind Speed, Mph	Rotor Diameter in Feet: 2	4	6	8	10	12	14	16	18	20
6	.001	.008	.018	.033	.051	.074	.100	.131	.166	.205
8	.005	.019	.044	.078	.121	.175	.238	.311	.393	.485
10	.009	.038	.085	.152	.237	.341	.464	.607	.768	.948
12	.016	.066	.148	.262	.409	.590	.802	1.048	1.327	1.638
14	.026	.104	.234	.416	.650	.936	1.274	1.665	2.107	2.601
16	.039	.156	.350	.622	.970	1.398	1.902	2.485	3.145	3.883
18	.055	.222	.498	.885	1.381	1.990	2.708	3.538	4.478	5.528
20	.076	.304	.683	1.214	1.895	2.730	3.714	4.853	6.142	7.583
22	.101	.405	.909	1.616	2.522	3.633	4.944	6.460	8.175	10.093
24	.131	.525	1.180	2.098	3.274	4.717	6.418	8.387	10.614	13.103

APPENDIX A—Continued

D = Rotor diameter, in ft.
V = Wind velocity, in mph
P = Available power in the wind, in kw.

Units:

P(ft. lb./sec.) = $(0.593)\frac{1}{2}$ (0.002378 lb. ft. $^{-4}$sec.^{2}) $(A)(V)^3$
where V is in ft./sec., since 1 mph = 0.000705 $(A)(V)^3$
= 1.466667 ft./sec.

Then:

P(ft. lb./sec.) = $(0.000705)(A)(V \times 1.466667)^3$
where A is in sq. ft. and V is in mph,

Then:

$P = 0.0022242\ (A)(V)^3$
since, 1 ft. lb./sec. = 1.3563636 w.

Then:

P(w.) = $(0.0022242)(A)(V)^3(1.3563636)$
= $0.0030168\ (A)(V)^3$
P(kw.) = $0.0000030168\ (A)(V)^3$

22	24	26	28	30	32	34	36	38	40
.248	.295	.346	.401	.461	.524	.592	.663	.739	.819
.587	.699	.820	.951	1.091	1.242	1.402	1.572	1.751	1.941
1.147	1.364	1.602	1.858	2.133	2.426	2.739	3.070	3.421	3.791
1.981	2.358	2.768	3.210	3.685	4.192	4.733	5.306	5.912	6.550
3.147	3.744	4.395	5.098	5.852	6.657	7.516	8.426	9.388	10.402
4.697	5.589	6.560	7.669	8.735	9.937	11.219	12.578	14.014	15.528
6.687	7.958	9.341	10.834	12.437	14.149	15.974	17.909	10.953	22.109
9.173	10.916	12.813	14.862	17.061	19.409	21.912	24.566	27.371	30.327
12.210	14.529	17.054	19.781	22.708	25.833	29.164	32.698	36.431	40.366
15.852	18.863	22.141	25.861	29.481	33.539	37.863	42.451	47.297	52.406

APPENDIX B

Average Wind Speeds, Monthly and Annual

Wind—Average Speed, Mph

Data Through 1979	Years	Jan.	Feb.	Mar.	Apr.
Birmingham, AL	36	8.5	9.0	9.4	8.5
Huntsville, AL	12	9.4	9.6	10.1	9.0
Mobile, AL	31	10.6	10.8	11.1	10.4
Montgomery, AL	35	7.8	8.4	8.5	7.4
Anchorage, AK	26	5.8	6.6	6.7	7.1
Annette, AK	28	12.1	12.3	11.1	11.2
Barrow, AK	47	11.3	10.9	11.1	11.4
Barter Island, AK	27	14.8	14.2	13.7	11.9
Bethel, AK	21	14.1	15.1	13.7	13.4
Bettles, AK	9	5.8	6.8	7.4	7.6
Big Delta, AK	20	10.8	9.9	8.2	7.6
Cold Bay, AK	24	17.8	17.8	17.4	18.2
Fairbanks, AK	28	2.9	3.9	5.1	6.5
Gulkana, AK	8	5.1	5.6	6.5	8.6
Homer, AK	9	7.8	7.7	7.4	7.3
Juneau, AK	36	8.4	8.8	8.8	8.8
King Salmon, AK	24	10.5	11.2	11.3	11.1
Kodiak, AK	26	12.4	12.1	11.8	10.9
Kotzebue, AK	34	14.8	12.7	12.5	12.9
Mc Grath, AK	30	2.8	4.1	5.1	6.3
Nome, AK	32	11.8	11.1	10.5	10.8
St. Paul Island, AK	13	20.8	21.8	20.1	19.2
Talkeetna, AK	6	6.3	5.0	4.9	4.4
Yakutat, AK	31	7.5	8.0	7.4	7.2
Flagstaff, AZ	16	7.7	7.5	8.3	8.7

SOURCE: *National Climatic Center, Comparative Climatic Data for the United States: Through 1979* (Asheville, N.C.: National Climatic Center, 1979).

May	June	July	Aug.	Sept.	Oct.	Nov.	Dec.	Annual
6.9	6.2	5.7	5.5	6.5	6.2	7.5	8.0	7.3
7.7	6.9	6.2	5.8	6.7	7.5	8.3	9.3	8.1
8.9	7.7	6.9	6.8	8.0	8.2	9.4	10.1	9.1
6.2	5.9	5.8	5.2	6.1	5.7	6.6	7.2	6.7
8.3	8.2	7.1	6.5	6.1	6.4	6.1	6.0	6.7
9.4	9.0	8.1	8.3	9.4	12.0	12.4	12.8	10.7
11.6	11.3	11.6	12.3	13.0	13.3	12.4	11.1	11.8
12.5	11.5	10.6	11.7	13.1	14.6	15.0	13.9	13.1
11.7	11.7	11.2	11.2	11.6	12.6	13.5	14.0	12.8
7.6	7.1	6.5	6.1	6.7	6.7	6.1	5.9	6.7
7.9	6.5	6.0	6.6	7.3	8.5	9.5	9.5	8.2
16.2	15.9	15.7	16.4	16.2	16.8	17.6	17.0	16.9
7.7	7.0	6.5	6.1	6.1	5.4	3.9	3.2	5.3
8.8	8.8	8.2	8.0	7.6	6.3	4.8	3.6	6.8
7.7	7.0	6.8	5.7	6.2	6.8	7.5	7.1	7.1
8.5	7.8	7.6	7.5	8.0	9.7	8.7	9.2	8.5
11.3	10.9	10.0	10.2	10.5	10.5	10.9	10.3	10.7
10.1	8.3	7.0	7.6	9.0	10.7	12.0	12.2	10.3
10.9	12.4	13.0	13.3	13.2	13.6	14.6	12.7	13.0
6.5	6.2	5.9	5.6	5.6	5.1	3.5	3.1	5.0
10.3	10.1	10.1	10.7	11.3	11.2	12.2	10.3	10.8
16.2	14.3	12.9	14.8	16.2	19.4	22.7	22.3	18.4
4.4	4.3	3.7	3.0	3.1	3.5	5.0	4.9	4.3
7.9	7.4	6.9	6.6	7.2	8.4	8.0	8.4	7.6
8.2	7.9	6.2	5.7	6.5	6.6	7.6	7.7	7.4

APPENDIX B—Continued

		Wind—Average Speed, Mph			
Data Through 1979	**Years**	**Jan.**	**Feb.**	**Mar.**	**Apr.**
Phoenix, AZ	34	5.2	5.9	6.6	7.0
Tucson, AZ	34	7.7	8.1	8.5	8.8
Winslow, AZ	35	7.1	8.5	10.6	11.3
Yuma, AZ	28	7.4	7.4	7.9	8.3
Fort Smith, AR	34	8.2	8.5	9.5	8.9
Little Rock, AR	37	8.8	9.2	10.0	9.4
Bakersfield, CA	40	5.3	5.8	6.6	7.2
Blue Canyon, CA	10	9.7	9.4	9.1	7.7
Eureka, CA	54	6.9	7.2	7.6	8.0
Fresno, CA	30	5.4	5.7	6.7	7.2
Long Beach, CA	20	5.5	6.1	6.9	7.4
Los Angeles, (Int'l. Airport), CA	31	6.7	7.3	8.0	8.4
Los Angeles (City), CA	24	6.8	6.9	7.0	6.6
Oakland, CA	32	6.8	7.3	9.0	9.6
Red Bluff, CA	35	9.0	9.2	9.7	9.6
Sacramento, CA	31	7.7	7.8	8.9	9.0
Sandberg, CA	30	17.0	16.4	16.7	16.4
San Diego, CA	39	5.7	6.3	7.2	7.7
San Francisco (Int'l. Airport), CA	52	7.1	8.5	10.3	12.1
San Francisco (City), CA	28	6.7	7.5	8.5	9.5
Santa Maria, CA	15	6.7	7.2	8.3	8.0
Stockton, CA	35	6.7	6.9	7.7	8.3
Alamosa, CO	3	5.3	5.6	9.2	10.7
Colorado Springs, CO	31	9.8	10.5	11.6	12.1
Denver, CO	31	9.0	9.3	10.0	10.4
Grand Junction, CO	33	5.4	6.6	8.3	9.6
Pueblo, CO	37	8.0	8.6	9.8	10.5
Bridgeport, CT	22	13.2	13.6	13.4	13.0
Hartford, CT	25	9.4	9.7	10.2	10.4
Wilmington, DE	31	10.0	10.5	11.1	10.5
Washington, DC (Dulles Airport)	17	8.4	9.0	9.2	9.0
Washington, DC (National Airport)	31	10.1	10.5	10.9	10.5
Apalachicola, FL	35	8.4	8.8	9.0	8.7
Daytona Beach, FL	34	9.1	9.9	10.1	9.9
Fort Myers, FL	34	8.5	9.1	9.4	9.0
Jacksonville, FL	30	8.5	9.4	9.3	9.0
Key West, FL	26	12.1	12.2	12.6	12.8

May	June	July	Aug.	Sept.	Oct.	Nov.	Dec.	Annual
7.1	7.0	7.3	6.7	6.4	5.9	5.4	5.2	6.3
8.6	8.6	8.3	7.7	8.2	8.1	8.0	7.8	8.2
10.9	10.6	9.1	8.4	8.2	7.7	7.3	6.7	8.9
8.3	8.5	9.5	8.9	7.3	6.6	6.9	7.2	7.8
7.8	6.7	6.3	6.4	6.6	6.8	7.7	8.0	7.6
8.0	7.5	6.9	6.6	6.9	7.0	8.2	8.5	8.1
8.0	8.0	7.2	6.8	6.3	5.6	5.1	5.0	6.4
7.7	7.5	6.9	6.8	7.3	8.1	8.2	7.9	8.0
7.9	7.4	6.8	5.8	5.5	5.6	6.0	6.4	6.8
8.0	8.1	7.2	6.7	5.9	5.2	4.7	5.0	6.3
7.3	6.9	6.7	6.5	6.1	5.8	5.6	5.2	6.3
8.2	7.9	7.6	7.6	7.2	6.8	6.5	6.4	7.4
6.3	5.7	5.4	5.3	5.3	5.7	6.4	6.6	6.2
10.0	10.0	9.4	9.0	7.8	6.8	6.3	6.5	8.2
9.3	9.2	8.0	7.6	7.9	8.2	8.4	8.3	8.7
9.4	9.9	9.1	8.7	7.7	6.7	6.3	7.0	8.2
15.9	15.1	12.8	12.6	12.9	14.9	16.1	16.4	15.3
7.7	7.6	7.1	7.1	6.8	6.4	5.7	5.4	6.7
13.2	13.9	13.5	12.8	11.0	9.3	7.2	6.8	10.5
10.4	10.9	11.2	10.5	9.1	7.6	6.3	6.5	8.7
8.3	7.9	6.5	6.2	5.9	6.2	6.6	6.4	7.0
9.2	9.2	8.2	7.7	7.1	6.4	5.8	6.2	7.4
11.5	10.0	8.1	8.1	8.2	7.7	7.1	6.3	8.0
11.6	10.9	9.6	9.3	9.7	9.9	9.7	9.9	10.4
9.5	9.1	8.5	8.2	8.2	8.2	8.7	8.9	9.0
9.6	9.8	9.3	9.0	9.0	7.9	6.6	5.8	8.1
9.8	9.4	8.7	8.0	8.0	7.5	7.6	7.9	8.7
11.7	10.4	9.9	10.1	11.1	11.8	12.6	13.0	12.0
9.2	8.4	7.7	7.4	7.5	8.0	8.6	8.9	8.8
9.0	8.4	7.7	7.4	7.9	8.2	9.1	9.4	9.1
7.7	6.8	6.2	6.0	6.3	6.6	7.7	7.9	7.6
9.3	8.8	8.2	8.0	8.3	8.6	9.2	9.5	9.3
7.8	7.2	6.6	6.7	8.1	8.1	8.1	8.0	8.0
9.2	8.4	7.6	7.3	8.7	9.5	8.8	8.8	8.9
8.2	7.4	6.8	6.8	7.7	8.5	8.2	8.2	8.2
8.5	8.3	7.4	7.2	8.1	8.5	8.1	8.1	8.4
10.8	9.8	9.9	9.6	10.1	11.3	12.0	12.1	11.3

APPENDIX B—Continued

Data Through 1979	Years	Wind—Average Speed, Mph Jan.	Feb.	Mar.	Apr.
Lakeland, FL	12	7.3	7.8	7.8	7.7
Miami, FL	30	9.5	10.1	10.5	10.7
Orlando, FL	31	9.1	9.8	9.9	9.5
Pensacola, FL	16	9.1	9.3	9.7	9.5
Tallahassee, FL	18	7.6	7.9	8.3	7.5
Tampa, FL	33	8.9	9.5	9.7	9.6
West Palm Beach, FL	37	9.9	10.3	10.7	11.0
Athens, GA	24	8.6	9.0	8.9	8.5
Atlanta, GA	41	10.5	10.9	10.9	10.0
Augusta, GA	29	7.1	7.7	8.0	7.6
Columbus, GA	21	7.5	8.1	8.2	7.4
Macon, GA	31	8.3	9.0	9.4	8.9
Savannah, GA	29	8.7	9.3	9.3	8.9
Hilo, HI	30	7.4	7.8	7.5	7.4
Honolulu, HI	30	10.0	10.7	11.6	12.1
Kahului, HI	18	11.2	11.3	12.3	13.1
Lihue, HI	29	10.5	11.3	11.8	12.3
Boise, ID	40	8.4	9.2	10.2	10.3
Pocatello, ID	27	11.2	10.9	11.5	11.8
Cairo, IL	22	9.8	9.8	10.6	10.2
Chicago (O'Hare Airport), IL	21	11.6	11.6	11.8	12.1
Chicago (Midway Airport), IL	37	11.5	11.5	11.9	11.8
Moline, IL	36	10.8	11.0	12.1	12.2
Peoria, IL	36	11.2	11.5	12.3	12.2
Rockford, IL	29	10.4	10.6	11.6	11.7
Springfield, IL	32	12.9	12.9	14.0	13.5
Evansville, IN	39	9.4	9.7	10.4	9.9
Fort Wayne, IN	33	11.7	11.5	12.2	11.9
Indianapolis, IN	31	11.1	11.1	11.8	11.4
South Bend, IN	31	12.1	11.7	12.3	12.0
Burlington, IA	38	11.3	11.5	12.3	12.1
Des Moines, IA	30	11.7	11.7	13.0	13.2
Sioux City, IA	39	11.2	11.3	12.3	13.3
Waterloo, IA	23	11.3	11.5	12.3	13.0
Concordia, KS	17	12.0	12.4	13.8	14.1
Dodge City, KS	37	13.6	14.1	15.8	15.7

May	June	July	Aug.	Sept.	Oct.	Nov.	Dec.	Annual
6.9	6.2	5.7	5.5	6.6	7.2	6.9	6.9	6.9
9.5	8.2	7.9	7.8	8.2	9.3	9.5	9.1	9.2
8.9	8.2	7.4	7.2	7.8	8.8	8.7	8.7	8.7
8.5	7.5	6.8	6.6	7.8	8.0	8.3	9.0	8.4
6.7	6.4	5.7	5.5	6.5	6.8	6.5	6.8	6.8
9.0	8.3	7.5	7.3	8.1	8.9	8.7	8.8	8.7
9.7	8.1	7.5	7.6	8.6	10.1	10.0	9.9	9.5
7.1	6.6	6.2	5.6	6.4	6.8	7.4	8.1	7.4
8.6	7.9	7.4	7.1	8.0	8.4	9.1	9.8	9.1
6.5	6.1	5.9	5.5	5.7	5.8	6.2	6.7	6.6
6.7	6.2	5.6	5.4	6.7	6.5	6.5	7.0	6.8
7.8	7.3	6.9	6.4	7.1	7.0	7.3	7.8	7.8
7.9	7.6	7.2	6.7	7.5	7.6	7.6	8.1	8.0
7.2	7.0	6.8	6.8	6.7	6.6	6.6	7.1	7.1
12.2	12.9	13.7	13.5	11.7	10.9	11.1	11.0	11.8
13.2	14.8	15.8	14.9	12.9	12.0	11.7	11.6	12.9
11.9	12.4	13.0	12.6	11.2	10.9	11.7	11.3	11.7
9.6	9.1	8.5	8.3	8.3	8.5	8.5	8.4	8.9
10.6	10.3	9.2	8.9	9.1	9.3	10.3	10.3	10.3
8.2	7.4	6.5	6.2	7.0	7.3	9.1	9.3	8.5
10.6	9.2	8.1	8.2	8.7	9.9	11.0	11.0	10.3
10.4	9.5	8.4	8.2	9.0	9.9	11.3	11.2	10.4
10.4	9.2	7.6	7.3	8.1	9.2	10.8	10.7	9.9
10.3	9.2	8.0	7.8	8.6	9.5	11.1	11.0	10.2
10.6	9.3	8.1	7.9	8.6	9.5	10.4	10.4	9.9
11.6	10.0	8.5	8.0	9.1	10.5	12.7	12.8	11.4
8.1	7.3	6.3	5.9	6.5	7.0	8.8	9.1	8.2
10.4	9.2	8.2	7.7	8.5	9.2	11.1	11.4	10.3
9.6	8.6	7.5	7.2	8.0	8.9	10.5	10.6	9.7
10.7	9.4	8.3	8.0	8.8	9.8	11.3	11.6	10.5
10.4	9.3	7.9	7.8	8.7	9.5	11.0	11.0	10.2
11.5	10.4	9.0	8.8	9.4	10.5	11.6	11.5	11.0
11.9	10.8	9.1	9.0	9.7	10.4	11.2	10.8	10.9
11.5	10.1	8.6	8.5	9.0	10.2	11.2	11.2	10.7
12.4	12.2	11.5	11.3	11.2	11.8	11.5	11.8	12.2
14.7	14.4	12.9	12.7	13.6	13.6	13.7	13.5	14.0

APPENDIX B—Continued

Data Through 1979	Years	Wind—Average Speed, Mph Jan.	Feb.	Mar.	Apr.
Goodland, KS	31	12.4	12.6	14.2	14.6
Topeka, KS	30	10.3	10.8	12.6	12.5
Wichita, KS	26	12.4	12.9	14.4	14.4
Cincinnati Airport–Covington, KY	32	10.7	10.6	11.2	10.8
Lexington, KY	32	11.4	11.4	11.8	11.2
Louisville, KY	32	9.7	9.8	10.4	9.9
Baton Rouge, LA	26	9.3	9.6	9.7	9.1
Lake Charles, LA	18	10.5	10.6	10.9	10.3
New Orleans, LA	31	9.5	9.9	10.1	9.5
Shreveport, LA	27	9.8	10.1	10.5	10.1
Caribou, ME	15	12.4	12.0	12.9	11.7
Portland, ME	39	9.3	9.5	10.1	10.0
Baltimore, MD	29	10.0	10.6	11.1	10.9
Blue Hill Observatory, MA	56	17.4	17.3	17.4	16.6
Boston, MA	22	14.2	14.1	13.9	13.3
Worcester, MA	21	12.4	12.2	11.8	11.5
Alpena, MI	19	8.4	8.0	8.5	8.8
Detroit (City Airport), MI	45	11.7	11.5	11.5	11.1
Detroit (Metro Airport), MI	21	11.7	11.6	11.7	11.6
Flint, MI	38	11.9	11.5	12.1	11.7
Grand Rapids, MI	16	11.7	10.9	11.2	11.2
Houghton Lake, MI	14	10.1	9.2	9.3	9.8
Lansing, MI	20	12.2	11.4	11.5	11.6
Marquette, MI	27	8.9	8.8	8.4	8.6
Muskegon, MI	19	12.8	12.0	12.2	12.0
Sault Ste. Marie, MI	38	10.1	9.8	10.3	10.6
Duluth, MN	30	11.9	11.6	12.0	13.0
International Falls, MN	27	9.0	9.0	9.5	10.4
Minneapolis–St. Paul, MN	41	10.4	10.5	11.3	12.3
Rochester, MN	19	14.2	13.4	13.9	14.3
Saint Cloud, MN	9	8.2	7.8	9.0	9.8
Jackson, MS	16	8.8	8.7	9.3	8.5
Meridian, MS	20	7.0	7.3	7.9	7.1
Columbia, MO	9	10.6	11.3	12.1	11.4
Kansas City, MO	7	10.7	11.6	12.4	12.2
Kansas City, MO	33	9.9	10.3	11.6	11.5
Saint Joseph, MO	22	10.6	10.8	12.5	12.9

May	June	July	Aug.	Sept.	Oct.	Nov.	Dec.	Annual
13.6	13.0	12.1	11.8	12.2	11.7	11.9	12.0	12.7
11.1	10.3	8.7	8.8	9.0	9.4	10.2	10.2	10.3
12.7	12.3	11.2	11.2	11.4	12.1	12.2	12.2	12.5
8.8	7.9	7.1	6.7	7.4	8.1	9.6	10.2	9.1
9.1	8.2	7.5	7.0	7.9	8.5	10.4	11.1	9.6
8.1	7.4	6.7	6.4	6.8	7.2	8.9	9.3	8.4
8.0	6.8	6.1	5.7	6.9	6.8	7.9	8.4	7.9
9.0	7.5	6.5	6.2	7.3	7.6	9.1	9.3	8.7
8.2	6.9	6.2	6.0	7.3	7.5	8.7	9.1	8.3
8.6	7.8	7.4	7.1	7.5	7.7	8.8	9.2	8.7
11.4	10.4	9.8	9.3	10.4	10.9	11.1	11.5	11.2
9.2	8.2	7.6	7.4	7.8	8.4	8.7	9.0	8.8
9.4	8.6	8.0	8.0	8.2	8.9	9.4	9.4	9.4
14.7	13.8	13.0	12.6	13.5	15.2	16.4	16.8	15.4
12.2	11.4	10.8	10.7	11.2	12.1	12.9	13.8	12.6
10.4	9.2	8.6	8.4	8.9	9.7	10.5	11.2	10.4
8.1	7.2	6.8	6.5	6.7	7.4	8.0	8.1	7.7
9.9	9.2	8.4	8.2	8.9	9.6	11.4	11.4	10.2
10.3	9.1	8.5	8.3	8.8	9.6	11.1	11.2	10.3
10.3	9.1	8.2	7.9	9.0	9.9	11.4	11.5	10.4
9.9	9.0	8.4	8.2	8.5	9.6	10.4	10.7	10.0
9.0	8.0	7.6	7.2	8.0	9.1	9.9	9.6	8.9
10.3	9.2	8.2	7.8	8.5	9.6	10.8	11.3	10.2
8.2	7.4	7.1	7.5	8.4	8.7	9.0	9.1	8.3
10.3	9.5	8.8	8.7	9.3	11.0	11.9	12.1	10.9
10.1	8.8	8.2	8.1	8.9	9.4	10.1	9.9	9.5
12.1	10.7	9.7	9.7	10.6	11.3	12.0	11.4	11.3
9.9	8.6	7.9	7.8	8.8	9.5	9.8	9.0	9.1
11.3	10.5	9.3	9.1	9.8	10.4	10.9	10.3	10.5
13.1	12.0	10.7	10.5	11.4	12.8	13.2	13.5	12.7
8.9	8.2	7.0	6.4	7.1	7.8	8.3	7.7	8.0
7.0	6.2	5.9	5.7	6.4	6.3	7.4	8.2	7.4
5.8	5.0	4.8	4.6	5.3	5.0	6.0	6.6	6.0
9.0	8.7	8.2	7.9	8.3	9.4	10.4	10.7	9.8
10.0	9.7	8.4	9.1	8.5	10.1	11.2	11.3	10.4
10.4	10.0	8.8	8.8	9.0	9.1	10.2	9.9	10.0
11.2	9.8	7.9	7.4	8.0	8.8	9.9	10.0	10.0

APPENDIX B—Continued

Wind—Average Speed, Mph

Data Through 1979	Years	Jan.	Feb.	Mar.	Apr.
St. Louis, MO	30	10.4	10.8	11.8	11.4
Springfield, MO	34	12.0	12.4	13.4	12.7
Billings, MT	40	13.1	12.5	11.7	11.7
Glasgow, MT	10	10.2	10.6	11.3	13.0
Great Falls, MT	38	15.5	14.7	13.5	13.0
Havre, MT	13	10.8	10.6	10.4	11.5
Helena, MT	39	7.0	7.6	8.6	9.3
Kalispell, MT	17	6.4	6.4	7.5	8.2
Miles City, MT	30	9.6	9.8	10.8	11.8
Missoula, MT	35	5.2	5.6	6.7	7.5
Grand Island, NB	30	11.8	11.9	13.6	14.2
Lincoln, NB	7	9.9	10.6	12.1	13.1
Norfolk, NB	3	12.8	12.2	14.1	14.6
North Platte, NB	27	9.2	10.0	11.7	12.8
Omaha, NB	43	11.1	11.4	12.6	13.1
Scottsbluff, NB	29	10.6	11.4	12.3	12.8
Valentine, NB	12	9.8	9.9	11.2	12.0
Elko, NV	30	5.4	5.9	6.7	7.2
Ely, NV	41	10.4	10.5	10.8	11.0
Las Vegas, NV	31	7.1	8.4	9.8	10.9
Reno, NV	37	5.8	6.1	7.6	8.1
Winnemucca, NV	30	7.6	7.8	8.6	8.6
Concord, NH	38	7.4	8.0	8.3	7.9
Mt. Washington, NH	45	45.9	44.6	42.0	36.5
Atlantic City, NJ	21	11.8	12.0	12.3	12.1
Newark, NJ	36	11.2	11.6	11.9	11.3
Trenton, NJ	32	9.8	10.2	10.7	10.4
Albuquerque, NM	40	8.0	8.8	10.1	11.0
Roswell, NM	7	7.8	8.7	10.6	10.6
Albany, NY	41	9.9	10.4	10.7	10.6
Binghamton, NY	28	11.8	11.8	11.8	11.5
Buffalo, NY	40	14.5	14.0	13.7	13.0
New York (Central Park), NY	57	10.7	10.9	11.0	10.5
New York (JFK Airport), NY	21	13.7	14.0	13.9	13.2
New York (La Guardia Airport), NY	31	13.9	13.9	14.0	12.9
Rochester, NY	39	11.9	11.7	11.2	10.8
Syracuse, NY	30	11.1	11.2	11.1	11.0

May	June	July	Aug.	Sept.	Oct.	Nov.	Dec.	Annual
9.4	8.7	7.8	7.5	7.9	8.7	9.9	10.3	9.5
10.8	10.0	8.8	8.9	9.6	10.4	11.5	11.8	11.0
11.0	10.4	9.7	9.7	10.4	11.1	12.3	13.2	11.4
11.9	11.1	10.5	10.9	10.9	10.6	9.5	10.1	10.9
11.5	11.3	10.1	10.4	11.5	13.5	14.8	16.0	13.0
10.5	9.9	9.5	9.3	9.7	10.2	10.1	10.5	10.2
9.0	8.7	7.9	7.6	7.5	7.3	7.2	7.1	7.9
7.8	7.4	6.7	6.7	6.4	5.4	5.6	6.0	6.7
11.1	10.4	9.7	9.7	9.9	9.8	9.8	9.8	10.2
7.3	6.9	6.7	6.6	5.9	5.0	4.9	4.8	6.1
12.8	12.1	10.7	10.7	11.1	11.4	11.8	11.7	12.0
10.8	10.3	10.2	10.2	9.6	10.1	10.1	10.1	10.6
12.7	11.9	11.1	10.4	10.8	11.5	11.9	12.5	12.2
11.8	10.6	9.5	9.4	9.7	9.5	9.5	9.1	10.2
11.3	10.4	9.0	9.1	9.6	10.0	11.1	10.8	10.8
11.9	10.5	9.3	9.0	9.4	9.6	10.3	10.6	10.6
11.7	10.4	9.6	9.8	10.5	9.9	10.6	9.8	10.4
6.9	6.7	6.2	6.0	5.4	5.1	5.1	5.1	6.0
10.9	10.7	10.3	10.5	10.4	10.3	10.1	10.1	10.5
10.9	10.8	10.0	9.4	8.7	7.9	7.3	7.1	9.0
7.7	7.3	6.7	6.3	5.5	5.3	5.2	5.0	6.4
8.6	8.5	8.3	7.8	7.5	7.3	7.1	7.1	7.9
7.1	6.4	5.6	5.3	5.5	6.0	6.5	7.1	6.8
30.0	27.1	24.9	25.1	28.4	33.4	38.3	44.1	35.0
10.6	9.7	8.9	8.5	9.0	9.6	10.9	11.2	10.5
10.0	9.3	8.8	8.6	8.9	9.3	10.1	10.8	10.2
9.0	8.4	7.8	7.6	7.9	8.3	9.2	9.3	9.0
10.5	10.0	9.1	8.2	8.6	8.3	7.9	7.7	9.0
10.2	9.5	8.5	7.9	7.9	7.9	7.5	7.4	8.7
9.1	8.2	7.4	7.0	7.4	8.0	9.0	9.3	8.9
10.1	9.2	8.4	8.3	8.7	9.8	10.8	11.4	10.3
11.7	11.2	10.5	10.0	10.6	11.4	12.9	13.5	12.3
8.8	8.1	7.7	7.6	8.1	8.9	9.9	10.4	9.4
11.9	11.0	10.7	10.4	10.7	11.4	12.4	13.0	12.2
11.5	10.8	10.2	10.1	10.8	11.6	12.5	13.5	12.1
9.3	8.5	8.0	7.7	8.1	8.8	10.3	10.9	9.8
9.4	8.6	8.3	8.1	8.6	9.2	10.4	10.7	9.8

APPENDIX B—Continued

Data Through 1979	Years	Wind—Average Speed, Mph Jan.	Feb.	Mar.	Apr.
Asheville, NC	15	9.8	9.9	9.5	9.0
Cape Hatteras, NC	22	12.6	12.9	12.3	12.3
Charlotte, NC	30	8.0	8.5	8.9	8.9
Greensboro, NC	51	8.2	8.7	9.3	9.0
Raleigh, NC	30	8.6	9.1	9.5	9.1
Wilmington, NC	28	9.3	10.2	10.4	10.6
Bismarck, ND	40	10.1	10.1	11.1	12.4
Fargo, ND	37	12.9	12.8	13.4	14.4
Williston, ND	15	9.9	9.9	10.3	11.4
Akron, OH	31	11.7	11.4	11.6	11.0
Cincinnati, OH	43	8.3	8.4	9.0	8.4
Cleveland, OH	38	12.5	12.2	12.4	11.8
Columbus, OH	30	10.4	10.3	10.8	10.1
Dayton, OH	36	11.8	11.8	12.3	11.8
Mansfield, OH	11	13.4	12.6	12.5	12.3
Toledo, OH	24	11.0	10.8	11.0	10.9
Youngstown, OH	31	11.8	11.6	11.6	11.2
Oklahoma City, OK	31	13.2	13.5	15.0	14.8
Tulsa, OK	31	10.8	11.3	12.7	12.4
Astoria, OR	26	9.2	8.9	9.0	8.6
Burns, OR	8	6.2	6.3	7.4	8.7
Eugene, OR	27	8.3	7.9	8.6	7.7
Medford, OR	30	4.1	4.5	5.3	5.7
Pendleton, OR	29	8.4	8.9	10.1	10.6
Portland, OR	31	10.1	9.0	8.3	7.2
Salem, OR	31	8.5	7.8	8.1	7.2
Sexton Summit, OR	6	12.8	12.5	11.7	10.6
Guam, Pacific	6	8.2	10.2	9.0	8.9
Johnston Island, Pacific	19	14.6	16.2	16.8	16.6
Koror Island, Pacific	19	8.6	8.8	8.6	7.7
Kwajalein Island, Pacific	15	16.7	17.1	16.2	15.3
Majuro, Marshall Is. Pacific	20	12.8	13.8	13.2	12.2
Pago Pago, American Samoa	12	8.5	7.8	7.7	7.5
Ponape Island, Pacific	14	8.6	9.5	8.4	7.3
Truk, Caroline Is. Pacific	19	10.8	11.5	10.8	9.8
Wake Island, Pacific	31	13.6	13.5	14.3	15.7
Yap Island, Pacific	16	10.2	10.6	10.0	9.0
Allentown, PA	30	10.9	11.2	11.7	11.0

May	June	July	Aug.	Sept.	Oct.	Nov.	Dec.	Annual
7.1	6.2	5.9	5.5	5.8	7.0	8.4	9.0	7.8
11.3	10.8	10.3	9.6	10.6	11.3	11.3	11.8	11.4
7.6	6.9	6.6	6.5	6.8	7.1	7.3	7.4	7.5
7.7	7.0	6.5	6.3	6.7	7.1	7.6	7.7	7.6
7.7	7.0	6.7	6.4	6.8	7.2	7.8	8.0	7.8
9.4	8.6	8.1	7.6	8.0	8.3	8.2	8.6	8.9
12.0	10.7	9.4	9.6	10.1	10.1	10.2	9.6	10.4
13.4	12.0	10.8	11.3	12.3	12.9	13.2	12.5	12.6
11.2	10.1	9.3	9.6	10.0	10.2	9.3	9.7	10.1
9.4	8.5	7.6	7.4	8.0	9.1	11.0	11.4	9.8
6.7	6.4	5.2	5.1	5.4	6.1	7.7	7.9	7.1
10.3	9.4	8.7	8.4	9.1	10.1	12.0	12.4	10.8
8.5	7.5	6.7	6.4	6.8	7.7	9.5	9.9	8.7
9.9	9.1	8.1	7.5	8.3	9.1	11.2	11.5	10.2
10.3	9.9	8.4	8.4	9.0	10.6	11.9	12.7	11.0
9.8	8.4	7.5	7.2	7.7	8.7	10.2	10.5	9.5
9.8	8.7	7.9	7.6	8.3	9.4	11.1	11.5	10.0
13.1	12.5	11.1	10.7	11.2	12.1	12.5	12.7	12.7
11.0	10.3	9.2	9.1	9.3	10.0	10.5	10.6	10.6
8.5	8.3	8.4	7.9	7.4	7.5	8.4	9.2	8.4
8.6	7.9	7.7	7.1	7.0	6.4	5.4	5.8	7.0
7.4	7.4	8.0	7.5	7.4	6.6	7.1	7.7	7.6
5.7	5.9	5.7	5.2	4.5	3.7	3.5	3.6	4.8
10.2	10.5	9.5	9.1	8.9	8.1	8.0	8.5	9.2
7.0	7.1	7.5	7.1	6.5	6.5	8.5	9.6	7.9
6.6	6.5	6.5	6.3	6.1	6.2	7.3	8.2	7.1
10.2	11.1	11.8	10.9	11.1	12.0	14.1	13.3	11.8
8.3	6.4	5.1	4.8	4.7	6.2	7.8	9.1	7.4
16.1	16.2	16.0	15.5	14.4	15.0	15.8	17.0	15.9
6.6	6.0	6.7	6.7	7.0	7.4	6.6	7.6	7.4
13.9	12.7	10.8	9.8	9.3	10.3	12.2	16.6	13.4
11.1	9.8	8.5	7.4	7.1	7.6	9.0	12.4	10.4
9.0	11.5	12.0	11.9	11.2	11.7	9.7	8.7	9.8
6.6	5.6	4.9	4.5	4.7	4.8	5.6	7.7	6.5
8.7	7.4	7.4	7.1	7.7	7.9	8.2	9.8	8.9
14.3	12.5	12.8	12.0	12.6	14.1	15.7	14.7	13.8
7.8	6.4	6.1	6.2	6.6	6.5	7.6	9.0	8.0
9.1	8.1	7.2	6.9	7.3	8.3	9.7	10.2	9.3

APPENDIX B—Continued

Data Through 1979	Years	Wind—Average Speed, Mph Jan.	Feb.	Mar.	Apr.
Erie, PA	26	13.5	12.4	12.3	11.7
Harrisburg, PA	40	8.4	9.2	9.7	9.4
Philadelphia, PA	39	10.5	11.1	11.5	11.0
Pittsburgh (Int'l. Airport), PA	27	10.8	10.9	11.1	10.7
Avoca, PA	24	9.0	9.1	9.3	9.6
Williamsport, PA	19	9.1	9.3	9.4	9.2
San Juan, PR	24	9.1	9.3	9.7	9.4
Providence, RI	26	11.5	11.7	12.2	12.3
Charleston, SC	30	9.3	10.1	10.2	9.9
Columbia, SC	31	7.1	7.6	8.3	8.3
Greenville–Spartanburg, SC	17	7.3	8.0	8.0	7.7
Aberdeen, SD	13	11.4	11.6	12.5	13.2
Huron, SD	40	11.6	11.7	12.7	14.0
Rapid City, SD	29	10.7	11.1	12.8	13.4
Sioux Falls, SD	31	11.0	11.1	12.6	13.5
Bristol–Johnson City, TN	25	6.6	6.9	7.4	7.1
Chattanooga, TN	39	7.2	7.6	8.1	7.7
Knoxville, TN	37	8.1	8.5	9.1	8.9
Memphis, TN	31	10.5	10.5	11.3	10.7
Nashville, TN	38	9.2	9.4	10.0	9.4
Oak Ridge, TN	16	4.8	5.0	5.3	5.7
Abilene, TX	35	12.1	12.8	14.2	14.2
Amarillo, TX	38	13.1	14.2	15.6	15.5
Austin, TX	38	9.9	10.2	11.0	10.7
Brownsville, TX	37	11.7	12.3	13.6	14.2
Corpus Christi, TX	37	12.1	12.9	14.0	14.4
Dallas–Fort Worth, TX	26	11.3	12.0	13.1	12.7
Del Rio, TX	16	8.8	9.5	10.9	11.0
El Paso, TX	37	8.9	9.8	11.7	11.8
Galveston, TX	93	11.6	11.8	11.9	12.1
Houston, TX	10	8.4	8.8	9.5	9.2
Lubbock, TX	30	12.3	13.6	15.1	15.1
Midland–Odessa, TX	26	10.2	11.3	12.7	12.9
Port Arthur, TX	26	11.3	11.7	12.1	12.2
San Angelo, TX	30	10.3	10.9	12.4	12.2
San Antonio, TX	37	9.2	9.9	10.6	10.6
Victoria, TX	18	10.8	11.1	11.8	12.0

May	June	July	Aug.	Sept.	Oct.	Nov.	Dec.	Annual
10.2	9.6	9.1	9.1	10.1	11.4	13.1	13.7	11.3
7.7	6.9	6.3	6.0	6.2	6.6	7.9	8.1	7.7
9.7	8.8	8.1	7.9	8.3	8.9	9.6	10.1	9.6
9.2	8.2	7.5	7.1	7.6	8.5	10.0	10.5	9.3
8.6	7.8	7.3	7.1	7.3	7.9	8.5	8.8	8.4
8.0	7.0	6.4	6.0	6.2	6.8	8.1	8.7	7.8
8.8	9.1	10.0	9.3	7.7	7.0	7.6	8.7	8.8
11.0	10.0	9.5	9.3	9.5	9.7	10.5	11.0	10.7
8.9	8.5	8.0	7.4	8.0	8.1	8.1	8.7	8.8
6.9	6.7	6.4	5.9	6.1	6.1	6.4	6.6	6.9
6.8	6.2	5.8	5.5	5.9	6.4	6.5	7.2	6.8
12.4	10.6	9.6	10.3	10.6	11.1	10.9	10.7	11.2
12.7	11.5	10.7	10.9	11.6	11.5	12.0	11.3	11.8
12.5	10.8	10.0	10.2	11.0	11.1	10.9	10.5	11.3
12.0	10.7	9.7	9.8	10.2	10.8	11.6	10.8	11.2
5.3	4.7	4.2	3.8	4.3	4.7	5.8	6.0	5.6
6.1	5.4	5.2	4.7	5.0	5.1	6.2	6.6	6.2
7.2	6.6	6.2	5.7	5.8	5.8	7.1	7.5	7.2
8.9	8.1	7.5	7.0	7.5	7.8	9.3	10.0	9.1
7.6	7.0	6.4	6.1	6.3	6.6	8.4	8.9	7.9
4.5	4.2	3.9	3.7	3.8	3.6	4.1	4.5	4.4
13.2	13.1	10.9	10.5	10.4	11.1	11.7	12.1	12.2
14.8	14.4	12.5	12.1	12.9	13.0	13.1	13.0	13.7
9.8	9.4	8.5	8.0	8.0	8.1	9.1	9.2	9.3
13.4	12.4	11.6	10.4	9.5	9.6	10.8	10.9	11.7
12.9	11.9	11.5	10.8	10.2	10.0	11.4	11.3	12.0
11.1	10.8	9.4	9.0	9.3	9.6	10.6	11.0	10.8
10.7	11.5	10.9	10.2	9.2	9.1	8.5	8.4	9.9
10.9	10.0	8.8	8.3	8.2	8.0	8.4	8.4	9.4
11.5	10.7	9.8	9.4	10.1	10.3	11.2	11.3	11.0
8.0	7.4	6.6	5.5	6.7	6.5	7.7	7.7	7.7
14.4	13.9	11.3	10.0	10.5	11.2	11.7	12.0	12.6
12.6	12.2	10.6	10.0	10.0	10.0	10.1	10.1	11.0
10.5	9.0	7.8	7.4	8.6	8.9	10.3	10.7	10.1
11.4	11.3	9.7	9.2	9.0	9.3	9.9	10.0	10.4
10.2	10.1	9.2	8.5	8.5	8.5	8.9	8.6	9.4
10.8	9.7	8.8	8.3	8.6	8.7	9.7	10.1	10.0

APPENDIX B—Continued

Wind—Average Speed, Mph

Data Through 1979	Years	Jan.	Feb.	Mar.	Apr.
Waco, TX	30	11.9	12.2	13.3	13.2
Wichita Falls, TX	31	11.3	11.9	13.5	13.3
Salt Lake City, UT	50	7.7	8.2	9.3	9.5
Burlington, VT	36	9.6	9.3	9.4	9.3
Lynchburg, VA	23	8.8	8.7	9.3	9.2
Norfolk, VA	31	11.6	12.0	12.3	11.7
Richmond, VA	31	8.0	8.5	8.9	8.8
Roanoke, VA	31	9.8	10.3	10.5	10.1
Olympia, WA	27	7.4	7.3	7.5	7.3
Quillayute, WA	13	7.3	7.2	7.3	6.7
Seattle (Int'l. Airport), WA	31	10.1	9.8	10.1	9.7
Spokane, WA	32	8.7	9.2	9.6	9.8
Walla Walla, WA	47	5.1	5.5	6.2	6.1
Yakima, WA	28	5.8	6.5	8.0	8.6
Beckley, WV	16	11.1	11.2	11.7	10.9
Charleston, WV	32	7.6	7.8	8.4	7.7
Elkins, WV	24	7.3	8.0	8.2	7.9
Huntington, WV	18	7.5	7.6	8.1	7.6
Parkersburg, WV	76	7.2	7.6	7.8	7.3
Green Bay, WI	30	11.2	10.8	11.0	11.6
La Crosse, WI	29	8.6	8.5	9.2	10.4
Madison, WI	33	10.5	10.4	11.2	11.4
Milwaukee, WI	39	12.9	12.7	13.2	13.1
Casper, WY	29	16.6	15.2	14.2	12.8
Cheyenne, WY	22	15.5	15.2	14.9	14.6
Lander, WY	33	6.2	6.2	7.2	8.0
Sheridan, WY	39	7.8	8.0	9.1	9.9

May	June	July	Aug.	Sept.	Oct.	Nov.	Dec.	Annual
12.1	11.8	10.8	9.9	9.5	10.0	10.9	11.3	11.4
12.2	12.1	10.9	10.4	10.4	10.6	11.4	11.3	11.6
9.4	9.3	9.4	9.6	9.1	8.5	7.8	7.5	8.8
8.8	8.3	7.8	7.5	8.1	8.7	9.5	9.8	8.8
7.7	7.0	6.6	6.3	7.0	7.4	8.0	7.9	7.8
10.3	9.6	8.8	8.8	9.6	10.3	10.6	11.0	10.6
7.7	7.2	6.6	6.3	6.5	6.8	7.3	7.5	7.5
8.1	7.1	6.7	6.3	6.3	7.1	8.7	9.1	8.3
6.8	6.6	6.1	5.8	5.6	5.9	6.6	7.6	6.7
6.5	6.2	5.9	5.4	5.4	5.8	6.6	7.2	6.5
9.1	8.9	8.3	8.0	8.2	8.8	9.2	9.9	9.2
9.0	8.9	8.3	8.1	8.1	8.0	8.3	8.8	8.7
5.7	5.5	5.4	5.1	4.7	4.5	4.8	5.2	5.3
8.5	8.3	7.7	7.4	7.4	6.7	5.8	5.3	7.2
9.3	7.9	7.0	6.7	7.5	8.9	10.1	11.1	9.4
6.2	5.6	5.1	4.5	4.8	5.3	6.8	7.2	6.4
6.7	4.9	4.3	4.1	4.4	5.0	6.9	6.9	6.3
6.0	5.4	4.9	4.8	4.9	5.6	6.9	7.4	6.4
6.0	5.4	5.1	4.9	5.1	5.5	6.7	6.7	6.3
10.5	9.4	8.5	8.1	9.2	10.1	11.2	10.8	10.2
9.6	8.3	7.6	7.4	8.2	9.2	9.7	8.7	8.8
10.3	9.2	8.1	8.0	8.7	9.5	10.7	10.2	9.9
11.9	10.5	9.7	9.6	10.6	11.5	12.7	12.5	11.7
11.8	11.1	10.1	10.4	11.0	12.2	14.4	16.2	13.0
13.0	11.7	10.5	10.6	11.3	12.3	13.5	15.0	13.2
7.9	7.8	7.7	7.5	7.0	6.2	5.7	5.9	6.9
9.1	8.1	7.2	7.3	7.5	7.5	7.8	7.7	8.1

APPENDIX C

Summary of "1/7 Power Rule" for Various Heights of Wind Systems

Height in Feet of Tower to be Used (H)	Height in Feet at Which Wind Was Measured (H_0): 15	20	25	30	35	40	45	50
20	1.04	1.00	0.97	0.94	0.92	0.91	0.89	0.88
30	1.10	1.06	1.03	1.00	0.98	0.96	0.94	0.93
40	1.15	1.10	1.07	1.04	1.02	1.00	0.98	0.97
50	1.19	1.14	1.10	1.08	1.05	1.03	1.02	1.00
60	1.22	1.17	1.13	1.10	1.08	1.06	1.04	1.03
70	1.25	1.20	1.16	1.13	1.10	1.08	1.07	1.05
80	1.27	1.22	1.18	1.15	1.13	1.10	1.09	1.07
90	1.29	1.24	1.20	1.17	1.14	1.12	1.11	1.09
100	1.31	1.26	1.22	1.19	1.16	1.14	1.12	1.10

APPENDIX C—Continued

To calculate wind speeds at a height greater than that of the actual wind measurements, look at the intersecting point on the table of the height at which the wind was measured and the height of the tower to be used. Multiply the corresponding factor by the wind speeds actually recorded at the lower height.*

*Based upon the formula:

$$\frac{H}{H_0}^{1/7} = \frac{V}{V_0}$$

Where:

H = Height at which wind speed V is to be estimated
H_0 = Height at which wind speed V_0 was measured
V = Wind speed at height H
V_0 = Wind speed at height H_0

APPENDIX D

Tables for Estimates of Financing a Wind System

Future Value of Electricity Table

To find the value of electricity in a specific future year at a specific percent of increase in rates, look under Column A where that year and that percent intersect. Multiply the corresponding factor by your utility company's present rate per kwh.

To find the cumulative value of electricity generated by a wind system over a specific period of years at a specific percent of increase in rates, look under Column B where the year ending the period and the percent of increase intersect. The figure you'll find will be the accumulated value in interest earned by each dollar saved after the designated period of years. Column C indicates the average dollars saved per year in interest earned over a particular period of years at a specific percent of increase in utility rates. (See Ch. 2 for examples.)

APPENDIX D—Continued

	1%			2%		
Year	A	B	C	A	B	C
1	1.00	1.00	1.00	1.00	1.00	1.00
2	1.01	2.01	1.01	1.02	2.02	1.01
3	1.02	3.03	1.01	1.04	3.06	1.02
4	1.03	4.06	1.02	1.06	4.12	1.03
5	1.04	5.10	1.02	1.08	5.20	1.04
6	1.05	6.15	1.03	1.10	6.31	1.06
7	1.06	7.21	1.03	1.13	7.43	1.06
8	1.07	8.29	1.04	1.15	8.58	1.07
9	1.08	9.37	1.04	1.17	9.75	1.08
10	1.09	10.46	1.05	1.20	10.95	1.09
11	1.10	11.57	1.05	1.22	12.17	1.11
12	1.12	12.08	1.06	1.24	13.41	1.12
13	1.13	13.81	1.06	1.27	14.68	1.13
14	1.14	14.95	1.07	1.29	15.97	1.14
15	1.15	16.10	1.07	1.32	17.29	1.15
16	1.16	17.26	1.08	1.35	18.64	1.16
17	1.17	18.43	1.08	1.37	20.01	1.18
18	1.18	19.61	1.09	1.40	21.41	1.19
19	1.20	20.81	1.10	1.43	22.84	1.20
20	1.21	22.02	1.10	1.46	24.30	1.21
21	1.22	23.24	1.11	1.49	25.78	1.23
22	1.23	24.47	1.11	1.52	27.30	1.24
23	1.24	25.72	1.12	1.55	28.84	1.25
24	1.26	26.97	1.12	1.58	30.42	1.27
25	1.27	28.24	1.13	1.61	32.03	1.28
30	1.33	34.78	1.16	1.78	40.57	1.35

APPENDIX D—Continued

Year	3% A	3% B	3% C	4% A	4% B	4% C
1	1.00	1.00	1.00	1.00	1.00	1.00
2	1.03	2.03	1.02	1.04	2.04	1.02
3	1.06	3.09	1.03	1.08	3.12	1.04
4	1.09	4.18	1.05	1.12	4.25	1.06
5	1.13	5.31	1.06	1.17	5.42	1.08
6	1.16	6.47	1.08	1.22	6.63	1.11
7	1.19	7.66	1.09	1.27	7.90	1.13
8	1.23	8.89	1.11	1.32	9.21	1.15
9	1.27	10.16	1.13	1.37	10.58	1.18
10	1.30	11.46	1.15	1.42	12.01	1.20
11	1.34	12.81	1.16	1.48	13.49	1.23
12	1.38	14.19	1.18	1.54	15.03	1.25
13	1.43	15.62	1.20	1.60	16.63	1.28
14	1.47	17.07	1.22	1.67	18.29	1.31
15	1.51	18.60	1.24	1.73	20.02	1.33
16	1.58	20.16	1.26	1.80	21.82	1.36
17	1.60	21.76	1.21	1.87	23.70	1.39
18	1.65	23.41	1.30	1.95	25.65	1.42
19	1.70	25.12	1.32	2.03	27.67	1.46
20	1.75	26.87	1.34	2.11	29.78	1.49
21	1.81	28.68	1.37	2.19	31.97	1.52
22	1.86	30.54	1.39	2.28	34.25	1.56
23	1.92	32.45	1.41	2.37	36.62	1.59
24	1.97	34.43	1.43	2.46	39.08	1.63
25	2.03	36.45	1.46	2.56	41.66	1.67
30	2.36	47.58	1.59	3.24	56.08	1.87

APPENDIX D—Continued

	5%			6%		
Year	A	B	C	A	B	C
1	1.00	1.00	1.00	1.00	1.00	1.00
2	1.05	2.05	1.03	1.06	2.06	1.03
3	1.10	3.15	1.05	1.12	3.18	1.06
4	1.16	4.31	1.08	1.19	4.37	1.09
5	1.22	5.53	1.11	1.26	5.64	1.13
6	1.28	6.80	1.13	1.34	6.98	1.16
7	1.34	8.14	1.16	1.42	8.39	1.20
8	1.41	9.55	1.19	1.50	9.90	1.24
9	1.49	11.03	1.23	1.59	11.49	1.28
10	1.55	12.58	1.26	1.69	13.18	1.32
11	1.63	14.21	1.29	1.79	14.97	1.36
12	1.71	15.92	1.33	1.90	16.87	1.41
13	1.80	17.71	1.36	2.01	18.88	1.45
14	1.89	19.60	1.40	2.13	21.02	1.50
15	1.98	21.58	1.44	2.26	23.28	1.55
16	2.08	23.66	1.48	2.40	25.67	1.60
17	2.18	25.84	1.52	2.54	28.21	1.66
18	2.29	28.13	1.56	2.69	30.90	1.72
19	2.41	30.54	1.61	2.85	33.76	1.78
20	2.53	33.07	1.65	3.03	36.79	1.84
21	2.65	35.72	1.70	3.21	39.99	1.90
22	2.79	38.51	1.75	3.40	43.39	1.97
23	2.93	41.43	1.80	3.60	47.00	2.04
24	3.07	44.50	1.85	3.82	50.82	2.12
25	3.23	47.73	1.91	4.05	54.86	2.19
30	4.12	66.44	2.21	5.42	79.06	2.64

APPENDIX D—Continued

	7%			8%		
Year	A	B	C	A	B	C
1	1.00	1.00	1.00	1.00	1.00	1.00
2	1.07	2.07	1.04	1.08	2.08	1.04
3	1.14	3.21	1.07	1.17	3.25	1.08
4	1.23	4.44	1.11	1.26	4.51	1.13
5	1.31	5.75	1.19	1.36	5.87	1.17
6	1.40	7.15	1.19	1.47	7.36	1.23
7	1.50	8.65	1.24	1.59	8.92	1.27
8	1.61	10.26	1.28	1.71	10.64	1.33
9	1.72	11.98	1.32	1.85	12.49	1.39
10	1.84	13.82	1.38	2.00	14.49	1.45
11	1.97	15.78	1.43	2.16	16.65	1.51
12	2.10	17.89	1.49	2.33	18.98	1.58
13	2.25	20.14	1.55	2.52	21.50	1.65
14	2.41	22.55	1.61	2.72	24.21	1.73
15	2.58	25.13	1.68	2.94	27.15	1.81
16	2.76	27.88	1.74	3.17	30.32	1.90
17	2.95	30.84	1.81	3.43	33.75	1.99
18	3.16	34.00	1.89	3.70	37.45	2.08
19	3.38	37.38	1.97	4.00	41.45	2.18
20	3.62	41.00	2.05	4.32	45.76	2.29
21	3.87	44.87	2.14	4.66	50.42	2.40
22	4.14	49.01	2.23	5.03	55.46	2.52
23	4.43	53.44	2.32	5.44	60.89	2.65
24	4.74	58.18	2.42	5.87	66.76	2.78
25	5.07	63.25	2.53	6.34	73.11	2.92
30	7.11	94.46	3.15	9.32	113.28	3.78

APPENDIX D—Continued

	9%			10%		
Year	A	B	C	A	B	C
1	1.00	1.00	1.00	1.00	1.00	1.00
2	1.09	2.09	1.05	1.10	2.10	1.05
3	1.19	3.28	1.09	1.21	3.31	1.10
4	1.30	4.57	1.14	1.33	4.64	1.16
5	1.41	5.98	1.20	1.46	6.11	1.22
6	1.54	7.52	1.25	1.61	7.72	1.29
7	1.68	9.20	1.31	1.77	9.49	1.36
8	1.83	11.03	1.38	1.95	11.44	1.43
9	1.99	13.02	1.45	2.14	13.58	1.50
10	2.17	15.19	1.52	2.36	15.94	1.59
11	2.37	17.56	1.60	2.59	18.53	1.68
12	2.58	20.14	1.68	2.85	21.38	1.78
13	2.81	22.95	1.77	3.14	24.52	1.89
14	3.07	26.02	1.86	3.45	27.97	2.00
15	3.34	29.36	1.96	3.80	31.77	2.12
16	3.67	33.02	2.06	4.18	35.95	2.25
17	3.97	36.99	2.18	4.59	40.54	2.38
18	4.32	41.31	2.30	5.05	45.60	2.53
19	4.71	46.03	2.42	5.56	51.16	2.69
20	5.14	51.17	2.56	6.12	57.28	2.86
21	5.51	56.68	2.70	6.73	64.00	3.05
22	6.10	62.78	2.85	7.40	71.40	3.23
23	6.65	64.43	2.80	8.14	79.54	3.46
24	7.25	76.68	3.20	8.95	88.50	3.69
25	7.91	84.58	3.38	9.85	98.34	3.93
30	12.16	136.14	4.54	14.92	163.55	5.45

APPENDIX D—Continued

	11%			12%		
Year	**A**	**B**	**C**	**A**	**B**	**C**
1	1.00	1.00	1.00	1.00	1.00	1.00
2	1.11	2.11	1.06	1.12	2.12	1.06
3	1.23	3.34	1.11	1.25	3.37	1.12
4	1.37	4.71	1.18	1.40	4.78	1.19
5	1.52	6.23	1.25	1.57	6.35	1.27
6	1.69	7.91	1.32	1.76	8.12	1.35
7	1.87	9.78	1.40	1.97	10.09	1.44
8	2.08	11.86	1.48	2.21	12.30	1.54
9	2.30	14.16	1.57	2.48	14.78	1.64
10	2.56	16.72	1.67	2.77	17.55	1.75
11	2.84	19.56	1.78	3.11	20.65	1.88
12	3.15	22.71	1.89	3.48	24.13	2.01
13	3.50	26.21	2.02	3.90	28.03	2.16
14	3.88	30.10	2.15	4.36	32.39	2.31
15	4.31	34.41	2.29	4.89	37.28	2.49
16	4.78	39.19	2.45	5.47	42.75	2.67
17	5.31	44.50	2.62	6.13	48.88	2.66
18	5.90	50.40	2.80	6.87	55.75	3.10
19	6.54	56.94	3.00	7.69	63.44	3.34
20	7.26	64.20	3.21	8.61	72.05	3.60
21	8.06	72.37	3.44	9.65	81.70	3.89
22	8.95	81.21	3.69	10.80	92.50	4.20
23	9.93	91.15	3.96	12.10	104.60	4.55
24	11.03	102.17	4.26	13.55	118.16	4.92
25	12.24	114.41	4.58	15.18	133.33	5.33
30	20.62	199.02	6.63	26.75	217.45	7.25

APPENDIX D—Continued

	13%			14%		
Year	A	B	C	A	B	C
1	1.00	1.00	1.00	1.00	1.00	1.00
2	1.13	2.13	1.07	1.14	2.14	1.07
3	1.28	3.41	1.14	1.30	3.44	1.15
4	1.44	4.85	1.21	1.48	4.92	1.23
5	1.63	6.48	1.30	1.69	6.10	1.32
6	1.84	8.32	1.39	1.93	8.54	1.42
7	2.08	10.40	1.49	2.14	10.73	1.53
8	2.35	12.76	1.59	2.50	13.23	1.65
9	2.66	15.42	1.71	2.85	16.09	1.79
10	3.00	18.42	1.84	3.25	19.34	1.93
11	3.39	21.81	1.98	3.71	23.04	2.09
12	3.84	25.65	2.14	4.23	27.27	2.27
13	4.33	29.98	2.31	4.82	32.09	2.47
14	4.90	34.88	2.49	5.49	37.58	2.68
15	5.53	40.42	2.69	6.26	43.84	2.92
16	6.25	46.67	2.92	7.13	50.98	3.19
17	7.07	53.74	3.16	8.14	59.12	3.48
18	7.99	61.73	3.43	9.28	68.39	3.80
19	9.02	70.75	3.72	10.58	78.97	4.16
20	10.20	80.95	4.05	12.06	91.53	4.58
21	11.52	92.47	4.40	13.74	105.27	5.01
22	13.02	105.49	4.89	15.67	120.94	5.50
23	14.71	120.20	5.23	17.86	138.80	6.03
24	16.63	136.83	5.70	20.36	159.16	6.63
25	18.79	155.62	6.22	23.21	182.37	7.29
30	34.62	243.20	9.77	44.69	357.28	11.91

APPENDIX D—Continued

Loan Repayment Table

To calculate the amount of principal and interest paid upon a loan for a wind power system, look on the table for the intersection of the number of years for the loan and its percent of interest. Multiply the corresponding factor by the amount of the loan. The resulting figure will be the total principal and interest to be paid upon the loan during each year.

Percent of Interest

Number of Years	1%	2%	3%	4%	5%	6%	7%
1	1.010	1.020	1.030	1.040	1.050	1.060	1.070
2	.508	.515	.523	.530	.538	.545	.553
3	.340	.347	.354	.360	.367	.374	.381
4	.256	.263	.269	.275	.282	.289	.295
5	.206	.212	.218	.225	.231	.237	.244
6	.173	.179	.185	.191	.197	.203	.210
7	.149	.155	.161	.167	.173	.179	.186
8	.131	.137	.142	.149	.155	.161	.167
9	.117	.123	.128	.134	.141	.147	.153
10	.106	.111	.117	.123	.130	.136	.142
11	.096	.102	.108	.114	.120	.127	.133
12	.089	.095	.105	.107	.113	.119	.126
13	.082	.088	.094	.100	.106	.113	.120
14	.077	.083	.089	.095	.101	.108	.114
15	.0072	.078	.084	.090	.096	.103	.110

8%	9%	10%	11%	12%	13%	14%
1.080	1.090	1.100	1.110	1.120	1.130	1.140
.561	.568	.576	.584	.592	.599	.607
.388	.395	.402	.409	.416	.424	.431
.302	.309	.315	.322	.329	.336	.343
.250	.257	.264	.271	.277	.284	.291
.216	.223	.230	.236	.243	.250	.257
.192	.199	.205	.212	.219	.226	.233
.174	.181	.187	.194	.201	.208	.216
.160	.167	.174	.181	.188	.195	.202
.149	.156	.163	.170	.177	.184	.192
.140	.147	.154	.161	.168	.176	.183
.133	.140	.147	.154	.161	.169	.177
.127	.134	.141	.148	.156	.163	.171
.121	.128	.136	.143	.151	.159	.167
.117	.124	.131	.139	.147	.155	.163

APPENDIX E

Characteristics of Manufacturers' Wind System Models

	Rotor Size		Power Rating	Speed Range, Mph	
Manufacturer/ Model	Rotor Dia-meter, Ft.	Rotor Area, Sq. Ft.	Rated Power, Kw., at Wind Speed, Mph, at Rotor Speed, Rpm	Cut-in, Cut-out, Survival	Number of Blades
Alcoa/ "634214"	42.0	1,385	57.0 kw. 30 mph 64 rpm	12 60 150	3
Mehrkam/ "445"	40.0	1,257	37.0 kw. 33 mph 90 rpm	5 40 120	6
Mehrkam/ "440"	38.0	1,134	20.0 kw. 25 mph 60 rpm	5 40 120	6
Wind Engineering/ "Windgen 25"	38.0	1,134	25.0 kw. 25 mph 80 rpm	15 50 . . .	3
D.A.F./ "VAWT 25 × 55"	36.7	1,307	37.0 kw. 33 mph 90 rpm	13 60 130	2
Windworks/ "10KW"	33.0	855	10.0 kw. 22 mph . . .	8	3
Tumac/ "10 Meter"	32.8	845	20.0 kw.	. . . .	3

NOTES: This information is based upon manufacturer's published data and may be subject to change. For current information, contact the wind system manufacturer.

Some wind power systems have not been included in this appendix because the relevant information was not available at the time of publication.

Rotor	Gears	Generator	Weight, Lbs.	
Type,* Material, Speed Control	**Gear Ratio**	**Type, Voltage(s)**	**Rotor**	**System**
VAWT-Darrieus Aluminum Brake	. . .	Induction . . .	. . .	. . .
HAWT-Down-wind Aluminum Brake	20:1	Alternator 120, 240 volts d.c.	1,250	9,500
HAWT-Down-wind Aluminum Brake	20:1	. . . . . .	925	7,000
HAWT-Up-wind . . . Flyball	22.5:1	Induction 220 volts a.c.	160	. . .
VAWT-Darrieus Aluminum Spoiler flaps	. . .	Induction . . .	7,000	8,000
HAWT-Down-wind Composite Coning	1:1	Alternator 120, 240 volts d.c.	. . .	. . .
VAWT-Darrieus	. . .	. . . . . .	. . .	. . .

* HAWT = Horizontal-Axis Wind Turbine
VAWT = Vertical-Axis Wind Turbine

APPENDIX E—Continued

	Rotor Size		Power Rating	Speed Range, Mph	
Manufacturer/ Model	Rotor Diameter, Ft.	Rotor Area, Sq. Ft.	Rated Power, Kw., at Wind Speed, Mph, at Rotor Speed, Rpm	Cut-in, Cut-out, Survival	Number of Blades
Wind Power/ "Storm Master 10"	32.8	845	18.0 kw. 24 mph . . .	9 150 160	3
Grumman/ "Windstream 33"	32.3	819	15.0 kw. 24 mph . . .	9	3
Jay Carter/ "Mod 25"	32.0	804	25.0 kw. 25 mph . . .	7.5	2
Alcoa/ "452011"	30.0	707	22.0 kw.	. . . .	3
Dakota/ "10 KW"	28.0	616	10.0 kw. 20 mph . . .	5	3
Astral/Wilcon/ "AW10-B"	26.0	531	10.0 kw. 22 mph 225 rpm	. . . .	3
Millville "10-3-IND"	25.0	491	10.0 kw. 25 mph 80 rpm	9 60 80	3
Env. Energies "HWT 15"	24.0	452	15.0 kw. 25 mph . . .	8.7	8
Env. Energies/ "HWT 12"	22.0	380	12.0 kw. 25 mph . . .	8.7	8
Elektro/ "WV120G"	21.7	370	10.0 kw. 31 mph . . .	7	3
D.A.F./ "VAWT 20 × 30"	20.0	389	14.0 kw. 30 mph . . .	12 60 130	2
Env. Energies/ "HWT 9"	20.0	314	9.0 kw. 25 mph . . .	8.7	8
Product Development/ "4500"	20.0	314	4.0 kw. 20 mph . . .	9	3
Whirlwind/ "Model AA"	19.5	299	4.0 kw. 22 mph . . .	8	2

Rotor	Gears	Generator	Weight, Lbs.	
Type,* Material, Speed Control	**Gear Ratio**	**Type, Voltage(s)**	**Rotor**	**System**
HAWT-Down-wind Composite Blade stall	. . .	Induction 120 volts a.c.	295	875
HAWT	. . .	Alternator 240, 480 volts d.c.	. . .	. . .
HAWT-Down-wind Composite Flyball, Blade stall, Brake	. . .	Induction 220, 440 volts a.c.	. . .	. . .
VAWT-Darrieus Aluminum Brake	. . .	Induction . . .	. . .	. . .
HAWT-Up-wind	1:1	Alternator 160 volts d.c.	. . .	. . .
HAWT-Up-wind Composite . . .	25:1	Alternator 220 volts d.c.	. . .	. . .
HAWT-Up-wind Aluminum Flyball	24:1	Induction 220 volts a.c.	230	850
HAWT	. . .	. . . . . .	. . .	. . .
HAWT	. . .	. . . . . .	. . .	. . .
HAWT-Up-wind . . . Flyball	. . .	. . . . . .	. . .	. . .
VAWT-Darrieus Aluminum Spoiler flaps	. . .	Induction . . .	800	1,000
HAWT	. . .	. . . . . .	. . .	. . .
HAWT-Up-wind	. . .	Alternator 220 volts d.c.	. . .	. . .
HAWT	. . .	Alternator 32, 120, 240 volts d.c.	. . .	. . .

* HAWT = Horizontal-Axis Wind Turbine
VAWT = Vertical-Axis Wind Turbine

APPENDIX E—Continued

	Rotor Size		Power Rating	Speed Range, Mph	
Manufacturer/ Model	**Rotor Diameter, Ft.**	**Rotor Area, Sq. Ft.**	**Rated Power, Kw., at Wind Speed, Mph, at Rotor Speed, Rpm**	**Cut-in, Cut-out, Survival**	**Number of Blades**
Bertoia/ "A.P.S."	18.0	254	2.0 kw. 18 mph 150 rpm	5 70 . . .	3
Env. Energies/ "HWT 6"	18.0	254	6.0 kw. 25 mph . . .	8.7	8
Elektro/ "WV50G"	16.4	211	5.4 kw. 27 mph . . .	7	3
North Wind/ "HR2"	16.4	211	2.2 kw. 20 mph 250 rpm	8 . . . 105	3
Tumac/ "5 Meter"	16.4	211	5.0 kw.	. . . .	3
Kedco/ "1620"	16.0	201	3.0 kw. 25 mph 250 rpm	11 60 110	3
Kedco/ "1610"	16.0	201	2.0 kw. 22 mph 250 rpm	10 60 110	3
Kedco/ "1605"	16.0	201	1.9 kw. 20 mph 250 rpm	7 60 110	3
Kedco/ "1600"	16.0	201	1.2 kw. 17 mph 250 rpm	7 60 110	3
American/ "16 Ft."	15.3	183	2.0 kw. 20 mph 90 rpm	10 35 100	48
Dynergy/ "5 Meter"	15.0	150	3.2 kw. 24 mph 200 rpm	10 . . . 100	3
Independent/ "Skyhawk IV"	15.0	177	4.0 kw. 27 mph . . .	8	3
Pinson/ "Cycloturbine C2E3"	15.0	120	5.0 kw. 30 mph . . .	8 30 . . .	3
Elektro/ "WV35G"	14.5	165	3.8 kw. 27 mph . . .	7	3

Rotor	Gears	Generator	Weight, Lbs.	
Type,* Material, Speed Control	**Gear Ratio**	**Type, Voltage(s)**	**Rotor**	**System**
HAWT-Down-wind Aluminum . . .	1:1	Alternator . . .	300	700
HAWT	. . .	. . . . . .	. . .	. . .
HAWT-Down-wind . . . Flyball	. . .	. . . . . .	. . .	. . .
HAWT-Up-wind Wood Tilt-up	1:1	Alternator 24, 32, 48, 110 volts d.c.	. . .	785
VAWT-Darrieus	. . .	. . . . . .	. . .	. . .
HAWT-Down-wind Aluminum Flyball	8.76:1	Alternator 120 volts d.c.	75.5	293
HAWT-Down-wind Aluminum Flyball	8.76:1	Alternator 120 volts d.c.	75.5	267
HAWT-Down-wind Aluminum Flyball	8.76:1	Alternator 120 volts d.c.	75.5	217
HAWT-Down-wind Aluminum Flyball	8.76:1	Alternator 14 volts d.c.	75.5	217
HAWT-Up-wind Aluminum . . .	30:1	Alternator . . .	135	420
VAWT-Darrieus Aluminum Blade stall, Brake	. . .	. . . . . .	312	850
HAWT-Up-wind Wood Blade-actuated	1:1	D.C. Generator 32, 120, 220 volts d.c.	75	600
VAWT-H-rotor . . . Centrifugal	. . .	. . . . . .	. . .	. . .
HAWT-Up-wind	. . .	Alternator . . .	. . .	. . .

* HAWT = Horizontal-Axis Wind Turbine
VAWT = Vertical-Axis Wind Turbine

	Rotor Size		Power Rating	Speed Range, Mph	
Manufacturer/ Model	**Rotor Diameter, Ft.**	**Rotor Area, Sq. Ft.**	**Rated Power, Kw., at Wind Speed, Mph, at Rotor Speed, Rpm**	**Cut-in, Cut-out, Survival**	**Number of Blades**
Dakota/ "BC4"	14.0	154	4.0 kw. 27 mph 300 rpm	8 40 90	3
Power Group/ "Hummingbird"	14.0	154	4.0 kw. 22 mph . . .	8	3
Wind Titan	14.0	154	1.2 kw. 25 mph . . .	7.5	2
Independent/ "Sky Hawk II"	13.6	145	2.0 kw. 27 mph . . .	8	3
Dunlite/ "2000"	13.5	143	2.0 kw. 25 mph 185 rpm	8 . . . 110	3
Enertech/ "1800"	13.2	137	1.5 kw. 21 mph 170 rpm	8 40 120	3
Aero Power/ "SL 1500"	12.0	113	1.4 kw 25 mph 500 rpm	6 100 100	3
Kedco/ "1210"	12.0	113	2.0 kw. 26 mph 300 rpm	11 70 110	3
Kedco/ "1205"	12.0	113	1.2 kw. 22 mph 300 rpm	8 70 110	3
Kedco/ "1200"	12.0	113	1.2 kw. 22 mph 300 rpm	7 70 110	3
Sencenbaugh/ "1000"	12.0	113	1.0 kw. 23 mph 290 rpm	6 60 80	3
Altos/ "BWP-12B"	11.5	100	2.0 kw. 28 mph 116 rpm	8 60 90	24
American/ "12 Ft."	11.5	100	1.0 kw. 20 mph 120 rpm	. . . .	36
Dunlite/ "High Winds"	10.0	79	1.0 kw. 37 mph . . .	14 . . . 110	3

Rotor	Gears	Generator	Weight, Lbs.	
Type,* Material, Speed Control	**Gear Ratio**	**Type, Voltage(s)**	**Rotor**	**System**
HAWT-Up-wind Wood Blade-actuated	1:1	D.C. Generator 12 volts d.c.	75	600
HAWT-Up-wind Composite Side-tilt	1:1	Alternator 120, 220 volts d.c.	25	500
HAWT	. . .	. . . . . .	. . .	. . .
HAWT-Up-wind Wood Blade-actuated	1:1	D.C. Generator 24, 32, 48, 120, 220 volts d.c.	75	485
HAWT-Up-wind Steel Flyball	. . .	Alternator 12, 24, 32, 48, 120 volts d.c.	130	500
HAWT-Down-wind Wood Brake, Tip brakes	11.4:1	Induction 120 volts a.c.	48	185
HAWT-Up-wind Wood Blade-actuated	. . .	Alternator 12, 24, 32, 48, 120 volts d.c.	50	160
HAWT-Down-wind Aluminum Flyball	8.76:1	Alternator 120 volts d.c.	71	252
HAWT-Down-wind Aluminum Flyball	8.76:1	Alternator 28 volts d.c.	71	202
HAWT-Down-wind Aluminum Flyball	8.76:1	Alternator 14 volts d.c.	71	202
HAWT-Up-wind Wood Side-tilt	3:1	Alternator 14, 28 volts d.c.	16	180
HAWT-Up-wind Aluminum Side-tilt	17:1	Alternator 120, 220 volts d.c.	111	300
HAWT-Up-wind Aluminum Side-tilt	30:1	. . . . . .	92	320
HAWT-Up-wind Steel . . .	5:1	Alternator 12, 24, 32, 48, 120 volts d.c.	130	500

* HAWT = Horizontal-Axis Wind Turbine
VAWT = Vertical-Axis Wind Turbine

APPENDIX E—Continued

	Rotor Size		Power Rating	Speed Range, Mph	
Manufacturer/ Model	**Rotor Diameter, Ft.**	**Rotor Area, Sq. Ft.**	**Rated Power, Kw., at Wind Speed, Mph, at Rotor Speed, Rpm**	**Cut-in, Cut-out, Survival**	**Number of Blades**
Elektro/ "WV15G"	10.0	79	1.2 kw. 27 mph . . .	7	2
TWR	10.0	79	0.5 kw. 25 mph . . .	8.5	2
Whirlwind/ "Model A"	10.0	79	2.0 kw. 25 mph 900 rpm	8 50 80	2
Winflo/ "Mod 2000"	10.0	79	2.0 kw. 25 mph . . .	7	2
Aerolectric/ "Wind Wizard"	9.0	64	0.6 kw. 26 mph 377 rpm	9 40 90	3
Bergey/ "650"	8.5	57	.65 kw. 25 mph . . .	10 40 120	2
Elektro/ "WV05G"	8.2	53	0.5 kw. 19 mph . . .	7	2
Altos/ "BWP-8A"	8.0	50	1.5 kw. 28 mph 165 rpm	8 40 100	24
Dragonfly	8.0	50	0.24 kw. 20 mph . . .	10	4
American/ "8 Ft."	7.6	46	0.5 kw. 20 mph 150 rpm	10 35 120	24
Sencenbaugh/ "400"	7.0	39	0.4 kw. 20 mph 1,000 rpm	9 60 80	3
Megatech/ "W1P-A1"	6.0	28	0.4 kw. 27 mph . . .	6	2
Sencenbaugh/ "500"	6.0	28	0.5 kw. 24 mph 1,000 rpm	10 60 120	3
Winco/ "1222H"	6.0	28	0.2 kw. 23 mph 900 rpm	7 70 70	2

Rotor	Gears	Generator	Weight, Lbs.	
Type,* Matérial, Speed Control	**Gear Ratio**	**Type, Voltage(s)**	**Rotor**	**System**
HAWT-Up-wind	. . .	. . . . . .	. . .	. . .
. . . .	. . .	. . . . . .	. . .	. . .
HAWT-Down-wind Wood Brake	1:1	Alternator 12, 24, 32, 48, 120 volts d.c.	. . .	71
. . . .	. . .	. . . . . .	. . .	. . .
HAWT-Up-wind Wood Side-tilt	. . .	Alternator 24 volt d.c.	22	50
HAWT-Up-wind Aluminum Side-tilt	1:1	Alternator 12, 24, 36, 48, 120 volts d.c.	. . .	. . .
HAWT-Up-wind	. . .	. . . . . .	. . .	. . .
HAWT-Up-wind Aluminum Side-tilt	11:1	Alternator 24 volts d.c.	59	250
. . . .	. . .	. . . . . .	. . .	. . .
HAWT-Up-wind Aluminum Side-tilt	. . .	. . . . . .	50	. . .
HAWT-Up-wind Wood Side-tilt	1:1	Alternator 14 volts d.c.	30	65
HAWT-Up-wind Wood . . .	. . .	. . . . . .	. . .	. . .
HAWT-Up-wind Wood Side-tilt	1:1	Alternator 14, 24 volts d.c.	9	140
HAWT-Up-wind Wood Spoiler flaps	1:1	D.C. Generator 12, 24 volts d.c.	20	134

* HAWT = Horizontal-Axis Wind Turbine
VAWT = Vertical-Axis Wind Turbine

APPENDIX E—Continued

	Rotor Size		Power Rating	Speed Range, Mph	
Manufacturer/ Model	**Rotor Dia-meter, Ft.**	**Rotor Area, Sq. Ft.**	**Rated Power, Kw., at Wind Speed, Mph, at Rotor Speed, Rpm**	**Cut-in, Cut-out, Survival**	**Number of Blades**
Winflo/ "Windgen Mod 500"	6.0	28	0.5 kw. 25 mph . . .	7	3
Winflo/ "Windgen Mod 30"	3.0	7	0.03 kw. 25 mph . . .	. . . .	10
Zephyr/ "Tetrahelix"	2.0	3	0.007 kw. 25 mph . . .	12 . . . 65	2
Sencenbaugh/ "24–14"	1.67	2	0.0024 kw. 21 mph 3,200 rpm	8 . . . 100	3

Rotor	Gears	Generator	Weight, Lbs.	
Type,* Material, Speed Control	**Gear Ratio**	**Type, Voltage(s)**	**Rotor**	**System**
. . . .	. . .	. . . . . .	. . .	. . .
. . . .	. . .	. . . . . .	. . .	. . .
VAWT	1:1	D.C. Generator 14 volts d.c.	3	5
VAWT-Up-wind Aluminum . . .	. . .	. . . 14 volts d.c.	. . .	18

* HAWT = Horizontal-Axis Wind Turbine
VAWT = Vertical-Axis Wind Turbine

APPENDIX F

Manufacturers' Estimated Energy Output for Their Wind Power Systems

Kilowatt Hours per Month

Manufacturer/Model	Rotor Diameter, Feet	Average Wind Speed, Mph 7	8	9	10
Alcoa/"634214"	42	. . .	. . .	. . .	. . .
Mehrkam/"445"	40	. . .	. . .	. . .	5,000
Mehrkam/"440"	38	. . .	. . .	. . .	4,333
Windworks/"10KW"	33	. . .	1,250	. . .	1,667
Wind Power/"Storm Master 10"	32.8	. . .	. . .	. . .	1,080
Grumman/"Windstream 33"	32.3	. . .	. . .	. . .	. . .
Jay Carter/"Mod 25"	32	. . .	. . .	. . .	2,160
Alcoa/"452011"	30	. . .	. . .	. . .	. . .
Alcoa/"271806"	30	. . .	. . .	. . .	. . .
Dakota Wind/"10 KW"	28	. . .	. . .	. . .	. . .
Astral/Wilcon/"AW10-B"	26	. . .	. . .	. . .	1,333
Millville/"10-3-IND"	25	. . .	. . .	. . .	504
Env. Energies/"HWT 15"	24	. . .	. . .	. . .	. . .

NOTES: This information is based upon manufacturer's published data and may be subject to change. For current information, contact the wind system manufacturer.

Some wind power systems have not been included in this appendix because the relevant information was not available at the time of publication.

11	12	13	14	15	16	17	18
. . .	4,167	. . .	. . .	8,333	. . .	. . .	13,333
. . .	8,750	. . .	. . .	. . .	. . .	. . .	. . .
. . .	7,853	. . .	. . .	. . .	. . .	. . .	. . .
. . .	2,500	. . .	. . .	3,333	4,167	. . .	. . .
. . .	1,667	. . .	. . .	2,580	. . .	. . .	. . .
. . .	2,667	. . .	. . .	. . .	. . .	. . .	. . .
. . .	3,528	. . .	. . .	5,400	. . .	. . .	. . .
. . .	2,083	. . .	. . .	3,750	. . .	. . .	5,833
. . .	667	. . .	. . .	1,333	. . .	. . .	2,083
1,000	. . .	. . .	4,000	. . .	. . .	. . .	8,000
1,583	1,916	2,167	2,667	. . .	. . .	. . .	. . .
. . .	612	. . .	. . .	1,584	. . .	. . .	. . .
. . .	1,956	. . .	. . .	. . .	. . .	. . .	. . .

APPENDIX F—Continued

Kilowatt Hours per Month

Manufacturer/Model	Rotor Diameter, Feet	Average Wind Speed, Mph 7	8	9	10
Env. Energies/"HWT 12"	22	. . .	. . .	. . .	. . .
Env. Energies/"HWT 9"	20	. . .	. . .	. . .	. . .
Product Development/"4500"	20	. . .	. . .	. . .	632
Enertech/"4000"	19.7	. . .	270	. . .	540
Whirlwind/"Model AA"	19.5	. . .	. . .	. . .	325–600
Bertoia/"A.P.S."	18	100–300	. . .	. . .	. . .
Env. Energies/"HWT 6"	18	. . .	. . .	. . .	. . .
Elektro/"WV50G"	16.4	. . .	200–300	. . .	. . .
North Wind/"HR2"	16.4	. . .	150	. . .	370
Dynergy/"5 Meter"	15	. . .	. . .	. . .	212
Independent/"Sky Hawk IV"	15	. . .	. . .	. . .	300–500
Pinson/"Cycloturbine C2E3"	15	. . .	. . .	. . .	250
Dakota/"BC4"	14	. . .	. . .	. . .	. . .
Independent/"Sky Hawk II"	13.6	. . .	. . .	. . .	150–300
Dunlite/"2000"	13.5	. . .	80–125	. . .	. . .
Enertech/"1800"	13.2	. . .	120	. . .	240
Aero Power/"SL 1500"	12	. . .	. . .	. . .	157
Sencenbaugh/"1000"	12	. . .	. . .	. . .	125
Whirlwind/"Model A"	10	. . .	. . .	. . .	130–240
Winco/"1222H"	6	. . .	. . .	. . .	20

11	12	13	14	15	16	17	18
. . .	1,565	. . .	. . .	. . .	. . .	. . .	. . .
. . .	1,174	. . .	. . .	. . .	. . .	. . .	. . .
. . .	964	1,125	. . .	. . .	. . .	. . .	. . .
. . .	800	. . .	1,000	. . .	. . .	. . .	. . .
. . .	525–975	. . .	775–1375	. . .	. . .	. . .	. . .
. . .	. . .	. . .	. . .	. . .	. . .	. . .	. . .
. . .	783	. . .	. . .	. . .	. . .	. . .	. . .
. . .	400–620	. . .	. . .	. . .	600–930	. . .	. . .
. . .	575	. . .	700	. . .	850	. . .	910
. . .	366	. . .	. . .	714	. . .	. . .	. . .
. . .	400–600	. . .	400–700	. . .	. . .	. . .	. . .
. . .	458	. . .	. . .	800	. . .	. . .	. . .
250	. . .	. . .	1,000	. . .	. . .	. . .	. . .
. . .	250–450	. . .	350–550	. . .	. . .	. . .	. . .
. . .	160–250	. . .	. . .	. . .	. . .	. . .	. . .
. . .	370	. . .	500	. . .	. . .	. . .	. . .
. . .	258	. . .	406	. . .	. . .	. . .	. . .
. . .	. . .	200	. . .	380	. . .	. . .	. . .
. . .	210–390	. . .	310–550	. . .	. . .	. . .	. . .
. . .	26	. . .	30	. . .	. . .	. . .	. . .

APPENDIX G

Power Curves for Various Commercial Wind Systems

Power, Kilowatts (Rated wind speed is in bold type)

Manufacturer/ Model	**Rotor Diameter, Feet**	**Wind Speed, Mph** 8	9	10	11	12	13	14	15	16	17
Alcoa/"634214"	42	...	...	...	...	1.0	3.0	5.0	7.0	9.0	14.0
Windworks/"10KW"	33	.2	.4	.8	1.1	1.9	2.0	3.0	4.0	5.0	6.0
Wind Power/ "Storm Master 10"	32.8	.2	.3	1.0	1.1	1.9	2.2	3.0	4.0	6.0	**6.9**
Jay Carter/ "Mod 25"	32	...	...	1.0	2.0	2.5	3.0	4.0	5.0	6.0	8.0
Astral/Wilcon/ "AW10-B"	26	...	.5	1.0	1.5	2.0	2.2	3.2	3.9	4.2	5.0
Millville/ "10-3-IND"	25	...	...	.4	.6	.8	1.3	1.7	2.2	2.6	3.2
Env. Energies/ "HWT 15"	24	...	.2	.5	1.0	1.5	1.8	2.0	3.0	4.0	5.0
Env. Energies/ "HWT 12"	22	...	.2	.5	1.0	1.5	1.8	2.0	3.0	3.9	5.0
Env. Energies/ "HWT 9"	20	...	.2	.5	1.0	1.5	1.8	2.0	2.1	2.2	3.2
Enertech/"4000"	19.7	...	.2	.5	.6	.6	1.0	1.2	1.5	1.8	2.2
Bertoia/"A.P.S."	18	.2	.2	.3	.4	.5	.8	1.0	1.2	1.3	1.6
Env. Energies/ "HWT 6"	18	...	.2	.5	1.0	1.5	1.8	2.0	2.1	2.2	2.3

NOTES: This information is based upon manufacturer's published data and may be subject to change. For current information, contact the wind system manufacturer.

Some models of wind power systems have not been included in this appendix because the relevant information was not available at the time of publication.

18	19	20	21	22	23	24	25	26	27	28	29	30	31	Power Coefficient
17.5	20.0	25.0	28.0	32.0	36.0	39.0	42.0	45.0	47.5	**51.0**	**54.0**	56.0	55.0	.530
8.0	8.5	9.0	**9.5**	**10.0**	10.0	10.0	10.0	10.0	10.0	10.0	10.0	10.0	10.0	.364
8.0	8.0	8.0	8.0	8.0	8.0	8.0	8.0	8.0	8.0	8.0	8.0	8.0	8.0	.387
10.0	12.0	14.0	16.0	18.0	20.0	**23.0**	**25.0**	26.0	27.0	28.0	29.0	30.1	30.2	.659
6.1	7.2	8.4	9.0	**9.7**	**10.0**	10.0	10.0	10.0	10.0	10.0	10.0	10.0	10.0	.513
3.6	4.0	4.8	5.6	6.4	8.0	**8.4**	**10.0**	10.0	10.0	10.0	10.0	10.0	10.0	.432
6.0	8.0	9.0	10.5	12.0	13.0	**14.0**	**14.5**	15.0	15.5	16.0	16.2	16.5	16.7	.680
5.5	6.0	7.0	8.0	9.0	10.0	**11.0**	**11.5**	12.0	12.5	13.0	13.2	13.5	13.7	.642
4.0	4.5	5.0	6.0	7.0	8.0	**8.5**	**9.0**	9.2	9.4	9.5	9.8	10.0	10.5	.608
2.4	2.7	2.9	3.2	3.6	**3.8**	**4.0**	4.1	4.4	4.5	4.6	4.6	4.7	4.6	.315
2.0	2.1	2.2	2.3	2.4	2.5	2.5	2.6	2.7	2.8	2.9	**3.0**	**3.0**	3.0	.145
2.5	3.0	4.0	4.5	5.0	5.3	**5.8**	**6.0**	6.3	6.4	6.5	6.8	7.0	7.1	.500

APPENDIX G—Continued

Power, Kilowatts (Rated wind speed is in bold type)

Manufacturer/ Model	Rotor Diameter, Feet	Wind Speed, Mph 8	9	10	11	12	13	14	15	16	17
North Wind/"HR2"	16.4	.1	.2	.4	.5	.7	.9	1.0	1.2	1.5	1.7
Kedco/"1620"	16	...	...	.1	.2	.2	.3	.5	.6	.8	1.1
Kedco/"1610"	16	...	...	.1	.2	.2	.3	.4	.6	.8	1.0
Kedco/"1605"	16	.2	.2	.4	.4	.6	.6	.8	1.0	1.2	1.5
Kedco/"1600"	16	.2	.2	.2	.4	.6	.7	.8	1.1	**1.2**	**1.2**
American/ "16 Ft."	15.3	...	...	.3	.4	.6	.8	.9	1.0	1.2	1.4
Dynergy/ "5 Meter"	15	...	...	...	.2	.4	.6	.8	1.0	1.2	1.4
Independent/ "Sky Hawk IV"	15	.2	.2	.3	.4	.6	.8	1.0	1.7	1.4	1.8
Pinson/ "Cycloturbine C2E3"	15	...	.1	.2	.3	.5	.6	.7	.8	.9	1.0
Dakota/"BC 4"	14	.2	.3	.4	.5	.6	.7	.8	1.0	1.2	1.3
Independent/ "Sky Hawk II"	13.6	.2	.2	.3	.4	.4	.5	.7	.8	.8	1.0
Dunlite/"2000"	13.5	...	.1	.1	.2	.3	.4	.5	.6	.7	.8
Enertech/"1800"	13.2	.0	.1	.2	.2	.3	.4	.5	.7	.8	.8
Aero Power/ "SL 1500"	12	.2	.2	.3	.4	.5	.6	.6	.8	.8	.9
Kedco/"1210"	12	...	...	...	.2	.2	.3	.3	.4	.4	.5
Kedco/"1205"	12	.0	.1	.2	.2	.3	.4	.4	.6	.6	.8
Kedco/"1200"	12	.1	.1	.1	.2	.2	.3	.3	.4	.5	.6
Sencenbaugh/ "1000"	12	.1	.1	.2	.2	.3	.3	.4	.4	.5	.6
Altos/ "BWP-12B"	11.5	...	...	.1	.1	.2	.2	.3	.4	.5	.6
American/ "12 Ft."	11.5	...	...	.2	.2	.3	.3	.4	.5	.6	.7
Chalk/"11.5 Ft."	11.5	.05	.1	.2	.2	.3	.3	.4	.5	.6	.8

18	19	20	21	22	23	24	25	26	27	28	29	30	31	Power Coefficient
1.8	**2.0**	**2.2**	2.2	2.3	2.4	2.4	2.4	2.5	2.6	2.6	2.6	2.7	2.8	.432
1.5	2.0	2.1	2.5	2.8	3.0	**3.0**	**3.0**	3.0	3.0	3.0	3.0	3.0	3.0	.304
1.2	1.8	2.0	**2.0**	**2.0**	2.0	2.0	2.0	2.0	2.0	2.0	2.0	2.0	2.0	.297
1.8	**1.9**	**1.9**	1.9	1.9	1.9	1.9	1.9	1.9	1.9	1.9	1.9	1.9	1.9	.376
1.2	1.2	1.2	1.2	1.2	1.2	1.2	1.2	1.2	1.2	1.2	1.2	1.2	1.2	.387
1.8	**1.9**	**2.0**	2.0	2.0	2.0	2.0	2.0	2.0	2.0	2.0	2.0	2.0	2.0	.435
1.8	2.0	2.2	2.4	2.8	**3.2**	**3.3**	3.8	4.0	4.4	4.8	5.0	5.4	6.0	.506
2.3	2.8	3.3	3.4	3.5	3.6	3.7	**3.8**	**4.0**	4.0	4.1	4.2	4.2	4.2	.427
1.1	1.3	1.5	1.6	1.8	2.0	2.5	2.8	3.2	3.4	3.7	**4.0**	**5.0**	5.0	.512
1.5	1.8	2.0	2.3	2.5	3.0	3.3	3.8	**3.9**	**4.0**	4.0	4.0	4.0	4.0	.419
1.4	1.5	1.6	1.7	1.8	1.8	1.9	**2.0**	**2.0**	2.0	2.1	2.2	2.2	2.2	.260
.9	1.0	1.2	1.3	1.5	1.6	**1.8**	**2.0**	2.2	2.4	2.6	2.7	2.8	2.9	.284
1.0	1.2	**1.3**	**1.5**	1.6	1.8	1.8	1.9	2.0	2.0	2.1	2.1	2.0	2.0	.360
1.0	1.1	1.2	1.2	1.3	1.4	**1.4**	**1.4**	1.5	1.5	1.5	1.6	1.6	1.6	.371
.6	.8	1.0	1.0	1.2	1.5	1.8	**2.0**	**2.0**	2.0	2.0	2.0	2.0	2.0	.320
.9	1.1	1.2	**1.2**	**1.2**	1.2	1.2	1.2	1.2	1.2	1.2	1.2	1.2	1.2	.317
.7	.8	1.0	**1.1**	**1.2**	1.2	1.2	1.2	1.2	1.2	1.2	1.2	1.2	1.2	.317
.7	.7	.8	.9	**1.0**	**1.0**	1.1	1.1	1.1	1.2	1.2	1.2	1.2	1.2	.231
.7	.8	.9	1.0	1.2	1.4	1.4	1.5	1.6	**1.7**	**1.8**	1.9	2.0	2.5	.289
.8	**1.0**	**1.0**	1.2	1.4	1.6	1.6	1.6	1.6	1.6	1.6	1.6	1.6	1.6	.386
.9	**1.0**	**1.2**	1.4	1.6	1.8	2.1	2.3	2.6	3.0	3.3	3.6	4.0	4.0	.471

APPENDIX G—Continued

Power, Kilowatts (Rated wind speed is in bold type)

Manufacturer/ Model	Rotor Diameter, Feet	Wind Speed, Mph 8	9	10	11	12	13	14	15	16	17
Dunlite/ "High Winds"	10	...	...	...	...	...	...	...	.1	.2	.3
Whirlwind/ "Model A"	10	...	.05	.1	.15	.2	.2	.3	.4	.5	.6
Aerolectric/ "Wind Wizard"	9	...	.06	.08	.1	.12	.14	.2	.28	.3	.4
Bergey/"1000"	8.5	...	...	...	.05	.1	.15	.25	.3	.35	.4
Bergey/"650"	8.5	...	...	...	...	.1	.15	.2	.23	.25	.3
Altos/"BWP-8A"	8	...	...	.05	.07	.1	.15	.25	.35	.4	.6
American/ "8 Ft."	7.6	...	...	.04	.06	.1	.14	.2	.22	.3	.3
Sencenbaugh/ "400"	7	...	.02	.05	.08	.1	.14	.16	.2	.23	.3
Sencenbaugh/ "500"	6	...	...	.02	.04	.06	.08	.10	.14	.16	.2
Winco/"1222H"	6	...	.005	.006	.01	.01	.015	.02	.03	.035	.05

18	19	20	21	22	23	24	25	26	27	28	29	30	31	Power Coefficient
.4	.4	.5	.6	.6	.6	.7	.7	.8	.8	.8	.9	**.9**	**.9**	.080
.8	.9	1.0	1.2	1.5	1.7	**1.8**	**2.0**	2.0	2.0	2.0	2.0	2.0	2.0	.519
.4	.5	.5	.5	.6	.6	.6	**.6**	**.7**	.7	.7	.7	.7	.8	.194
.5	.6	.6	.7	.8	.8	**.9**	**1.0**	1.1	1.2	1.2	1.3	1.4	1.5	.374
.3	.4	.4	.4	.5	.6	**.6**	**.6**	.7	.7	.8	.8	.9	1.0	.243
.8	.8	1.0	1.0	1.1	1.2	1.2	1.3	1.4	**1.4**	**1.5**	1.6	1.6	1.6	.476
.4	**.5**	**.5**	.6	.7	.8	.8	.8	.8	.8	.8	.8	.8	.8	.457
.3	**.4**	**.4**	.4	.5	.5	1.5	1.5	.5	.5	.5	.5	.4	.4	.381
.2	.3	.3	.4	.4	**.5**	**1.5**	.5	.6	.6	.6	.6	.6	.2	.407
.1	.1	.1	.2	**.2**	**.2**	1.2	.2	.2	.2	.2	.2	.2	.2	.185

Appendix H

Energy Output Estimates of Various Commercially Available Wind Systems Based upon Rayleigh Wind Distribution

Kilowatt Hours per Month

Manufacturer/ Model	Rotor Dia- meter, Feet	Average Wind Speed, Mph 8	9	10	11	12	13	14	15	16
Alcoa/"634214"	42	968	1,742	2,755	3,974	5,354	6,846	8,404	9,987	11,561
Windworks/"10KW"	33	645	967	1,330	1,712	2,097	2,471	2,828	3,162	3,470
Wind Power/ "Storm Master 10"	32.8	664	979	1,319	1,664	1,999	2,315	2,608	2,875	3,116
Jay Carter/ "Mod 25"	32	854	1,344	1,947	2,649	3,428	4,258	5,117	5,983	6,840
Astral/Wilcon/ "AW10-B"	26	641	939	1,273	1,627	1,989	2,345	2,689	3,014	3,318
Millville/ "10-3-IND"	25	326	512	737	996	1,280	1,580	1,886	2,190	2,488
Env. Energies/ "HWT 15"	24	517	820	1,194	1,626	2,099	2,596	3,102	3,606	4,097

NOTES: This information is calculated from manufacturers' power curves, which are not uniformly derived, and may be subject to change. For current information, contact the wind system manufacturer.

Some models of wind power systems have not been included in this appendix because the relevant information was not available at the time of publication.

APPENDIX H—Continued

Kilowatt Hours per Month

Manufacturer Model	Rotor Diameter, Feet	Average Wind Speed, Mph 8	9	10	11	12	13	14	15	16
Env. Energies/ "HWT 12"	22	488	751	1,064	1,416	1,796	2,193	2,596	2,997	3,389
Env. Energies/ "HWT 9"	20	408	606	838	1,099	1,383	1,679	1,982	2,283	2,579
Enertech/ "4000"	19.7	253	369	500	640	786	933	1,077	1,217	1,351
Bertoia/ "A.P.S."	18	209	295	388	486	585	682	777	866	951
Env. Energies/ "HWT 6"	18	374	532	707	895	1,094	1,297	1,501	1,703	1,899
North Wind/ "HR2"	16.4	223	311	405	500	594	685	772	854	930
Kedco/"1620"	16	101	168	251	346	448	552	656	756	852
Kedco/"1610"	16	93	149	217	290	366	442	515	583	648
Kedco/"1605"	16	197	271	350	428	504	576	642	703	758
Kedco/"1600"	16	174	233	292	348	398	444	484	520	552
American/ "16 Ft."	15.3	175	253	336	419	500	577	648	713	772
Dynergy/ "5 Meter"	15	138	222	323	440	570	710	858	1,010	1,163
Independent/ "Sky Hawk IV"	15	205	303	419	547	683	821	958	1,091	1,217
Pinson/ "Cycloturbine C2E3"	15	137	200	272	355	449	553	664	781	902
Dakota/"BC4"	14	208	289	381	483	594	709	826	924	1,056
Independent/ "Sky Hawk II"	13.6	154	213	278	347	418	488	556	622	685
Dunlite/"2000"	13.5	100	147	202	264	331	403	478	554	631
Enertech/"1800"	13.2	113	164	221	284	349	414	479	542	602
Aero Power/ "SL 1500"	12	157	208	261	314	366	417	465	511	554
Kedco/"1210"	12	110	166	225	286	347	408	468	526	582
Kedco/"1205"	12	98	140	185	232	279	323	366	405	441
Kedco/"1200"	12	80	113	151	191	233	275	316	354	391

APPENDIX H—Continued

Kilowatt Hours per Month

Manufacturer Model	Rotor Dia-meter, Feet	Average Wind Speed, Mph 8	9	10	11	12	13	14	15	16
Sencenbaugh/ "1000"	12	96	130	166	203	241	279	316	351	385
Altos/"BWP-12B"	11.5	59	93	134	181	233	290	349	411	473
American/ "12 Ft."	11.5	85	126	174	225	279	333	386	437	485
Chalk/"11.5 Ft."	11.5	99	145	202	270	347	434	528	626	727
Dunlite/ "High Winds"	10	14	28	46	68	93	120	148	177	205
Whirlwind/ "Model A"	10	73	111	158	212	270	332	395	457	517
Aerolectric/ "Wind Wizard"	9	45	64	86	109	132	156	179	201	223
Bergey/"1000"	8.5	39	63	91	123	158	196	235	274	314
Bergey/"650"	8.5	28	45	64	85	108	132	157	183	208
Altos/"BWP-8A"	8	50	82	120	164	210	259	308	356	403
American/ "8 Ft."	7.6	36	55	78	104	131	158	185	211	235
Sencenbaugh/ "400"	7	33	48	65	83	101	118	134	150	164
Sencenbaugh/ "500"	6	21	32	46	62	79	97	116	134	152
Winco/"1222H"	6	5	9	14	19	26	32	39	45	52

Source List

Wind-Measuring Instruments

Aeolian Kinetics P.O. Box 100 Providence, RI 20901	Wind odometers, wind frequency analyzers.
Approach Fish, Inc. 314 Jefferson St. Clifton Forge, VA 24422	Tethered kites.
Bendix Instruments Division 1400 Taylor Ave. Baltimore, MD 21204	Strip-chart recorders.
Davis Instrument Mfg. Co. 513 E. Thirty-sixth St. Baltimore, MD 21218	Strip-chart recorders.
Dwyer Instruments P.O. Box 737 Michigan City, IN 46360	Meter anemometers.
Enertech Corp. P.O. Box 420 Norwich, VT 05055	Wind odometers.

Helion, Inc. Box 445 Brownsville, CA 95919	Wind frequency analyzers.
Kahl Scientific Instrument Corp. Box 1166 El Cajon, CA 92022	Wind odometers, strip-chart recorders.
Maximum, Inc. 42 South Ave. Natick, MA 01760	Meter anemometers.
Natural Power, Inc. Francestown Turnpike New Boston, NH 03070	Hand-held anemometers, wind odometers, wind monitors, strip-chart recorders, wind frequency analyzers, amp.-hour meters, tachometers.
Science Associates, Inc. 230 Nassau St. Princeton, NJ 08540	Hand-held anemometers, meter anemometers, wind odometers, strip-chart recorders.
Sencenbaugh Wind Electric P.O. Box 11174 Palo Alto, CA 94306	Wind odometers.
Sign X Laboratories, Inc. Essex, CT 06426	Wind frequency analyzers.
Simerl, R. A., Instruments Division 238 West St. Annapolis, MD 21401	Meter anemometers.
Spectrex Co. Bragg Hill Rd. Waitsfield, VT 05673	Watt-hour meters.

Stewart, M.C. Co.
Ashburnham, MA 01430

Wind odometers.

Texas Electronics
P.O. Box 7225
Inwood Station
Dallas, TX 75209

Strip-chart recorders.

Weather Measure Corp.
P.O. Box 41257
Sacramento, CA 95841

Hand-held anemometers, strip-chart recorders, wind frequency analyzers.

Generators

The following companies sell generators suitable for use with experimental or home-built wind systems.

Georator Corp.
9016 Prince William St.
Manassas, VA 22110

The model "36-011" produces 5 kw. of power at 1,800 rpm.

Leece Neville Co.
Cleveland, OH 44103

Low-speed alternators useful for wind applications.

Natural Power, Inc.
Francestown Turnpike
New Boston, NH 03070

The N.P.I. alternator is a low-cost, rewound heavy-duty alternator that generates 1.5 kw. at 1,800 rpm.

Zephyr Wind Dynamo Co.
P.O. Box 241
21 Stamwood St.
Brunswick, ME 04011

Zephyr's model "VLS-PM," permanent magnet alternator, has a rated power of 15 kw. at only 300 rpm.

Also, Zephyr's OBT student windmill is a miniature wind system suitable for schools.

Rotors

The following companies manufacture rotors suitable as replacements or for experimental use.

Aero Power Systems, Inc. 2398 Fourth St. Berkeley, CA 94710	Machine-fabricated, Sitka spruce blades including restored pre-R.E.A. Jacobs system replacements.
Aerowind, Inc. Rt. 2, Box 74 Milaca, MN 56353	Restored pre-R.E.A. Jacobs system replacement blades and governors.
Astral/Wilcon P.O. Box 291 Milbury, MA 01527	Composite glass-fiber, foam-filled blades including restored pre-R.E.A. Jacobs system replacements.
Neil Harman 705 E. Iona Fountain, CT 80817	Wooden rotor blades.
Propellor Engineering Duplicating 403 Avenida Teresa San Clemente, CA 92672	Laminated wooden blades including restored pre-R.E.A. Jacobs system replacements.
Santa Rosa Machine Works P.O. Box 1541 1270 Airport Blvd. Santa Rosa, CA 95402	Machine-carved, wooden rotors.
Sencenbaugh Wind Electric P.O. Box 11174 Palo Alto, CA 94306	Replacements for Sencenbaugh and Electro systems of laminated Sitka spruce. Also custom-built.

Towers

Aermotor Division of Valley Industries P.O. Box 1364 Conway, AR 72032	Water-pumping windmill towers.
Dempster Industries P.O. Box 848 Beatrice, NE 68310	Water-pumping windmill towers.
Heller-Aller Co. Perry and Oakwood St. Napoleon, OH 43545	Water-pumping windmill towers.
Natural Power, Inc. Francestown Turnpike New Boston, NH 03070	Octahedron towers.
Solargy Corp. 17914 E. Warren Ave. Detroit, MI 48224	Tubular steel towers.
Tri-Ex Tower Corp. 7182 Rasmussen Ave. Visalia, CA 93277	Self-supporting towers.
Unarco-Rohn 6718 W. Plank Rd. Peoria, IL 61601	Self-supporting and guyed towers with tower-top adapter kits for many brands of wind systems.
Valmont Industries, Inc. Valley, NE 68064	Tubular steel towers.

Batteries

This is a partial list of industrial and deep-cycle (including golf cart) battery manufacturers. Consult your local Yellow Pages for names of local manufacturers and suppliers. Battery manufacturers and suppliers are also sources for hydrometers, battery chargers, and battery stands. *The Starting, Lighting, Ignition, and Generating* (S.L.I.G.) *Buyers' Guide,* a worldwide guide to battery manufacturers, is available from the Independent Battery Manufacturers Association, Inc., 100 Larchwood Dr., Largo, FL 33540.

C & D Batteries
3043 Walton Rd.
Plymouth Meeting, PA 19462

Delatron Systems Corp.
553 Lively Blvd.
Elk Grove Village, IL 60007

E.S.B., Inc.
2000 E. Ohio Bldg.
Cleveland, OH 44101

Globe-Union
5757 N. Green Bay Ave.
Milwaukee, WI 53201

Gould, Inc.
485 Calhoun St.
Trenton, NJ 08618

Mule Battery Co.
325 Valley St.
Providence, RI 02908

SGL Batteries Mfg. Co.
14650 Dequindre
Detroit, MI 48212

Surrette Storage Battery Co., Inc.
P.O. Box 711
Salem, MA 01970

Trojan Batteries
1125 Mariposa St.
San Francisco, CA 94107

Battery Accessories

E-Z Red Co.
Deposit, NY 13754

Temperature-compensated hydrometer.

Perfection Electronic Products
1530 Rochester Rd.
P.O. Box 28UA
Royal Oak, MI 48068

Battery indicator gauges.

Inverters

Several inverter manufacturers are listed below. Each company makes several models and not all of the features listed in the chart are available on each model. For names of additional inverter manufacturers, check with your local electronics or wind products distributor.

	Type			Power Range, Watts	A.C. Wave Form		D.C. Input, Volts						Induction Start
	Vibrator	Rotary	Electronic		Square-Wave	Sine-Wave	12	24	32	36	48	120	
ATR Electronics, Inc. 300 E. Fourth St. St. Paul, MN 55101	X			275-1,100	X		X	X	X		X		
Best Energy Systems for Tomorrow (Best), Inc. Rt. 1, Box 106 Necedah, WI 54646			X	1,000–5,000		X	X	X	X	X	X		X
Carter Motor Co. 2711 W. George St. Chicago, IL 60618		X		250–750		X	X	X	X		X	X	
Delatron Systems, Corp. 553 Lively Blvd. Elk Grove Village, IL 60007			X	3,000–6,000		X					X	X	X
Dynamote Corp. 1130 N.W. Eighty-fifth St. Seattle, WA 98117			X	120–1,000	X		X	X	X				X
Elgar Corp. 8225 Mercury Ct. San Diego, CA 92111			X	600–1,000	X			X			X		
Independent Energy Systems, Inc. 6043 Sterrettania Rd. Fairview, PA 16415			X	3,000–6,000	X	X						X	X
Nova Electric Mfg. Co. 263 Hillside Ave. Nutley, NJ 07110			X	125–10,000		X	X	X			X	X	
Soleq Corp. 5969 Elston Ave. Chicago, IL 60646			X	1,500–6,000		X	X	X	X		X	X	X
Topaz Electronics 3855 Ruffin Ct. San Diego, CA 92123			X	250–10,000		X	X	X			X	X	X
Tripp-Lite 133 N. Jefferson Chicago, IL 60606			X	100–1,000	X		X	X					
Wilmore Electronics P.O. Box 2973 Durham, NC 27705			X	300–1,500	X	X	X	X	X	X	X	X	

Synchronous Inverters

Real Gas & Electric Co., Inc.
P.O. Box "F"
Santa Rosa, CA 95402

Sine-Sync synchronous Inverter and Inverter.

Windworks, Inc.
Box 329, Rt. 3
Mukwonago, WI 53149

Gemini synchronous inverter.

Control Panels

Montana Natural Power
P.O. Box 393
Basin, MT 59631

Control panel for 12-volt systems.

Natural Power, Inc.
Francestown Turnpike
New Boston, NH 03070

Manufactures control panels for 32 and 110-volt wind systems as well as the dynamic load switch for load control.

West Wind
Box 1465
Farmington, NM 87401

Manufactures the voltage control switch for load control.

Standby Power and Auxiliary Power

Battery Chargers

Sencenbaugh Wind Electric
P.O. Box 11174
Palo Alto, CA 94306

Gas Engine Driven Generators

Empire Electric Co.
5200-02 First Ave.
Brooklyn, NY 11232

Homelite Division of Textron, Inc.
P.O. Box 7047
Charlotte, NC 28217

Ideal Electric
615 First St.
Mansfield, OH 44903

Kato Engineering
3201 Third Ave. N.
Mankato, MN 56001

McCulloch
989 S. Brooklyn Ave.
Wellsville, NY 14895

Onan Corp.
1400 Seventy-third Ave., NE.
Minneapolis, MN 55432

Winco Division of Dyna
Technology
225 S. Cordova St.
LeCenter, MN 56053

Winpower
1225 First Ave., East
Newton, IA 50208

Electrical Wind Machine Manufacturers

Aerolectric
13517 Winter Lane
Cresaptown, MD 21502

Aero Power
2398 Fourth St.
Berkeley, CA 94710

Aerowatt, S.A.
c/o Automatic Power, Inc.
P.O. Box 18738
Houston, TX 77023

Alcoa Technical Center
Aluminum Co. of America
Alcoa Center, PA 15069

Altos: The Alternate Current
P.O. Box 905
Boulder, CO 80302

American Wind Turbine, Inc.
1016 E. Airport Rd.
Stillwater, OK 74074

Astral/Wilcon
P.O. Box 291
Milbury, MA 01527

Bergey Wind Power Co.
2001 Priestly Ave.
Norman, OK 73069

Bertoia Studio
644 Main St.
Bally, PA 19503

Jay Carter Enterprises
P.O. Box 684
Burkburnett, TX 76354

Chalk Wind Systems
P.O. Box 446
St. Cloud, FL 32769

Coulson Wind Electric
RFD 1, Box 225
Polk City, IA 50226

Dakota Wind & Sun, Ltd.
P.O. Box 1781
811 First Ave., NW
Aberdeen, SD 57401

Dominion Aluminum
Fabricating Indal, Ltd.
(D.A.F.)
3570 Hawkestone Rd.
Mississauga, Ontario
Canada L5C 2U4

Dragonfly Wind Electric
33810 Navarro Ridge Rd.
Albion, CA 95410

Dunlite Electrical Products Co.
c/o Enertech Corporation
P.O. Box 420
Norwich, VT 05055

Dynergy Corp.
P.O. Box 428
1269 Union Ave.
Laconia, NH 03246

Elektro GMBH
Winterhur-Schweiz
St. Galler-Strasse 27
Switzerland

Enertech Corp.
P.O. Box 420
Norwich, VT 05055

Environmental Energies, Inc.
Front St.
Copemish, MI 49625

Grumman Energy Systems
4175 Veterans Memorial Highway
Ronkonkoma, NY 11779

Hinton Research
417 Kensington
Salt Lake City, UT 84115

Independent Energy Systems, Inc.
6043 Sterrettania Rd.
Fairview, PA 16415

Jacobs Wind Electric Co., Inc.
2180 W. First St., Suite 410
Fort Myers, FL 33901

Kaman Aerospace
Old Windsor Rd.
Bloomfield, CT 06002

Kedco, Inc.
9016 Aviation Blvd.
Inglewood, CA 90301

McDonnell Aircraft Corp.
P.O. Box 516
St. Louis, MI 63166

Megatech Corp.
29 Cook St.
Billerica, MA 01866

Mehrkam Energy Development Co.
179 East Rd. #2
Hamburg, PA 19526

Millville Windmills & Solar Equipment Co.
P.O. Box 32
10335 Old Dr.
Millville, CA 96062

North Wind Power Co.
P.O. Box 315
Warren, VT 05674

Pinson Energy Corp.
P.O. Box 7
Marstons Mills, MA 02648

Power Group International Corp.
Suite 106
13315 Stuebener-Airline Rd.
Houston, TX 77014

Product Development Institute
4440 Secor Rd.
Toledo, OH 43623

Sencenbaugh Wind Electric
P.O. Box 11174
Palo Alto, CA 94306

Tumac Industries
650 Fort St.
Colorado Springs, CO 80915

TWR Enterprises
Sun-Wind-Home Concepts
72 W. Meadow Lane
Sandy, UT 84070

Whirlwind Power Company
2458 W. Twenty-ninth Ave.
Denver, CO 80211

Winco Division of Dyna Technology
225 S. Cordova St.
LeCenter, MN 56053

Wind Engineering Co.
P.O. Box 5936
Lubbock, TX 79417

Wind Power Systems, Inc.
P.O. Box 17323
San Diego, CA 92117

Windworks, Inc.
Box 329, Rt. 3
Mukwonago, WI 53149

Winflo Power Ltd.
15-90 Esna Park Drive, Unit 15
Markham, Ontario
Canada L3R 2R7

W.T.G. Energy Systems, Inc.
251 Elm St.
Buffalo, NY 14203

Zephyr Wind Dynamo Co.
P.O. Box 241
21 Stamwood St.
Brunswick, ME 04011

Mechanical Wind Machine Manufacturers

Aermotor Division of Valley Industries
P.O. Box 1364
Conway, AR 72032

American Wind Turbine, Inc.
1016 E. Airport Rd.
Stillwater, OK 74074

Bowjon
2829 Burton Ave.
Burbank, CA 91504

Dempster Industries, Inc.
P.O. Box 848
Beatrice, NE 68310

Dynergy Corp.
P.O. Box 428
1269 Union Ave.
Laconia, NH 03246

Heller-Aller Co.
Perry and Oakwood St.
Napoleon, OH 43545

Sparco (Denmark)
c/o Enertech Corporation
P.O. Box 420
Norwich, VT 05055

Wadler Manufacturing Co., Inc.
Rt. 2, Box 76
Galena, KS 66739

Installation Equipment

Lightning Arrestors

Approved Lightning Protection Co., Inc.
439 Meacham Ave.
Elmont, NY 11003

Zeus surge protector for lightning and a.c. surges.

Sencenbaugh Wind Electric
P.O. Box 11174
Palo Alto, CA 94306

Lightning protection for wind systems.

Thompson Lightning Protection, Inc.
901 Sibley Highway
St. Paul, MN 55118

Lightning protection engineering.

Safety Belts

Adirondack Wind Electric Co.
The Glen, Rt. 28
Warrensburg, NY 12885

Troll sit-harnesses.

Klein Tools, Inc.
7200 McCormick Rd.
Chicago, IL 60645

Linesman's safety belts.

Unarco-Rohn
6718 W. Plank Rd.
Peoria, IL 61601

Rohn-Loc safety climbing system.

Tackle Blocks

Sauerman Bros., Inc.
620 S. Twenty-eighth Ave.
Bellwood, IL 60104

Winches

Dutton-Lainson Co.
Hastings, NE 68901

Jeamar Winches Ltd.
52 Maple Ave.
Thornhill, Ontario
Canada L3T 3S7

Surplus Center
P.O. Box 82209
Lincoln, NE 68501

Appliances

D.C. Appliances, Retail

Allied Electronics
401 E. Eighth St.
Fort Worth, TX 76012

Relays, d.c.

James Bliss & Co., Inc.
Rt. 128
Dedham, MA 02026

Marine supplies.

W. W. Grainger, Inc.
5959 N. Howard St.
Chicago, IL 60648

Power tools, a.c.-d.c.

Montgomery Ward Co.

Norcold refrigerators, manual water softeners, 12-volt lights and accessories.

Newark Electronics
500 N. Pulaski Rd.
Chicago, IL 60624

Relays, d.c.

Prairie Sun and Wind Co. 4408 Sixty-second St. Lubbock, TX 79414	Fluorescent lights, 12 and 32-volt.
Sears, Roebuck and Co.	Piston pumps, 12-volt fans, air compressors, fluorescent lights, Norcold refrigerators, 12-volt freezer.
J. C. Whitney 1917 Archer Ave. Chicago, IL 60680	RV 12-volt fluorescent lights.
The Wilderness Home Power Co., Inc. P.O. Box 732 Laytonville, CA 95454	Lighting and accessories, 12-volt.

Fans

Merrin Electric Co. 1120 Clinton St. Hoboken, NJ 07030	32 and 120-volt d.c. fans.

Lighting

Iota Engineering Co. 1735 E. Ft. Lowell Rd. Tucson, AZ 85719	110-volt d.c. fluorescent.
McLean Electronics, Inc. 101-B Suburban Rd. San Luis Obispo, CA 93401	Fluorescent fixtures, 15-watt, 12-volt and 32-volt.
Triad-Utrad Division Litton Systems, Inc. 305 N. Briant St. Huntington, IN 46750	Model "PS-12," "PS-24," and "PS-32" operate on 12, 24, and 32-volt d.c. fluorescent ballasts, 70 w.

Motors

Bodine Electric Co. 2500 W. Bradley Pl. Chicago, IL 60618	Motors, d.c.

Carter Motor Co.
2711 W. George St.
Chicago, IL 60618

Motors, d.c. (shunt wound), 12 to 120-volt d.c. up to ¼ h.p., 1,800 and 3,600 rpm.

Universal Electric Co.
300 E. Main St.
Owosso, MI 48867

Wound filed d.c. motors to 1 h.p.

Pump Jacks

Aermotor Division of Valley Industries
P.O. Box 1364
Conway, AR 72032

Model "JE" pump jacks, pump cylinders, pump rod and leathers.

Dempster Industries
P.O. Box 848
Beatrice, NE 68310

Simpson model "BB-33" pump jacks, cylinders, pump rod and leathers.

Heller-Aller Co.
Perry and Oakwood St.
Napoleon, OH 43545

Freeze-proof windmill pumps.

Radios

Vintage Radio
Box 2045
Palos Verdes Peninsula, CA 90274

Tube radios.

Refrigeration

Dometic Sales Corp.
2320 Industrial Parkway
Elkhart, IN 46514

Large RV refrigerators that run on gas or 120-volt a.c. or d.c.

Koolatron Industries
56 Harvester Ave.
Batavia, NY 14020

Portable refrigerator uses thermoelectric modules to cool 40 lbs. of food. 12-volt.

Montgomery Ward Co.

Norcold refrigerators.

Norcold Co.
1501 Michigan Ave.
Sidney, OH 45365

RV refrigerators that run on 120-volt a.c. or 12-volt d.c. The 6-cubic foot model operates on 60 w. of power.

Stereo Receivers

Advent Corp.
195 Albany St.
Cambridge, MA 02139

Home-size stereo receiver, 12-volt d.c.

Publications and Organizations

Alternative Sources of Energy, 107 S. Central Ave., Milaca, MN 56353 (bimonthly, $15/yr.).

American Wind Energy Association, 1609 Connecticut Ave., NW, Washington, DC 20009 (Wind Energy Society and Trade Association).

Wind Energy Industry Association, P.O. Box 41157, Minneapolis, MN 55447 (Wind Trade Association).

Wind Energy Report, P.O. Box 14, Rockville Centre, NY 11571 (monthly, $95/yr.).

Wind Engineering, Multi-Science Publishing Co., Ltd., The Old Mill, Dorset Pl., London, E15 105 England (quarterly, $10/yr.).

Wind Power Digest, 115 E. Lexington, Elkhart, IN 46516 (quarterly, $8/yr.).

Bibliography

Altman, Alan. "Designing Windmill Blades." *Alternative Sources of Energy,* May 1974, pp. 10–13.

Anderson, Edwin. *Home Appliance Servicing.* Indianapolis, Ind.: Theodore Audel & Co., Division of Howard Sams & Co., Inc., 1976.

Annual Energy Requirements of Electrical Household Appliances, EEI-PUB #75-61. New York: Edison Electric Institute.

Bates, Donald and Cloud, Harold. "Energy Requirements of Electrical Equipment." *Agricultural Engineering Fact Sheet No. 1.* St. Paul, Minn.: University of Minnesota, 1969.

Bedford, Burnice and Hoft, R.G. *Principles of Inverter Circuits.* Somerset, N.J.: John Wiley & Sons, Inc., 1964.

Blanford, Percy. *Knots and Splices.* New York: Arco Publishing Co., 1965.

The Building Officials and Code Administrators International, Inc., Chicago, Ill.: The Building Officials and Code Administrators International, Inc., 1978.

Comparative Climatic Data for the United States: Through 1979. Asheville, N.C.: National Climatic Center, 1979. Also available for local areas: *Local Climatological Data* and *Percentage Frequency of Wind Direction and Speed.*

Croft, T. et al. *American Electricians' Handbook.* New York: McGraw-Hill Book Co., 1970.

Cullen, Jim. *The Wilderness Home Powersystem and How to Do It.* Laytonville, Calif.: The Wilderness Home Publishing Co., 1978.

DeRenzo, D. J. *Wind Power: Recent Developments.* Park Ridge, N.J.: Noyes Data Corp., 1979.

Dodd, D. W. *Lightning Protection for the Vertical-Axis Wind Turbine,* Report #SAND-77-1241. Springfield, Va.: National Technical Information Service (N.T.I.S.), 1977.

Dokol, Dan et al. "Using Existing House Wiring for Computer Remote Control." *Popular Electronics,* December 1977, pp. 60–65.

Eldridge, Frank. *Wind Machines,* Stock No. 038–000—27204. Washington, D.C.: U.S. Government Printing Office (U.S. G.P.O.).

Energy Storage Systems. Madison Heights, Mich.: Gulf & Western Energy Development Associates, June 1980.

Energy Task Force. *Windmill Power for City People,* Community Services Administration (C.S.A.) Pamphlet 6145-8. Washington, D.C.: U.S. G.P.O. 1977.

Exide Stationary Lead-Acid Battery System, Section 50.00. Philadelphia, Pa.: E.S.B. Brands, Inc., 1972.

Facts about Storage Batteries. Cleveland, Ohio: E.S.B. Brands, Inc., 1965.

Federal Energy Regulatory Commission. "Small Power Production and Co-Generation Facilities; Regulations Implementing Section 210 of the Public Utility Regulatory Policies Act of 1978," Docket no. RM79-55. *Federal Register,* vol. 45, no. 38, February 25, 1980. Washington, D.C.: U.S. G.P.O.

———. "Small Power Production and Co-Generation Facility Qualifying Status," Docket no. RM79-54. *Federal Register,* vol. 45, no. 56, March 20, 1980. Washington, D.C.: U.S. G.P.O.

Gipe, Paul. "VAWT Review," *Wind Power Digest.* Spring 1980, pp. 9–50.

"Going with the Wind," *EPRI Journal,* March 1980, pp. 9–17.

Golding, E. W. *The Generation of Electricity by Wind Power,* 1955. Reprint. New York: John Wiley & Sons, Inc., 1976.

Hackleman, Michael. *Electric Vehicles: Design and Build Your Own.* Mariposa, Calif.: Earthmind, 1977.

———. *The Homebuilt, Wind-Generated Electricity Handbook.* Mariposa, Calif.: Earthmind, 1975.

———. *Wind and Windspinners.* Mariposa, Calif.: Earthmind, 1974.

Hoffert, M. and Miller, G. "Augmented Vertical Axis Wind Energy System Evaluation." New York: New York University, Department of Applied Science, 1978.

Inglis, David R. *Wind Power and Other Energy Options.* Ann Arbor, Mich.: The University of Michigan Press, 1978.

International Association of Plumbing and Mechanical Officials and International Conference of Building Officials (ICBO). Whittier, Calif.: ICBO, 1976.

Johnson, Stephen B. "State Approaches to Solar Legislation: A Survey." *Solar Law Reporter,* vol. 1, no. 1, May/June 1979, pp. 55–217.

Kelly, Henry. *Application of Solar Technology to Today's Energy Needs.* Washington, D.C.: Office of Technology Assessment, Congress of the United States.

Lightning Protection Institute Installation Code LPI-175. St. Paul, Minn.: Thompson Lightning Protection, Inc., 1975.

Lindsay, T. J. *Power Inverter Technology: Technical Report.* Manteno, Ill.: Lindsay Publications, 1978.

McGeorge, John, "The Volt Box." *Alternative Sources of Energy,* no. 40, November 1979, pp. 41–43.

McGuigan, Dermot. *Harnessing the Wind for Home Energy.* Charlotte, Vt.: Garden Way Publishing, 1978.

Meyer, Hans. "Some Approaches to Power Conditioning." *Solar Engineering,* June 1978, pp. 31–33.

Mileaf, Harry. *Electricity Six.* New York: Hayden Book Co., 1966.

———. *Electricity Seven.* New York: Hayden Book Co., 1966.

The National Building Code. New York: American Insurance Association, 1976.

National Electrical Code 1978. Boston, Mass.: National Fire Protection Association, 1978.

National Science Foundation. *Legal-Institutional Implications of Wind Energy Conversion Systems,* NSF/RA-770203. Washington, D.C.: U.S. G.P.O., 1977.

Nissley, W. "The Wind Turbine/Water Twister Combination." Wilmington, Del.: All American Engineering Co., March 1978.

Park, Jack. *Simplified Wind Power Systems for Experimenters.* Brownsville, Calif.: Helion Co., 1975.

Park, Jack and Schwind, Dick. *Wind Power for Farms, Homes, and Small Industry,* Document #RFP-2841/1270/78/4. Springfield, Va.: N.T.I.S., 1978.

Planning a Wind-Powered Generating System. Norwich, Vt.: Enertech Corp., 1977.

Proceedings of the United Nations Conference on New Sources of Energy, Volume 7: Wind Power. New York: United Nations, 1964.

Reed, Jack. *Wind Power Climatology in the United States,* Document #SAND 74-0348. Springfield, Va.: N.T.I.S., 1975.

Rice, M. S. *Handbook of Airfoil Sections for Light Aircraft.* Appleton, Wis.: Aviation Publications, 1971.

Richter, H. P. *Wiring Simplified.* St. Paul, Minn.: Park Publishing, 1974.

Rossnagel, W. E. *Handbook of Rigging.* New York: McGraw-Hill Book Co. (Out of print).

Schacket, Sheldon. *The Complete Book of Electric Vehicles.* Northbrook, Ill.: Domus Books, 1979.

Simplified Electric Wiring Handbook. Chicago, Ill.: Sears Roebuck and Co., 1968.

Smith, Phil. *Knots for Mountaineering.* Redlands, Calif.: Citrograph Printing Company, 1975.

Solar Legislation. Rockville, Md.: National Heating and Cooling Center, 1979.

The Standard Building Code. Birmingham, Ala.: Southern Building Code Congress International, 1976.

The Storage Battery, Section 50.10. Philadelphia, Pa.: E.S.B. Brands, Inc., 1969.

U.S. Department of Energy, *First Semiannual Report: Rocky Flats Small Wind Systems Test Center Activities,* RFP# 2920/3533/78/6-1. Springfield, Va.: N.T.I.S., September 28, 1978.

———. *A Guide to Commercially Available Wind Machines,* Report #RFP-2836/3533/78/3. Springfield, Va.: N.T.I.S., 1978.

———. "Residential Conservation Service Program." Docket no. CAS-RM79-101, *Federal Register,* vol. 44, no. 247, December 21, 1979. Washington, D.C.: U.S. G.P.O.

Valkenburgh, Van. *Basic Electricity.* Rochelle Park, N.J.: Hayden Book Co., 1954.

Vinal, George. *Storage Batteries.* New York: John Wiley & Sons, Inc., 1940.

Walker, J. G. *The Automatic Operation of a Medium-Sized Wind-Driven Generator Running in Isolation.* Surrey, England: The British Electrical and Allied Industries Research Association, 1959.

Wegley, Orgell, and Drake. *A Siting Handbook for Small Wind Energy Conversion Systems,* Document #PNL 2521, Revision #1. Springfield, Va.: N.T.I.S., 1978.

Wheelock, Walt. *Ropes, Knots, and Slings for Climbers.* Glendale, Calif.: La Siesta Press, 1967.

The Windcyclopedia. Genesee Depot, Wis.: The Power Company-Midwest, Inc., 1980.

"Wind Turbines and Boats." *The Boat Builder,* no. 7. Bay City, Mich.: Gouegon Brothers, Inc., February 1980.

Index

C

D

E

Z